History of Life

AM BRYAN COLLEGE

WITHDRAWN

Askham Bryan College

069862

History of Life

Fifth Edition

Richard Cowen PhD

Professor Emeritus
Department of Geology, University of California, Davis, CA, USA

WITHDRAWN

ASKHAM BRYAN
COLLEGE
LEARNING RESOURCES

WILEY-BLACKWELL

A John Wiley & Sons, Ltd., Publication

This edition first published 2013 © 1990, 1995, 2000, 2005, 2013 by Richard Cowen.

Wiley-Blackwell is an imprint of John Wiley & Sons, formed by the merger of Wiley's global Scientific, Technical and Medical business with Blackwell Publishing.

Registered office: John Wiley & Sons, Ltd, The Atrium, Southern Gate, Chichester, West Sussex, PO19 8SQ, UK

Editorial offices: 9600 Garsington Road, Oxford, OX4 2DQ, UK
                     The Atrium, Southern Gate, Chichester, West Sussex, PO19 8SQ, UK
                     111 River Street, Hoboken, NJ 07030-5774, USA

For details of our global editorial offices, for customer services and for information about how to apply for permission to reuse the copyright material in this book please see our website at www.wiley.com/wiley-blackwell.

The right of the author to be identified as the author of this work has been asserted in accordance with the UK Copyright, Designs and Patents Act 1988.

All rights reserved. No part of this publication may be reproduced, stored in a retrieval system, or transmitted, in any form or by any means, electronic, mechanical, photocopying, recording or otherwise, except as permitted by the UK Copyright, Designs and Patents Act 1988, without the prior permission of the publisher.

Designations used by companies to distinguish their products are often claimed as trademarks. All brand names and product names used in this book are trade names, service marks, trademarks or registered trademarks of their respective owners. The publisher is not associated with any product or vendor mentioned in this book.

Limit of Liability/Disclaimer of Warranty: While the publisher and author(s) have used their best efforts in preparing this book, they make no representations or warranties with respect to the accuracy or completeness of the contents of this book and specifically disclaim any implied warranties of merchantability or fitness for a particular purpose. It is sold on the understanding that the publisher is not engaged in rendering professional services and neither the publisher nor the author shall be liable for damages arising herefrom. If professional advice or other expert assistance is required, the services of a competent professional should be sought.

Library of Congress Cataloging-in-Publication Data

Cowen, Richard, 1940-
    History of life / Richard Cowen. – Fifth edition.
          pages cm
    Includes bibliographical references and index.
    ISBN 978-0-470-67173-3 (hardback) – ISBN 978-0-470-67172-6 (paper)   1. Paleontology.   I. Title.
    QE711.2.C68 2013
    560–dc23
                              2012033617

A catalogue record for this book is available from the British Library.

Wiley also publishes its books in a variety of electronic formats. Some content that appears in print may not be available in electronic books.

Cover image: Cretaceous period fauna. Christian Jegou Publiphoto Diffusion/Science Photo Library.

An artist's reconstruction of marine and terrestrial fossils from various stages of the Cretaceous period (144 to 65 million years ago). Marine reptiles depicted include Mosasaurus, a large mosasaur (centre), and the giant pliosaur Kronosaurus (green, upper left). Between these are two ichthyosaurs. Also shown is the plesiosaur Elasmosaurus (blue, bottom, and rearing out of the water in a dramatic but unlikely fashion, top). The marine turtle Archelon is shown climbing onto the shore (to lay eggs?), watched by an oversized mammal. There are two cephalopods: a belemnite at left (orange) and an ammonite at lower right (brown). A group of duckbill dinosaurs (Ouranosaurus) patrols the shore, and pterosaurs watch from overhead.

Cover design by: Design Deluxe

Set in 10/11.5 pt Minion by Toppan Best-set Premedia Limited
Printed and bound in Malaysia by Vivar Printing Sdn Bhd

3   2014

Contents

Preface

For Everyone

For 36 years I taught a course called "History of Life" at the University of California, Davis. This book, now entering its fifth edition, was originally written for that course. But it is meant not just for students, but for everyone interested in the history of life on our planet. Fortunately, paleontology (=paleobiology), is accessible to the average person without deep scientific training. My aim is ambitious: I try to take you to the edges of our knowledge in paleontology, showing you how life has evolved on Earth, and how we have reconstructed the history of that evolution from the record of rocks and fossils.

However, there is a snag. Human history is never simple, even when we try to describe events that happened last year. It's even worse when we ask **why** events happened. It's not likely that any account of the history of life is going to be simple or easy either. The living world today contains all kinds of creatures that do unexpected things. There are frogs that fly and birds that can't. There are mammals that lay eggs, reptiles that have live birth, and amphibians that suckle their young. There are fishes that breathe air and mammals that never touch the land. We have to expect that there were complex and unusual ways of life in the past, and that evolution took some unexpected turns at times.

The challenge of teaching paleontology, and the challenge of writing a book like this, is to present a complex story in a way that is simple enough to grasp, yet true enough to real events that it paints a reasonable picture of what happened and why. I believe it can be done, and done so that you can learn enough to appreciate what's going on in current research projects.

Paleontologists can reconstruct how evolution happened and how the creatures of the past lived. We can't always prove it, any more than we can prove what really motivated George Washington. But we can state clearly what we know and don't know, we can suggest why certain events happened, and we can describe the evidence we used and the thoughts behind our suggestions. Then people can accept the ideas or not, as they wish.

Paleontologists have been collecting fossils, studying the rocks they came from, and assembling those data into a coherent framework, for over 200 years. At this point I think we are limited more by a lack of good ideas than by the facts available about the fossil record. I have not been shy about offering explanations of events as well as descriptions of them. Mostly they are other people's explanations, but now and again I've suggested some of my own. You can accept these or not, as you wish. The question you face is that facing a jury member: is this idea sound "beyond all reasonable doubt?" If you don't accept an explanation for an event, you can leave it as an abominable mystery, with no explanation at all, or you can suggest a better explanation yourself.

There is one caution, however. No one is allowed to dream up any old explanation for past events. A scientific suggestion (a hypothesis) has to fit the available evidence, and it has to fit with the laws of physics and chemistry, and with the principles of biology, ecology, and engineering that have been pieced together over the past 200 years of scientific investigation.

There's yet another wrinkle. A jury decides on a case, once and for all, with the evidence available. But in science, the jury is always out, and new evidence comes in all the time. You may have to change a verdict—without regret, because you made the best (wrong) decision you could based on the old evidence. Some of the ideas in the earlier editions are wrong, and you won't find them here; you'll find better ones. Sometimes the new answer is more complex, sometimes it is simpler. Always, however, the new idea fits the evidence better. That's the way science works: not on belief, not on emotional clinging to a favorite idea (even if it is your own), but on evidence.

I never expect to be able to write the final solution to the major questions about the history of life, but I do expect to be able to provide better answers this year than I could last year. If my lectures are the same this year as they were last year, then something is wrong with our science, or something is wrong with me. Paleontology is exciting because it is advancing so quickly.

Since paleontology is so fast-moving as a science, this book has changed too. I have radically re-written the sections on the origin of life, on extinctions, and on the origin and radiation of metazoan animals, of dinosaurs, of birds, and of humans. I have recognized more clearly that life evolved on an evolving planet, with changing chemistry, changing geography, and changing climate, and have tried to weave these threads into the tapestry as well. Every other section has been fine-tuned to reflect new research.

As I write, the full development of the Internet in publishing and educating and informing is in the early stages of a revolution. The Web site for this book, at http://mygeologypage.ucdavis.edu/cowen/HistoryofLife/ and at www.wiley.com/go/cowen/historyoflife allows me to add enriching material for which I could not find space in the printed page. It also allows me to connect you to other Internet sites for more illustrations, more detailed accounts of research projects, and snappy news articles. The Web site is meant to add real background and further details and perspectives for those who are interested. The essays and mini-essays are written in the same style as the book; the trivia are irresistible; and the Web links lead you to sites that are often richly illustrated in a way that no book can match.

I have written the book so that it stands alone without the Web site; but the Web pages are extensively linked back to the book. Over time, the Web page will also contain updates of material, new references, and new information.

To My Teaching Colleagues

The course for which this book was written serves three audiences at the same time: it is an introduction to paleontology; it is a "general education" course to introduce nonspecialists to science and scientific thought; and it can serve as an introduction to the history of life to biologists who know a lot about the present and little about the past. Therefore, the style and language of this book are aimed at accessibility. I do not use scientific jargon unless it is useful. I have tried to show how we reason out our conclusions— how we choose between bad ideas and good ones. I have not diluted the English language down to pidgin to make my points. In short, I have aimed this book at the intelligent nonspecialist.

I have not covered the fossil record evenly. I have tried to write compact essays on what I think are the most important events and processes that have molded the history of life. They illustrate the most important ways we go about reconstructing the life of the past. I've used case studies from vertebrates more than from other groups simply because those are the animals with which we are most familiar. Most fossils are marine invertebrates, and most paleontologists, including myself, are invertebrate specialists. I have tried to write briefly about invertebrates at an introductory level. They are not the easiest vehicles to use at this level, and that thought has controlled my choice of subject matter.

I'm pleased with the text of this book: I believe it communicates a lot about our science in the space available. But it's impossible to communicate paleontology well without a much greater visual component than can be included in a relatively inexpensive book format. So this book is better illustrated than previous editions, yet I use a lot more images in my classes than I can include in a book, in an attempt to bring fossil and living organisms into the classroom, and to give life to the words and names. The Web pages contain many sources for on-line illustrations that can be downloaded into your favorite presentation medium.

The references are a careful mixture of important books, primary literature, news reports, and review articles that bring the latest work into this edition as it went to press. I have deliberately skewed the lists to include items likely to be found in small college and city libraries.

If this book contained nothing controversial, it would be very dull and far from representing the state of paleontology as it stands today. I have tried to present arguments for and against particular ideas in case studies that are presented in some detail, such as major extinctions and major evolutionary innovations. Often, however, space or conviction has led me to present only one side of an argument. Please share your dissatisfaction and/or more complete knowledge with your students, and tell them why my treatment is one-sided or just plain wrong. That way everyone wins by exposure to the give and take of scientific argument as it ought to be practiced between colleagues.

To Students

Several thousand people like you have voted with their comments, questions, body language, and formal written evaluations on the content of my course. Students had more influence on the style and content of this book than anyone else. So you and your peers at the University of California, Davis, can take whatever credit is due for the style in which the material is presented.

After all the thanks, however, I do have another point to make. You don't have to take any of the interpretations in this book at face value. Facts are facts, but ideas are only suggestions. If you can come up with a better idea than one of those I've included here, then work on it, starting with the literature references. It would make a great term paper, and (more important) you might be right. The 1960s slogan "Question Authority!" is still valid. Your suggestion wouldn't be the first time that a student found a new and better idea for interpreting the fossil record.

Why do I, and why should you, bother with the past? If we don't understand the past, how can we deal intelligently with the present? We and our environment are reaching such a state of crisis that we need all the help we can get. Nature has run a series of experiments over the last 3.5 billion years on this planet, changing climate and geography, and introducing new kinds of organisms. If we can read the results of those experiments from the fossil record, we can perhaps define the limits to which we can stretch our present biosphere before a biological disaster happens.

The real pay-off from paleontology for me is the fun involved in reconstructing extinct organisms and ancient communities, but if one needs a concrete reason for looking at the fossil record, the future of the human race is surely important enough for anyone.

This Book

I begin this book with the formation of Earth and the great unsolved problem of the origin of life. Then I describe an early Earth populated entirely by bacteria, so strange in its chemistry and ecology that it might well be another planet. Eventually, living things so alter their world that we begin to recognize environments and organisms that seem much more familiar. I describe the evolution of animals and begin to worry about their physical and ecological environment. By now, we are dealing with a world whose geography we can begin to reconstruct, which leads to chapters on plate tectonics and the climates of the past, and how they might have affected living things.

The vast record of invertebrates allows us to measure the diversity of life through time, which shows that there have been times of high diversity, and times of dramatic extinctions. I deal with extinction, mainly to look at the crises or "mass extinctions" that have occurred sporadically through time. Then I turn largely to the history of vertebrates, following some of the great anatomical, physiological, and ecological innovations by which fishes gradually evolved into the major classes of tetrapods on the land, including ourselves.

Some of my colleagues are dubious about evolutionary "progress," but I regard the evidence for it as overwhelming, and present many examples. I have not tried to write a simple historical catalog of fossils. Instead, I have tried to set interesting episodes in Earth history into a global picture. For example, the tragedy that overtook Mesozoic communities 65 million years ago has to be seen in terms of their success until that time, and the subsequent radiation of the mammals can only be appreciated against a background of changing planetary geography and climate. Finally, the rise to ecological dominance of humans has its counterpart in the massive changes in land faunas that accompanied it, all set in the context of the ice ages.

Further Reading

I have tried to list widely sold paperbacks and articles in journals such as *Nature*, *Science*, *Discover*, *Scientific American*, *National Geographic Magazine*, and *American Scientist*, perhaps the six most widely distributed journals that deal with all aspects of science. I also list books and articles in specialized journals: generally the writing is more detailed and more technical in such journals.

Important earlier work is often summarized in more recent articles I have selected. Always, however, you should be able to work quickly backward to older papers from the references in recent articles.

An increasing number of scientists publish in open-access journals that are available on the Web, and/or they place their publications in Web-accessible sites associated with their labs or their courses. I have tried to refer to these often.

For more references, references available freely on the Web, further reading, notes, extra stories and mini-essays by RC, sources for classroom images, and updates, see the Web site for this book: http://www.wiley.com/go/cowen/historyoflife, mirrored at http://mygeologypage.ucdavis.edu/cowen/HistoryofLife/

Thanks

I thank all those reviewers who have given careful and calm advice over the years. I hope they recognize their contribution, and I apologize for churlishly ignoring some of them. I owe a great deal to the people at Blackwell/Wiley who have encouraged and helped me over the years. For this edition, it has been a delight to work with Ian Francis, Kelvin Matthews, and Delia Sandford. None of these people should be blamed for deficiencies: please complain directly to me at rcowen@ucdavis.edu.

Finally, I thank once again my wife Jo for tolerating my neglect of our home and property while all this was in process.

Winters, California   June 2012

About the Companion Website

This book is accompanied by a companion website:

www.wiley.com/go/cowen/historyoflife

The website includes:
- Powerpoints of all figures from the book for downloading
- PDFS of tables from the book
- Additional web resources

ONE

The Origin of Life on Earth

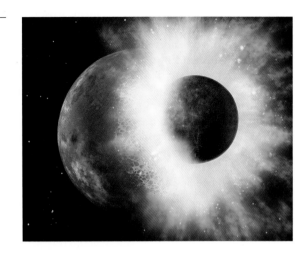

In This Chapter

First, I describe what geology and paleontology aim to study. But in dealing with the history of life, we have to face the most difficult question first: where did Earth's life come from? Astronomers find that organic compounds exist almost everywhere in space, yet we only know of life on one planet: Earth. I discuss the planets and moons of our solar system, and there are good reasons why none of them (except Earth) have life. Life exists in cells, so I discuss at length how complex organic molecules might have come together inside cells which survived, reproduced, and evolved on the early Earth. Laboratory experiments have already mimicked many of the steps in that process in the laboratory, but there is still a lot of work to be done.

How Geology Works

Geology is the study of the Earth we live on. It draws on methods and principles of its own, but also draws from many other sciences: physics, chemistry, biology, mathematics, and statistics are just a few. Geologists cannot be narrow specialists, because geology is a broad science that works best for people who think broadly. So geologists cannot be successful if they are geeks (though a few seem to manage it). Above all, geology deals with the reality of the Earth: its rocks, minerals, its rivers, lakes and oceans, its surface and its deep structure. Always, the reality of evidence from field work controls what can and what cannot be said about the Earth. Geological ideas are tested against evidence from rocks, and many beautiful ideas have failed that demanding test.

Some geologists deal with Earth as it is now: they don't need to look at the past. Deep Earth history doesn't matter much to a geologist trying to deal with ecological repair to an abandoned gold mine. But many geologists do study Earth history, and they find that our planet has changed, at all scales of space and time, and sometimes in the

History of Life, Fifth Edition. Richard Cowen.
© 2013 by Richard Cowen. Published 2013 by Blackwell Publishing Ltd.

most surprising ways. For 200 years, fossils have provided direct and solid proof of change through time. Life began and evolved on a planet that is changing too. Fossils often provide insight into Earth's environmental changes, whether or not they survived those changes. Paleontology is not just a fascinating side branch of geology, but a vital component of it.

As they run their life processes, organisms take in, alter, and release chemicals. Given enough organisms and enough time, biological processes can change the chemical and physical world. Photosynthesis, which provides the oxygen in our atmosphere, is only one of these processes. In turn, physical processes of the earth such as continental movement, volcanism, and climate change affect organisms, influencing their evolution, and, in turn, affecting the way they affect the physical earth. This gigantic interaction, or *feedback mechanism*, has been going on since life evolved on Earth. Paleontologists and geologists who ignore this interaction are likely to get the wrong answers as they try to reconstruct the past.

How Paleontology Works

Traces of Earth's ancient life have been preserved in rocks as fossils. Paleontology is the science of studying these fossils. Paleontology aims to understand fossils as once-living organisms, living, breeding, and dying in a real environment on a real but past Earth that we can no longer touch, smell, or see directly. We perceive a virtual Earth through our study of fossils and the rocks they are preserved in.

Most paleontologists don't study fossils for their intrinsic interest, although some of us do. Their greater value lies in what they tell us about ourselves and our background. We care about our future, which is a continuation of our past. One good reason for trying to understand ancient life is to manage better the biology of our planet today, so we need to use some kind of reasonable logic for clear interpretation of the life of the past.

Some basic problems of paleontology are much like those of archaeology and history: how do we know we have found the right explanation for some past event? How do we know we are not just making up a story?

Anything we suggest about the biology of ancient organisms should make sense in terms of what we know about the biology of living organisms, unless there is very good evidence to the contrary. This rule applies throughout biology, from cell biochemistry to genetics, physiology, ecology, behavior, and evolution.

But suggestions are only suggestions until they are tested against real evidence from fossils and rocks. Since fossils are found in rocks, we have access to environmental information about the habitat of the extinct organism: for example, a rock might show clear evidence that it was deposited under desert conditions, or on a shallow-water reef. Thus fossils are not isolated objects but parts of a larger puzzle. For example, it is difficult to interpret the biology of the first land animals unless we consider environmental evidence preserved in the Devonian sediments in which they are preserved (Chapter 7).

An alert reader should be able to recognize four levels of paleontological interpretation. First, there are *inevitable conclusions* for which there are no possible alternatives. For example, there's no doubt that extinct ichthyosaurs were swimming marine reptiles. At the next level, there are *likely interpretations*. There may be alternatives, but a large body of evidence supports one leading idea. For example, there is good evidence that suggests ichthyosaurs gave birth to live young rather than laying eggs. Almost all paleontologists view this as the best hypothesis, and would be surprised if contrary evidence turned up.

Then there are *speculations*. They may be right, but there is not much real evidence one way or another. Paleontologists are allowed to accept speculations as tentative ideas to work with and to test carefully, but they should not be surprised or upset to find them wrong. For example, it seems reasonable to me that ichthyosaurs were warm-blooded, but it's a speculative idea because it's difficult to test. If new evidence showed that the idea was unlikely, I might be personally disappointed but I would not be distressed scientifically.

Finally, there are *guesses*. They may be biologically more plausible than other guesses one might make, but for one reason or another they are untestable and must therefore be classified as nonscientific. For example, if I asked an artist to draw me an ichthyosaur (Fig. 1.1), I might suggest bold black-and-white color patterns, like those of living orcas, but another paleontologist might opt for more muted patterns like those of living fishes. Both ideas are reasonable, and are surely better science than one might find in a piece of art, however pleasing it may be (Fig. 1.1c). But all these are guesses: there is no evidence at all.

You will find examples of all four kinds of interpretation in this book. Often it's a matter of opinion in which category to place different suggestions, and this problem has caused many controversies in paleobiology. Were dinosaurs warm-blooded? For most paleobiologists this is an inevitable conclusion from the evidence. Some think it's likely, some think it's only speculative, some think it's unlikely, and a few think it is plain wrong. New evidence almost always helps to solve old questions but also poses new ones. Without bright ideas and constant attempts to test them against evidence, paleontology would not be so exciting.

The fossil record gradually gets poorer as we go back in time, for two reasons. Biologically, there were fewer types of organisms in the past. Geologically, relatively few rocks (and fossils) have survived from older times, and those that have survived have often suffered heating, deformation, and other changes, all of which tend to destroy fossils. Earth's early life was certainly microscopic and soft-bodied, a very unpromising combination for fossilization. So direct evidence about early life on Earth is very scanty, though speculation and guesses are abundant.

Figure 1.1   Guesses about ichthyosaur color patterns. a) ichthyosaur painting by Heinrich Harder 1916. b) art by Nobu Tamura, with muted colors, placed into Wikimedia. c) stylistic art work © Danny Anduza, used by permission. See more of Danny's work at http://www.cafepress.com/dannysdinosaurs

Figure 1.2   Edgar Rice Burroughs published the fourth in his series of Martian stories, *Thuvia, Maid of Mars*, as a book in 1920. a) cover art by P. J. Monahan; b) scene for black-and-white inside art, by J. Allen St John. The beast is a thoat, based on the real Earth fossil *Thoatherium* from South America (wait for Chapter 16!).

The Origin of Life

The fact of observation is that there is no evidence of life, let alone evidence of intelligence or civilization, anywhere in the universe except on our planet, Earth (for example, Smith 2011). This fact comes in the face of strenuous efforts by science fiction writers, tabloid magazines, movie directors, and NASA publicists to persuade us otherwise (Fig. 1.2). However, we have to face up to its implications. Most important, it implies (but does not prove) that Earth's life evolved here on Earth. How difficult would that have been?

We can test the idea that life evolved here on Earth, from nonliving chemicals, by observation and experiment. Geologists and astronomers look for evidence from the Earth, Moon, and other planets to reconstruct conditions in the early solar system. Chemists and biochemists determine how complex organic molecules could have formed in such environments. Geologists try to find out when life appeared on Earth, and biologists design experiments to test whether these facts fit with ideas of the evolution of life from non-living chemicals.

Complex organic molecules have been found in interstellar space, in the dust clouds around newly forming stars, on comets and asteroids and interplanetary dust, and on the meteorites that hit Earth from time to time. These compounds form naturally in space, generated as gas clouds, dust particles, and cometary and meteorite surfaces are bathed in cosmic and stellar radiation. Laboratory experiments designed to mimic such conditions in space have yielded organic molecules. Probably any solid surface near any star in the universe received organic molecules at some point in its history (Ciesla and Sandford 2012). Analyses of meteorites that have hit the Earth show they were carrying many of the basic organic molecules needed in the evolution of life.

But life as we know it is not just made of organic compounds: life consists of cells, composed mostly of liquid water that is vital to life. It is almost impossible to imagine the formation of any kind of water-laden cell in outer space: that can only happen on a planet that had oceans and therefore an atmosphere.

Planets have organic compounds delivered to them from space, especially from comets or meteorites, but this process by itself is unlikely to lead to the evolution of life. For example, organic molecules must have been delivered everywhere in the solar system, including Mercury, Mars, Venus, and the Moon, only to be destroyed by inhospitable conditions on those lifeless planets.

If conditions on a planet's surface were mild enough to allow organic molecules to survive after they arrived on comets, it is very likely that organic molecules were also forming naturally on that planet too. Space-borne molecules may have added to the supply on a planetary surface, but they are unlikely to have been the only source of organic molecules there.

Planets in Our Solar System

Scientists reconstructed the process of star and planet formation long before we could check it by observing stars forming out in the universe. Stars form from collapsing clouds of dust and gas, and in the process, planets and smaller bodies often form in orbit around the new stars. Now that we have telescopes powerful enough, the theories have been confirmed. In 2010 a spectacular new star, surrounded by dust and gas, was discovered in the process of forming in the constellation Centaurus (Fig. 1.3). Astronomers have now found hundreds of planets around other stars, most of them large ones because they are easier to detect.

Our star the Sun formed with Earth as one of four terrestrial (rocky) planets in the inner part of our solar system. Venus and Earth are about the same size, and Mars and Mercury are significantly smaller. They all formed from dust and gas in the same way, about 4570 Ma (million years ago) (Lin 2008).

Most likely, all the planets were largely complete by 4500 Ma, though they were bombarded heavily for hundreds of millions of years afterwards as stray asteroids struck their surfaces. The heat energy released as the planets formed would have made them partly or totally molten. Clearly, a very young planet is not a place where life could evolve. Earth in particular was struck by a huge Mars-sized body late in its formation. That impact probably melted the entire Earth, while most of the debris collected close to Earth to form the Moon (Fig. 1.4).

All the inner planets melted deeply enough to have hot surfaces that gave off gases to form atmospheres. But there

Figure 1.3 A new star forms in the constellation Centaurus. a) a bright new star (left side of the image) with a dust cloud around it. NASA/JPL-Caltech/ESO/ S. Kraus image. b) artist's impression of the new star. NASA/JPL-Caltech/R. Hurt (SSC) image.

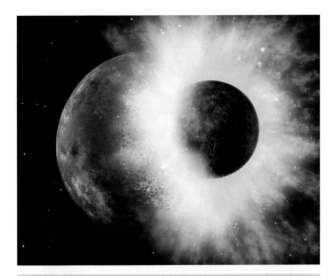

Figure 1.4 The early Earth was hit by a Mars-sized asteroid, and the debris that was blasted into space quickly collected to form the Moon. NASA/JPL-Caltech image.

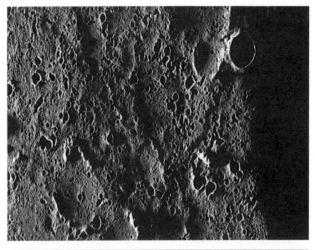

Figure 1.5 Image of the surface of Mercury: airless and lifeless. NASA image.

Figure 1.6 Image of the far side of the Moon: airless and lifeless. NASA image.

the similarity ended, and each inner planet has had its own later history.

Once a planet cools, conditions on its surface are largely controlled by its distance from the Sun and by any volcanic gases that erupt into its atmosphere from its interior. From this point onward, the geology of a planet greatly affects the chances that life might evolve on it.

Liquid water is vital for life as we know it, so surface temperature is perhaps the single most important feature of a young planet. Surface temperature is mainly determined by distance from the Sun: too far, and water freezes to ice; too close, and water evaporates to form water vapor.

But distance from the Sun is not the only factor that affects surface temperature. A planet with an early atmosphere that contained gases such as methane, carbon dioxide, and water vapor would trap solar radiation in the "greenhouse" effect, and would be warmer than an astronomer would predict just from its distance from the Sun.

In addition, distance from the Sun alone does not determine whether a planet has water, otherwise the Moon would have oceans like Earth's. The size of the planet is important, because gases escape into space from the weak gravitational field of a small planet. Gas molecules such as water vapor are lost faster from a small planet than from a larger one, and heavier gases as well as light ones are lost from a small planet. Thus Mars has only a thin atmosphere, and Mercury (Fig. 1.5) and the Moon (Fig. 1.6) have practically none.

Gases may be absorbed out of an atmosphere if they react chemically with the surface rocks of the planet. As they do so, they become part of the planet's geology, but may be released again if those rocks are melted in volcanic activity. But a small planet cools faster than a large one, so any volcanic activity quickly stops as its interior freezes. After that, no more eruptions can return or add gases to the atmosphere. Therefore, a small planet quickly evolves to have a very thin atmosphere or no atmosphere at all, and no chance of gaining one.

Volcanoes typically erupt large amounts of water vapor and CO_2, and these are both powerful greenhouse gases.

Earth would have been frozen for most of its history without volcanic CO_2 and water vapor in its atmosphere. Together they add perhaps 33°C to Earth's average temperature.

With these principles in mind, let's look at the prospects for life on other planets of our solar system. The brief story is that there is none. Both Mercury and the Moon had active volcanic eruptions early in their history, but they are small. They cooled quickly and are now solid throughout. Their atmospheric gases either escaped quickly to space from their weak gravitational fields or were blown off by major impacts. Today Mercury and the Moon are airless and lifeless.

Venus is larger than the Moon or Mercury, almost the same size as Earth. Volcanic rocks cover most of its surface. Like Earth, Venus has had a long and active geological history, with a continuing supply of volcanic gases for its atmosphere, and it has a strong gravitational field that can hold most gases. But Venus is closer to the Sun than Earth is, and the larger amount of solar radiation hitting the planet was trapped so effectively by water vapor and CO_2 that water molecules may never have been able to condense to become liquid water. Instead, water remained as vapor in the atmosphere until most of it was dissociated, broken up into hydrogen (H_2), which was lost to space, and oxygen (O_2), which was taken up chemically by reacting with hot surface rocks (Fig. 1.8).

Today Venus has a dense, massive atmosphere made largely of CO_2. Volcanic gases react in the atmosphere to make tiny droplets of sulfuric acid (H_2SO_4), forming thick clouds that hide the planetary surface. Water vapor has vanished completely. Although the sulfuric acid clouds reflect 80% of solar radiation, CO_2 traps the rest, so the surface temperature is about 450°C (850°F). We can be sure that there is no life on the grim surface of Venus under its toxic clouds.

Mars is much more interesting than Venus from a biological point of view. It is smaller than Earth (Fig. 1.7), and farther from the Sun. But it is large enough to have held on to a thin atmosphere, mainly composed of CO_2. Mars today is cold, dry, and windswept: dust storms sometimes cover half the planet.

No organic material can survive now on the surface of Mars. There is no liquid water, and the soil is highly oxidizing. But while Mars was still young, and was actively erupting volcanic gases from a hot interior, the planet may have had a thicker atmosphere with substantial amounts of water vapor. The crust still contains ice that could be set free as water if large impacts heated the surface rocks deeply enough to melt it, or if climatic changes were to melt it briefly.

So Mars does have water, but it is ice, frozen as part of the ice-caps, or under the surface sediment, where it is shielded from the sun. Ice can sublimate off the Martian surface, changing directly into water vapor. This blows around, sometimes being lost to space, sometimes freezing out again in the Martian winter.

Figure 1.7 Earth and Mars at the same scale. North poles at left. NASA/JPL-Caltech image.

Mars occasionally had surface water in the distant past. Canyons, channels, and plains look as if they were shaped by huge floods (Fig. 1.9), and other features look like ancient sandbars, islands, and lake beds. Ancient craters on Mars, especially in the lowland plains, have been eroded by gullies, and sheets of sediment lap around and inside the old craters, sometimes reducing them to ghostly rims sticking out of the flat surface.

Mars was too small to sustain geological activity for long. As the little planet cooled, its volcanic activity stopped (Figs 1.8, 1.10). Its atmosphere was largely lost, blasted off by impacts, or by slow leakage to space, or by chemical reactions with the rocks and soil. There may never have been oceans, and even lakes would have lasted a very short time. The surface is a dry frozen waste, and likely has been for well over three billion years. Even floods generated by a large meteorite impact would drain away or evaporate very quickly: they could not have lasted long enough to sustain life. In short, Mars is a lifeless ice-ball, and has been for billions of years.

In 1996, researchers reported they had found fossil bacteria in a meteorite that originated on Mars. (It was blasted into space by an asteroid impact, and fell on to Earth's Antarctic ice cap after spending thousands of years in space.) The researchers suggested that the bacteria were Martian. By now the report has been discounted: the objects are not bacteria and they are not evidence for life on Mars.

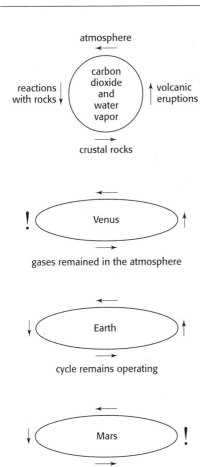

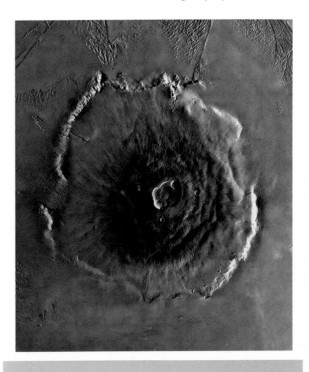

Figure 1.10 Olympus Mons, an enormous but long-extinct volcano on Mars, standing 27 km (17 miles) higher than the average crust of Mars, and over 600 km (370 miles) across at the base. NASA image.

Figure 1.8 An idealized rocky planet, with surface reactions. Earth is like this, but Venus and Mars are not. This has made all the difference in their history. Mars is frozen and dead, Venus is hot and toxic. See text for details.

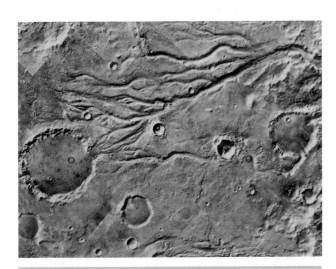

Figure 1.9 Ancient channels on the surface of Mars. NASA image.

The asteroid belt lies outside the orbit of Mars. Some asteroids have had a complex geological history, but here is no question of life in the asteroid belt now. Outside the asteroid belt, Jupiter and Saturn have ice-rich moons. But no planet or moon outside the orbit of Mars could trap enough solar radiation to form liquid water on its surface to provide the basis for life. Complex hydrocarbon compounds can accumulate and survive on asteroids, or in the atmospheres of the outer planets or on some of their satellites, but those bodies are frigid and lifeless.

Looking further afield, there is absolutely no evidence of life anywhere else in the universe. Many scientists argue that the universe is so vast that there must be other life out there, but that is speculation, not science. As we discover more planets around other stars, we find that many of them are in orbits that would make life impossible.

The Early Earth

So we return to Earth as the only known site of life. Gases released by eruptions and impacts formed a thick atmosphere around the early Earth, consisting mainly of CO_2 but with small amounts of nitrogen, water vapor, and sulfur gases. By about 4.4 billion years ago (4400 Ma or 4.4 Ga), Earth's surface was cool enough to have a solid crust, and liquid water accumulated on it to form oceans. Ocean water in turn helped to dissolve CO_2 out of the atmosphere and deposit it into carbonate rocks on the seafloor. This absorbed so much CO_2 that Earth did not develop runaway greenhouse heating as Venus did (Fig. 1.8). Large shallow oceans probably covered most of Earth, with a few crater rims and volcanoes sticking out as islands. The evidence for a cool watery Earth early in its history comes from a few zircon crystals that survived as recycled grains in later rocks. Some of the zircon crystals are dated close to 4.4 Ga.

We know from crater impacts and lunar samples that the Earth and Moon suffered a heavy late bombardment of asteroids around 3900 Ma, and the same event probably affected all the inner planets. Those catastrophic impacts must have destroyed almost all geological evidence of the early Earth's structure. Earth must have been hit by 100 or more giant asteroids and many smaller ones. At the same time, huge craters and basins filled with basalt lava were formed on the Moon (Fig. 1.11). The incoming asteroids seem to have been dislodged from their original orbits by changes in the orbits of Jupiter and perhaps Saturn as well, as those giant planets went through final gravitational adjustments in the complex dynamics of the solar system. The heat from the asteroid impacts probably melted the Earth's surface, boiled the ocean, and wiped out any life that might have evolved earlier. The life forms that were

Figure 1.11   The Late Heavy Bombardment hits the Moon (top), leaving scars that are still visible today (bottom). The effect on Earth would have been even greater because of Earth's greater mass. Image by Tim Wetherell of the Australian National University, and placed into Wikimedia.

our ancestors could not have evolved and survived until after the last sterilizing impact.

As the great bombardment died away, small late impacts may have encouraged the evolution of life on Earth. All comets and a few meteorites carry organic molecules, and comets in particular are largely made of ice. These bodies could have delivered organic chemicals and water to Earth. But Earth already had water, and processes here on Earth also formed organic chemicals. Intense ultraviolet (UV) radiation from the young Sun acted on the atmosphere to form small amounts of very many gases. Most of these dissolved easily in water, and fell out in rain, making Earth's surface water rich in carbon compounds. The compounds included ammonia (NH_3), methane (CH_4), carbon monoxide (CO), ethane (C_2H_6), and formaldehyde (CH_2O). They could have formed at a rate of millions of tons a year. Nitrates built up in water as photochemical smog and nitric acid from lightning strikes also rained out. But the most important chemical of all may have been cyanide (HCN). It would have formed easily in the upper atmosphere from solar radiation and meteorite impact, then dissolved in raindrops. Today it is broken down almost at once by oxygen, but early in Earth's history it built up at low concentrations in lakes and oceans. Cyanide is a basic building block for more complex organic molecules such as amino acids and nucleic acid bases. Life probably evolved in chemical conditions that would kill us instantly!

We have a good idea of the conditions of the early Earth, and of the many possible organic molecules that might have been present in its atmosphere and ocean. But how did that result in the evolution of life? First we look at the biology and the laboratory experiments that help us to solve the question, and then we look at real world environments to help us to work out where it happened.

Life Exists in Cells

The simplest cell alive today is very complex: after all, its ancestors have evolved through many billions of generations. We must try to strip away these complexities as we wonder what the first living cell might have looked like and how it worked.

A living thing has several properties: it has organized structure, and the capacity to reproduce (replicate itself), and to store information; and it has behavior and energy flow (metabolism). Mineral crystals have the first two but not the last two.

A living thing has a boundary that separates it from the environment. It operates its own chemical reactions, and if it did not have a boundary those reactions would be unable to work: they would be diluted by outside water, or compromised by outside contaminants. So a living "cell" has some sort of protective membrane around it.

A cell, like a computer, has hardware, software, and a protective case, all working well together. The case, or **cell membrane**, is made from molecules called **lipids**. The software that contains the information for running a cell is coded on **nucleic acids** (DNA and RNA), which use a four-character code rather than the two-character code (0 and 1) that our computers all use. The hardware consists largely of **proteins**, long molecules made from strings of **amino acids**. All those components had to become parts of a functioning organism.

A living thing can grow, and it can **replicate**, that is, it can make another structure just like itself. Both processes require complex chemistry. Growth and replication use materials that must be brought in from outside, through the cell wall.

A living thing interacts with its environment in an active way: it has **behavior**. The simplest behavior is the chemical flow of substances in and out of the cell, which can be turned on and off. The chemical flow will change the immediate environment, and the presence or absence of the desired chemicals will decide whether the cell turns the flow on or off. Temperature and other outside conditions also affect the behavior of even the simplest cell.

The chemical activity of the cell includes an energy flow that is called **metabolism** in living things. The cell must make molecules from simpler precursors, or break down complex molecules into simpler ones. If a cell grows or reproduces, it builds complex organic molecules, and those reactions need energy. The cell obtains that energy from outside, in the form of radiation or "food" molecules that it breaks down.

These attributes of a living cell are not different things: they are all intertwined, connected with gathering and processing energy and material into new chemical compounds (tissues), and continuing those processes into new cells. Any reconstruction of the evolution of life, as opposed to its creation by a Divine Being, must include a period of time during which lifeless molecules evolved the characters listed above and thereby became living. The phrase for this process is **chemical evolution**. We have to be able to argue that every step in the process could reasonably have happened on the early Earth in a natural, spontaneous way. It's easy to see that a protocell could grow effectively, given the right conditions. The critical turning point that defines life comes when relatively accurate replication evolves.

Even with a time machine, it would be very difficult to pick out the first living thing from the mass of growing organic blobs that must have surrounded it. But that cell survived and replicated accurately, and as time went by, its descendant cells that were more efficient remained alive and replicated, while those that were less efficient died or replicated more slowly. So as living things slowly emerged, chemical evolution slowly changed into *biological* evolution as we understand it today, subject to natural selection and extinction. Some lines of cells flourished, others became extinct. So living cells today do not exactly have the same genetic and biochemical machinery their ancestors had: they have long had major upgrades of their original software.

That brings one other concept into our discussion: *improvement* or *progress*. There is no question that the simplest living cells today are more efficient than their distant ancestors. Arguments rage about the politically correct word to use to describe this. The fossil record shows many examples of improved performance that can be analyzed mechanically. Living horses and living humans run far more efficiently, living whales swim more efficiently, and living birds fly more efficiently than their ancestors did. No doubt similar trends have occurred in physiology, biochemistry, reproduction, and so on. I can't think of a better word to describe this than **progress**.

We turn now to experiments that help us to see how life evolved from nonliving chemicals. The only life we know is on Earth, so we are testing the hypothesis that ingredients for the first cells were available on Earth, and that the first cells could have evolved along reasonable pathways.

The first stages in reconstructing the evolution of life were experiments in making the different necessary chemical components in plausible conditions. Now with success in that first stage, research has moved on to find how the components were successfully assembled into working units, getting closer to objects we might call "protocells".

Making Organic Molecules

In 1953 Stanley Miller, a young graduate student at the University of Chicago, passed energy (electric sparks) through a mixture of hydrogen, ammonia, and methane in an attempt to simulate likely conditions on the early Earth (Fig. 1.12). Any chemical products fell out into a protected flask. Among these products, which included cyanide and formaldehyde, were amino acids. This result was surprising at the time because amino acids are complex compounds, and are also vital components of all living cells.

The experiment that Miller published used a rather unlikely mixture of starting gases, but he also did a number of other experiments that gave similar results. Some were not published at the time, but Miller stored all his lab notes and experimental vials. When they were discovered after his death and analysed with 21st century techniques, it turns out that the best results came when Miller added volcanic gases to his mixtures (Parker et al. 2011).

It is now clear that almost all the amino acids found in living cells today could have formed naturally on the early Earth, from a wide range of ingredients, over a wide range of conditions. They form readily from mixtures that include the gases of Earth's early atmosphere. The same amino acids that form most easily in laboratory experiments are also the most common in living cells today. The only major condition is that amino acids do not form if oxygen is present.

Miller's experiments made amino acids in sterile glass flasks. But in later experiments, it was found that amino acids form even more easily on the surfaces of clay particles. Clay minerals are abundant in nature, have a long linear crystal structure, and are very good at attracting and adsorbing organic substances: cat litter is made from a natural clay and works on this principle.

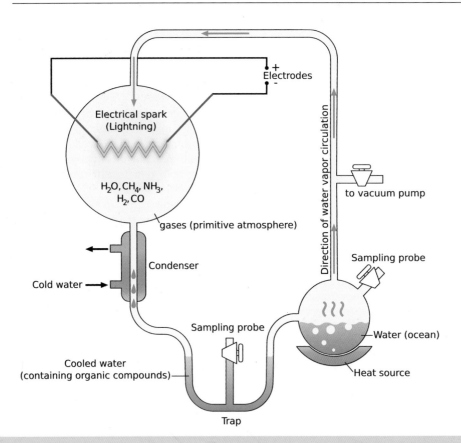

Figure 1.12 Stanley Miller's classic 1953 experiment, designed to simulate conditions on the early Earth. An atmosphere largely of water vapor, methane, and ammonia was subjected to lightning discharges. The reaction products cooled, condensed, and rained out to collect in the ocean. Those reaction products included amino acids. Diagram by Yassine Mrabet, and placed into Wikimedia.

People used to talk about "primordial soup", with the idea that interesting organic molecules would have been present throughout Earth's oceans. Everyone recognizes now that for the later stages of complex organic chemistry, organic molecules need to be concentrated, which allows them to react faster and more efficiently. Life may have begun in a rather unusual local environment.

For example, linking sequences of amino acid molecules into chains to form protein like molecules involves the loss of water, so scientists have tried evaporation experiments in simulated early Earth conditions. Four natural concentration mechanisms are evaporation; freezing; being enclosed inside membranes in scums, droplets, or bubbles; and concentration by being absorbed on to the surfaces of mineral grains. High temperatures help evaporation, but organic molecules tend to break down if they are heated too much. The longer the molecule, the more vulnerable it is to heat damage. However, experiments at low temperature form large molecules rather well. As water freezes into ice, other chemicals present are greatly concentrated. If they react to form larger organic molecules, the new molecules survive well.

Nucleic acids (**RNA** and **DNA**) have structures made of nucleic-acid bases or **nucleobases**; sugars; and phosphates. All the nucleobases have now been made in reasonable laboratory experiments. Sugars form in experiments that simulate water flow from hot springs over clay beds. Sugars and nucleobases could have formed in reactions powered by lightning. Naturally occurring phosphate minerals are associated with volcanic activity. Thus all the ingredients for nucleic acids were present on the early Earth, and the cell fuel ATP could also have formed easily.

Linking sugars, phosphates, and nucleobases to form fragments of nucleic acid called **nucleotides** also involves the loss of water molecules, and the phosphates themselves can act as catalysts here. Long nucleotides form much more easily on phosphate or clay surfaces than they do in suspension in water.

Many organic membranes are made of sheets of molecules called **lipids**. A lipid molecule has one end that attracts water and one end that repels water. Lipid molecules line up naturally with heads and tails always facing in opposite directions (Fig. 1.13); a bilayer sheet of lipid molecules therefore repels water. If a single or a double

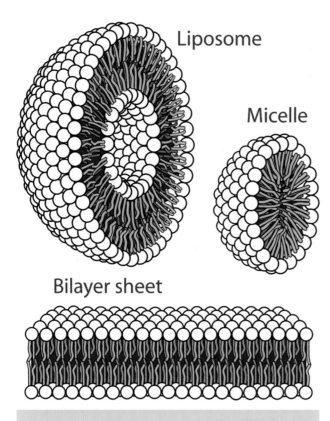

Liposome

Micelle

Bilayer sheet

Figure 1.13 The different shapes that lipid layers can form. Liposomes are also called vesicles. Vesicles can enclose mixtures of chemicals in a central cavity, and are very important in origin-of-life experiments. Image by Lady of Hats, Mariana Ruiz Villarreal, and placed into Wikimedia.

Figure 1.14 Sea foam, formed by waves on a South Australian beach. The dog is for scale. Photo taken by Bahudhara, and placed into Wikimedia.

Figure 1.15 A fragment of the Murchison meteorite yielded fatty acids that readily form into vesicles. Image from the US Department of Energy.

sheet of lipids happens to fold around to meet itself, it forms globular waterproof membranes (micelles) or hollow pills (liposomes or **vesicles**). Such shapes form spontaneously in lipid mixtures. Whipping up an egg in the kitchen produces lipid globules as the contents are frothed around. In the real world, lipid foams can form in the scum on wave surfaces (Fig. 1.14).

A breakthrough came when David Deamer's research group found that fatty acid molecules occur in the Murchison meteorite (Fig. 1.15), which fell in Australia in 1969. Those fatty acids could be extracted and formed into lipid vesicles by drying them out and then rewetting them (Fig. 1.16). Vesicles can also form from mixtures of molecules that would have been present on the early Earth. Deamer shook mixtures of lipids, amino acids and nucleic acids, and found that they formed spontaneously into many vesicles with organic molecules trapped inside them. They became tiny reaction chambers, inside which complex chemical changes could and did happen.

Nature has done experiments on making organic molecules. The meteorites and comets that strike Earth often carry organic compounds, and we can analyze them

knowing that they formed somewhere in space. The most common organic compounds in meteorites are also the most abundant in experiments that try to simulate chemistry in space. Many forms of amino acids, sugars, and nucleobases are found in meteorites, and so are fatty acids that easily form lipid membranes. Thousands of different organic compounds could have been supplied to the early Earth (Schmitt-Kopplin et al. 2010). We do not know how much organic matter was formed in natural processes on Earth and how much was delivered on comets and

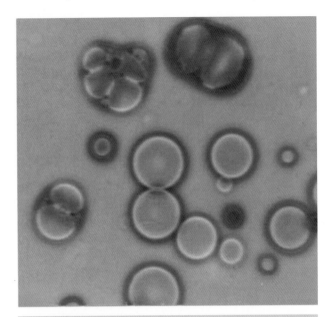

Figure 1.16 Lipid vesicles made in David Deamer's laboratory from fatty acids extracted from the Murchison meteorite. NASA image.

meteorites before and after the Late Bombardment. Either way, the right materials were present on the early Earth to encourage further reactions.

Toward the First Living Cell

How did basic organic molecules evolve into a cell that could reproduce itself? Deamer's early experiments began a new style of prebiotic experiments, using vesicles rather than test-tubes. After all, vesicles with cell-like contents could have formed in great numbers as waves thrashed around lipids on water surfaces (Fig. 1.14), or as lipid scums washed up on a muddy shore with clays in the water, or in the turbulent convection in and around hot springs. These vesicles would have had very variable contents (some with amino acids, primitive forms of nucleic acid, and so on). The "best" ones would have operated chemical reactions much more efficiently than the "worst". They would have done this because they had "better" nucleic acids, coded to produce "better" sets of protein enzymes to run efficient reactions.

Researchers have now found that vesicles can form 100 times as fast as usual if clay is added to the experimental mixtures. Some vesicles can take in substances from outside, through the lipid walls, and use them to build new walls and new contents: that is, they can grow. Irene Chen found that an active vesicle can "steal" (attract and absorb) part of the membrane from a less active neighbor and use it to grow! Vesicles can display a kind of "reproduction" in the sense that a large vesicle may divide into two, each keeping

some of the original vesicle contents (Chen et al. 2004, Chen 2009).

So we can imagine some watery environment where vesicles were growing and dividing more and more efficiently as their nucleic acids, their proteins, and their vesicle walls came to work well together.

In living cells today, information for making proteins is coded on long sequences of nucleic acid. The molecules of DNA that specify these protein structures are difficult to replicate, and replication requires many proteins to act as enzymes to catalyze the reactions. In living cells today, protein synthesis and DNA replication are interwoven: they depend on one another. So how could DNA and proteins have been formed independently, then evolved to depend on each other?

The answer lies with the simpler nucleic acid, RNA. Some RNA sequences called **ribozymes** can act as enzymes and make more RNA, even when no proteins are present. Other RNA sequences speed up the assembly of proteins. Perhaps the first living things were efficient vesicles that contained ribozymes with the right structure to replicate themselves accurately. (Such ribozymes have come to be called **naked genes**, but in reality they were inside vesicles.) Ribozymes would also have coded for the proteins needed to grow the vesicle and divide. In theory, RNA ribozymes on the early Earth could have replicated themselves with minimal proteins, in vesicles that we can now call **protocells**. Increasingly successful protocells would very quickly have outcompeted their neighbors. At some point, a successful protocell became the ancestor of all later life on Earth. The scenario that begins with ribozymes in an RNA world is currently the best hypothesis for the origin of life on Earth.

Where Did Life Evolve?

Most theories of the origin of life suggest surface or shoreline habitats in lakes, lagoons, or oceans. But it's unlikely that life evolved in the open sea. Complex organic molecules are vulnerable to damage from the sodium and chlorine in seawater. Most likely life evolved in lakes, or in seashore lagoons that were well supplied with river water. We have come to think of lagoons as tropical: the very name conjures up blue water and palm trees. Warm temperatures promote chemical reactions, and an early tropical island would most likely have been volcanic and therefore liable to have interesting minerals. But RNA bases are increasingly unstable as temperatures rise: normal tropical water, at 25°C, is about as warm at it could be for the origin of life.

So perhaps lakes or lagoons on cold volcanic islands were the best environments favoring organic reactions on the early Earth. In the laboratory, cyanide and formaldehyde reactions occur readily in half-frozen mixtures. Volcanic eruptions often generate lightning storms (Fig. 1.17), so eruptions, lightning, fresh clays, and near-freezing temperatures (ice, snow, hailstones) could all have been present

Figure 1.17 Volcanic lightning in an eruption cloud, at Rinjani volcano in Indonesia, 1995. Photograph by Oliver Spalt and placed into Wikimedia.

Figure 1.18 A volcanic island set in a cold climate: Onekotan, in the Kurile Islands on the Russian East coast. The southern volcano, on an island in a large crater, is Krenitzyn Peak. Image from NASA Earth Observatory.

on the shore of a cold volcanic island (Fig. 1.18). Note that if this environment is the correct one, there had to have been land and sea when life evolved: fresh water can only occur on Earth if it is physically separated from the ocean.

Solar radiation or lightning are likely energy sources for the reactions leading toward life. But deep in the oceans are places where intense geothermal heating generates hot springs on the sea floor. Most of these lie on the mid ocean ridges, long underwater rifts where the sea floor is tearing apart and forming new oceanic crust. Enormous quantities of heat are released in the process, much of it through hot water vents, and myriads of bacteria flourish in the hot water. Perhaps life began nowhere near the ocean surface, but deep below it, at these **hydrothermal vents** (Fig. 1.19).

Laboratory experiments have implied that amino acids and other important molecules can form in such conditions, even linking into short protein-like molecules, and currently the deep-sea hypothesis is popular. But if life evolved by way of naked genes, then it did not do so in hot springs. RNA and DNA are unstable at such high temperatures. Naked genes could not have existed (for long enough) in hot springs.

The deep-sea hypothesis, even though it looks unlikely (to me), has led to speculation that life might have evolved deep under the surface layers of other planets or satellites. (For example, Jupiter's moon Europa probably has liquid water under its icy crust, and Saturn's moon Enceladus has been seen to erupt water vapor "geysers".) The speculation helps to generate money for NASA's planetary probes. But the internal energy of such planets and moons is very low,

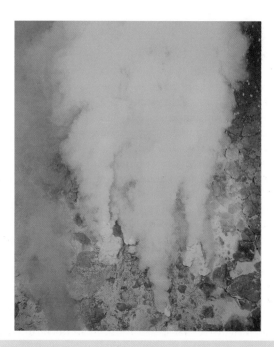

Figure 1.19 Hydrothermal vents on the Pacific Ocean floor. Image from NOAA.

and water-borne organic reactions are much less likely to work deep under the icy crust of Europa or Enceladus than in Earth's oceans. In any case, the under-ice oceans of icy moons are salty (that's how they were detected), so an origin of life is very unlikely in such environments.

Energy Sources for the First Life

Living things use energy. Much of biology consists of studying metabolism and ecology: how living things acquire and use the energy they need to grow and reproduce.

As we have seen, reactions powered by solar power, volcanic heat, lightning, or delivery from outside, built up a reservoir of simple organic chemicals on the early Earth. Protocells likely evolved in a watery environment that contained easily available chemical energy in naturally formed organic molecules such as ATP, amino acids, sugars, and other organic compounds.

So the first protocells had energy, fuel for cell growth and replication. But as they became more numerous and more effective in attracting and using organic molecules, there must have come a time when demand for energy exceeded supply. As simple organic molecules became scarcer and scarcer, protocells encountered the world's first energy crisis. This crisis would have happened first in those environments where protocells were most successful and abundant.

The energy that humans use so carelessly today comes from only two sources: solar energy and geothermal energy. Solar energy is in the form of direct radiation (heat and light); or as indirect energy, since solar energy powers wind and ocean currents, and evaporates water vapor that eventually falls as rain that runs hydroelectric plants. Even more indirect solar energy came from the sunlight that powered plant growth in the past, now found as fossil fuels in the earth: peat, coal, oil and gas. Geothermal energy can be tapped by drilling into steam vents or hot rocks, or by mining and concentrating radioactive minerals for fuel in nuclear power plants. Of the two sources, solar energy is by far the largest and easiest to manage.

Early cells found two very different solutions to their energy crisis that can still be seen among living organisms nearly 4 billion years later. Both depend on harnessing solar energy, but they occur in two very different kinds of organisms, using two very different processes.

Living organisms take in outside energy in two ways: **heterotrophy** and **autotrophy**. Heterotrophs obtain their metabolic energy by breaking down organic molecules they obtain from the environment: hummingbirds sip nectar and humans eat doughnuts. Heterotrophs do not pay the cost of building the organic molecules. They simply have to operate the reactions that break them down. But they must live where they can find "food" molecules. The first cells, living on the organic molecules around them, were heterotrophs.

Autotrophs do not need food molecules from outside: they make them inside the cell, paying the cost of building them by absorbing energy from outside. Autotrophy was evolved by some early cells, but not by all of them.

Heterotrophy

The simplest reaction used by cells to break down organic molecules is **fermentation**, to break down sugars such as glucose. This is what early heterotrophs must have done. Glucose is often called the universal cellular fuel for living organisms, and it was probably the most abundant sugar available on the early Earth. [Today, humans use fermenting microorganisms to produce beer, cheese, vinegar, wine, tea, and yogurt, and to break down much of our sewage.]

As heterotrophs used up the molecules that were easiest to break down, there would have been intense competition among them to break down more complex ones. One can imagine a huge advantage for cells that evolved enzymes to break down molecules that their competitors could not use. New sets of fermentation reactions would quickly have evolved, and different lineages of heterotrophic cells would have come to be specialists in their chemistry.

In becoming more efficient heterotrophs, some early cells found a way to import energy to make their internal chemistry run faster at no extra cost. In the last ten years, microbiologists have found that billions of heterotrophic microbes living in the world's shallow waters, in seas and lakes, and even in the ice around Antarctica, can absorb light energy and use it to help their internal chemical reactions. The molecules that can absorb light in this fashion are called **rhodopsins**.

We and many other creatures now use rhodopsins in our eye cells as light sensors. Light hitting a rhodopsin molecule activates it, and after a cascade of reactions, a nerve impulse is sent to the brain. Rhodopsins are the universal molecules in biological visual systems, allowing bacteria and fungi as well as humans to detect and react to light.

But the first rhodopsin molecules probably did something else. Rhodopsin is triggered by light to add electric charges to protons, and those protons can then be taken off to power chemical reactions inside the cell. Light-powered chemistry thus gives an advantage to rhodopsin-bearing heterotrophs over their competitors. Much of the biology in the ocean's surface waters is powered by rhodopsin reactions, and we knew nothing about them ten years ago! This system is called **phototrophy** ("feeding by light") because the rhodopsin reactions help to break down molecules, but do not build them up. Rhodopsin reactions aid heterotrophs, not autotrophs.

The first rhodopsin systems probably evolved only once, in some lucky mutant cell. The genes that code for rhodopsin are not large, and they seem to have passed easily from one cell to another, so that now, after billions of years, many different lineages of heterotrophic cells now use rhodopsin to save energy. Of course, rhodopsin is useful only in water that is shallow enough to receive sunlight. Heterotrophs living in dark environments must run at lower energy levels.

Autotrophy

Autotrophs generate their own energy, but in two completely different ways. Some extract chemical energy from inorganic molecules (**lithotrophy**), while others gain energy by trapping solar radiation (**photosynthesis**).

Lithotrophy can occur when a microorganism rips an oxygen molecule off one inorganic compound and transfers it to another, making an energy profit in the process. That energy is then used to build organic food molecules. For example, microorganisms called **methanogens** gain energy from lithotrophy by breaking up carbon dioxide and transferring the oxygen to hydrogen, forming water and methane as by-products:

$$4H_2 + CO_2 \rightarrow CH_4 + 2H_2O + energy$$

Methanogens are as different from true bacteria as bacteria are from us, and are part of a special group of microorganisms, the Archaea. Since carbon dioxide and hydrogen would have been available in the early ocean, it is reasonable to suggest that this reaction could have been used by very early cells. Indeed, based on their molecular genetics, Archaea were among the first living things on Earth.

If lithotrophy evolved very early, it may have been the first time (but not the last) that living things modified Earth's chemistry and climate. By replacing the greenhouse gas carbon dioxide with the even more powerful greenhouse gas methane, the activity of methanogens might have warmed the early Earth (Chapter 2).

Photosynthesis is simple in concept: energy from light is absorbed into specific molecules called **chlorophylls**. The process is biochemically more complex than lithotrophy or phototrophy. Chlorophylls (and the genes that code for making them) seem to have evolved only once.

The evolution of photosynthesis produced major ecological changes on Earth. Light energy trapped by chlorophyll was used to build more *biomass* (biological substance), giving photosynthetic cells an energy store, a buffer against times of low food supply, that could be used when needed. It's easy to see how such cells could come to depend almost entirely on photosynthesis for energy. In doing so, they did not have to compete directly with heterotrophs. In addition, as photosynthesizers died, and their cell contents were released into the environment, they inadvertently provided a dramatic new source of nutrition for heterotrophs. Photosynthesis greatly increased the energy flow in Earth's biological systems, and for the first time considerable amounts of energy were being transferred from organism to organism, in Earth's first true ecosystem.

The earliest photosynthetic cells probably used hydrogen from H2 or H2S. For example, the reaction

$$H_2S + CO_2 + light \rightarrow (CH_2O) + 2S$$

released sulfur into the environment as a by-product of photosynthesis. Later, photosynthetic bacteria began to break up the strong hydrogen bonds of the water molecule.

Bacteria that successfully broke down H_2O rather than H_2S, like this:

$$2H_2O + CO_2 + light \rightarrow (CH_2O) + 2O$$

immediately gained access to a much more plentiful resource. There was a penalty, however. The waste product of H_2S photosynthesis is sulfur (S), which is easily disposed of. The waste product of H_2O photosynthesis is an oxygen radical, monatomic oxygen (O), which is a deadly poison to a cell because it can break down vital organic molecules by oxidizing them. Even for humans, it is dangerous to breathe pure oxygen or ozone-polluted air for long periods.

Cells needed a natural antidote to this oxygen poison before they could operate the new photosynthesis consistently inside their cells. **Cyanobacteria** were the organisms that made the first breakthrough to oxygen photosynthesis using water. A lucky mutation allowed them to make a powerful antioxidant enzyme called **superoxide dismutase** to prevent O from damaging them: essentially, the enzyme packaged up the O into less dangerous O2 that was ejected out of the cell wall into the environment.

From then on, we can imagine early communities of microorganisms made up of autotrophs and heterotrophs, each group evolving improved ways of gathering or making food molecules.

Photosynthesizers need nutrients such as phosphorus and nitrogen to build up their cells, as well as light and CO_2. In most habitats, the nutrient supply varies with the seasons, as winds and currents change during the year. Light, too, varies with the seasons, especially in high latitudes. Since light is required for photosynthesis, great seasonal fluctuations in the primary productivity of the natural world began with photosynthesis. Seasonal cycles still dominate our modern world, among wild creatures and in agriculture and fisheries.

We can now envisage a world with a considerable biological energy budget and large populations of microorganisms: Archaea, photosynthetic bacteria, and heterotrophic bacteria. So there is at least a chance that a paleontologist might find evidence of very early life as fossils in the rock record. In Chapter 2 we shall look at geology, rocks, and fossils, instead of relying on reasonable but speculative arguments about Earth's early history and life.

Further Reading

Chen, I. A. et al. 2004. The emergence of competition between model protocells. *Science* 305: 1474–1476. Available at http://isites.harvard.edu/fs/docs/icb.topic459133.files/Papers/Chen-et-al-2004.pdf

Chen, I. A. 2009. Cell division: breaking up is easy to do. *Current Biology* 19: R327–328.

Chen, I. A. 2010. An RNA whirl. *Science* 330: 758. Available at http://www.sysbio.harvard.edu/csb/chen_lab/publications/Chen_2010_LifeFromRNAWorld.pdf

Chen, I. A. and P. Walde. 2010. From self-assembled vesicles to protocells. *Cold Spring Harbor Perspectives in Biology*. Available at http://www.cshperspectives.net/content/2/7/a002170.full

Ciesla, F. J. and S. A. Sandford. 2012. Organic synthesis via irradiation and warming if ice grains in the solar nebula. *Science* 336: 452–454.

Deamer, D. 2011. *First Life: Discovering the Connections Between Stars, Cells, and How Life Formed*. Berkeley: University of California Press.

Lin, D. N. C. 2008. The genesis of planets. *Scientific American* 298 (5): 50–59.

Mansy, S. S. and J. W. Szostak. 2009. Reconstructing the emergence of cellular life through the synthesis of model protocells. *Cold Spring Harbor Symposia Quantitative Biology* 74: 47–54. Available at http://genetics.mgh.harvard.edu/szostakweb/publications/Szostak_pdfs/Mansy_et_al_2009_CSHS.pdf

Martins, Z. et al. 2008. Extraterrestrial nucleobases in the Murchison meteorite. *Earth & Planetary Science Letters* 270: 130–136.

Parker, E. T. et al. 2011. Primordial synthesis of amines and amino acids in a 1958 Miller H2S-rich spark discharge experiment. *PNAS* 108: 5526–5531. Available at http://www.pnas.org/content/108/14/5526.long

Ricardo, A. and J. Szostak. 2009. Life on Earth. *Scientific American* 301 (3): 54–61. Available at http://origins.harvard.edu/OOLOE%20Scientific%20Am%209-2009.pdf

Schmitt-Kopplin, P. et al. 2010. High molecular diversity of extraterrestrial organic matter in Murchison meteorite revealed 40 years after its fall. *PNAS* 107: 2763–2768. Available at http://www.pnas.org/content/107/7/2763.long

Smith, H. A. 2011. Alone in the Universe. *American Scientist* 99: 320–328.

Szostak, J. W. 2009. Systems chemistry on early Earth. *Nature* 459: 171–172. Available at http://genetics.mgh.harvard.edu/szostakweb/publications/Szostak_pdfs/Szostak_2009_Nature.pdf

Questions for Thought, Study, and Discussion

1. It is clear that after Earth had cooled, comets and meteorites added important ingredients to its surface: ice (= water), and a great variety of organic molecules. Many scientists think that this "late accretion" gave Earth the ingredients for the formation of life. However, the same ingredients must have been added to Mars and Venus and the Moon also, with no sign that they ever evolved life. So why did Earth evolve life while the others did not?

2. Many movies have portrayed extinct animals. Suppose I said to you that none of the portrayals were scientific. Give a careful response to this assertion.

3. Where on Earth did life first evolve? When you decide where it was, give a careful summary of the evidence that helped you to come to your answer.

TWO
Earliest Life on Earth

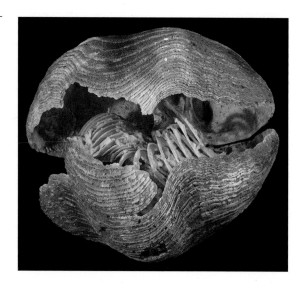

In This Chapter

We turn now to geological and paleontological evidence for Earth's early life. First I explain what fossils are and how we can find out how old they are. Since organisms run chemical reactions, they change Earth's chemistry, particularly in the ocean surface, on land, and in the atmosphere; and as they do so, they leave clues about ancient life process as chemical traces in ancient rocks. Ancient rocks may carry subtle chemical markers of ancient life, but in very special circumstances they can carry traces of ancient cells. As life expanded, its chemical influence on Earth's processes widened, and we see in particular ancient iron-bearing rocks that mark a transition from an anoxic atmosphere to the oxygen-bearing atmosphere you and I breathe today.

Introduction

When we move from astronomy and the laboratory to the Earth itself to search for evidence about early life, we look for **fossils**. A fossil is the remnant of an organism preserved in the geological record. There are three kinds of fossils, body fossils, trace fossils, and chemical fossils. We are most familiar with body fossils, in which part or all of an organism is preserved. If an organism had body parts that were made of resistant materials, such as shells, bones, or wood, it is much more likely than a "soft-bodied" creature to be preserved in the geological record. Such fossils may look more or less unchanged after death. Minerals may crystallize out of ground water to fill up large or small cracks, crevices, and cavities in the original substance, so body fossils may be denser and harder than they were in life. Sometimes the original shell or bone may be replaced by another mineral, making the fossil easier to recognize or to extract from the rock (Fig. 2.1).

Obviously, the hard parts of an organism are far more likely to be preserved than more fragile parts. But occasionally soft parts may leave an impression on soft sediment

History of Life, Fifth Edition. Richard Cowen.
© 2013 by Richard Cowen. Published 2013 by Blackwell Publishing Ltd.

before they rot. Even more rarely, a complete organism may be encased in soft sediment that later hardens into a rock. Bees, ants, flies, and frogs have been preserved as fossils in amber (fossilized tree resin) (Fig. 2.2a), and individual cells have been preserved in chert, a rock formed from silica gel that impregnated the cells and retained their shapes in three dimensions.

A **trace fossil** is not part of an organism at all, but it was made by an organism and therefore may tell us something about that creature. Trace fossils may be marks left by active organisms (footprints, trails, or burrows, Fig. 2.2b), or fecal masses (Fig. 2.3), or even a spider web. Trace fossils may give us insight into behavior that would not be available from a body fossil. For example, although dinosaur skeletons suggest that they could have run, trace fossils of dinosaur footprints tell us that they certainly did run [see Chapter 12].

Chemical fossils are compounds produced by organisms and preserved in the rock record. They may be molecules that were originally part of the organism, or molecules that were produced in the metabolic processes the organism operated. They may provide information about the organisms that produced them. In special cases where an organism absorbs one isotope of an atom over another in its food or water, the chemical fossils of these isotopes can be used to give information too, as described later in this chapter.

All kinds of agents may destroy or damage organisms beyond recognition before they can become fossils or while they are fossils. After death the soft parts of organisms may rot or be eaten. Any hard parts may be dissolved by water,

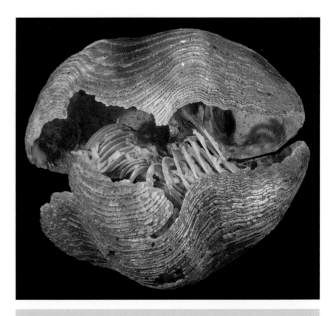

Figure 2.1 A brachiopod whose original calcite shell was replaced by silica. This made it fairly easy to dissolve the shell out of rock for study. This is the brachiopod *Spiriferina*, from the early Jurassic of France. In life, the spiral structure supported soft tissue that filtered sea water for plankton and oxygen. Photograph by Didier Descouens, who placed it into Wikimedia.

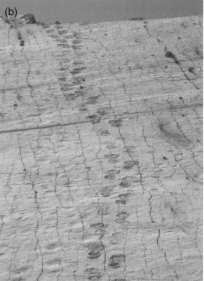

Figure 2.2 a) a fossil ant, preserved in the famous amber found on the shores of the Baltic Sea. Image by Anders Damgaard, and placed into Wikimedia. See www.amber-inclusions.dk for more of his images. b) the trackway of a dinosaur with very big feet, preserved on a tilted rock face in Bolivia. Image by Jerry Daykin, and placed into Wikimedia.

Figure 2.3 A trace fossil: a coprolite or fossil dung ball from a carnivorous dinosaur. This is a trace fossil because it was not part of the original organism but is evidence that it once existed (during the late Cretaceous in Saskatchewan, Canada). Scale is 15 cm (about 6 inches). Image from the United States Geological Survey.

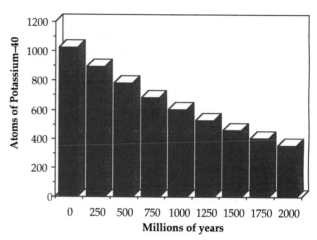

Figure 2.4 The radioactive decay of an isotope proceeds on a logarithmic time-table that is constant under all known conditions. If the decay is recorded in a rock or mineral, we can infer the date when the decay began. Often, but not always, that tells us the age of the rock. This graph shows the atoms of potassium-40 remaining in a crystal on a time scale measured in millions of years, compared to a starting value of 1024 atoms at time zero. For more details, see Hazen (2010).

or broken or crushed and scattered by scavengers or by storms, floods, wind, and frost. Remains must be buried to become part of a rock, but a fossil may be cracked or crushed as it is buried. After burial, groundwater seeping through the sediment may dissolve bones and shells. Earth movements may smear or crush the fossils beyond recognition or may heat them too much. Even if a fossil survives and is eventually exposed at the Earth's surface, it is very unlikely to be found and collected before it is destroyed by weathering and erosion.

Even when they are studied carefully, fossils are a very biased sample of ancient life. Fossils are much more likely to be preserved on the sea floor than on land. Even on land, animals and plants living or dying by a river or lake are more likely to be preserved than those in mountains or deserts. Different parts of a single skeleton have different chances of being preserved. Animal teeth, for example, are much more common in the fossil record than are tail bones and toe bones. Teeth are usually the only part of sharks to be fossilized. Large fossils are usually tougher than small ones and are more easily seen in the rock. Spectacular fossils are much more likely to be collected than apparently ordinary ones. Even if a fossil is collected by a professional paleontologist and sent to an expert for examination, it may never be studied. All the major museums in the world have crates of fossils lying unopened in the basement or the attic.

When we look at museum display cases, it seems that we have a good idea of the history of life. But most of the creatures that were living at any time are not in a museum. They were microscopic or soft-bodied, or both, or they were rare or fragile and were not preserved, or they have not been discovered. We do have enough evidence to begin to put together a story. But that story is always changing as we discover new fossils and look more closely at the fossils we have found already.

How to Find the Age of a Fossil

Fossils are found in rocks, and usually geologists try to establish the age of the containing rock or of a layer of rock that is not far under or over the fossil (so might be close to it in age). The age of rocks is measured in two different ways, known as relative and absolute dating.

Age dating of rocks can only work if one identifies components of the rocks that change with time or are in some way characteristic of the time at which the rocks formed. The same principles are used in dating archeological objects. Coins may bear a date in years (**absolute dating**), and one can be certain that a piece of jewelry containing a gold coin could not have been made before the date stamped on the coin. The age of waste dumps can be gauged by the type of container thrown into them: bottles with various shapes and tops, steel cans, aluminum cans, and so on. The age of old photographs, movies or paintings can often be judged by the dress or hairstyle of the people, or the cars or appliances shown in them.

Absolute geological ages can be determined because newly formed mineral crystals sometimes contain unstable, radioactive, atoms. Radioactive isotopes break down at a rate that no known physical or chemical agent can alter (Fig. 2.4), and as they do so they may change into other elements. For example, potassium-40, 40K, breaks down to form 40Ar, argon-40. By measuring the amount of radioactive decay in a mineral crystal, one can calculate the time since it was newly formed, just as one reads the date from a coin. The principle is simple, though the techniques are often laborious. For example, 40K breaks down to form 40Ar at a rate such that half of it has gone in about 1.3 billion years (Fig. 2.4). If we measure the 40Ar in a potassium feldspar crystal today, and find that half the original amount of 40K has gone, then the age of the crystal is 1300 Ma. Other dating methods use this same principle. (By convention, absolute ages in millions of years are given in megayears [Ma], while time periods or intervals are expressed in millions of years [m.y.].) Ages in billions of years are gigayears [Ga].)

Absolute dating must be done carefully. Crystals may have been reheated or even recrystallized, re-setting their radioactive clocks back to zero well after the time the rock originally formed. Chemical alteration of the rock may have removed some of the newly produced element, also giving a date younger than the true age. Geologists are familiar with these problems, and go to immense trouble to find and use fresh clean crystals.

Most elements used for radioactive age dating are not used by animals to build shells or bones, so usually we cannot date fossils directly. Instead, we have to measure the age of a lava flow or volcanic ash layer as close to the fossil-bearing bed as possible (Fig. 2.5), which does contain crystal we can use.

Paleontologists more often deal with a **relative time scale**, in which one says "Fossil A is older than Fossil B" (as in Fig. 2.5) without specifying the age in absolute years. This is much the same way that archeologists date Egyptian artifacts. We know which Pharaoh followed which, though we do not know the calendar years for some earlier dynasties. So Egyptian history is scaled according to the reigns of individual Pharaohs, rather than recorded in absolute years. One can work this way with fossils, because it is a fact of observation that fossils preserved in the rock record at particular times are almost always different from those preserved at other times. These principles have been firmly established over the past two centuries by geologists working in rock sequences to define successive layers, each layer lying on and thus being younger than the one underneath.

With the occasional check from absolute methods, the geological record has been arranged into a standard sequence: the **geological time scale** (Fig. 2.6). The time scale is divided into a hierarchy of units for easy reference, with the divisions between major units often corresponding to important changes in life on Earth. The names of the eras and periods are often unfamiliar and have bizarre historical roots. For example, the Permian period was named after Sir Roderick Murchison and the Comte d'Archiac took a stagecoach tour of Russia in 1841, and discovered unfamiliar new rocks near the city of Perm. After a while, however, the names and their sequence become not a matter for laborious memorization but the key to a vivid set of images of ancient life.

Life Alters a Planet

For too long, paleontologists thought of life as a set of passengers on a planet that had a certain geology, chemistry, and climate. Evolution took place as creature interacted with creature, or as a response to the physical environment. But we know now that biological processes dramatically affect the physical Earth, in a mutual interaction that has complex patterns. One can no longer study any component of the Earth system on its own because the interplay is so important. This may make life difficult for Earth scientists, but we do our humble best.

For example, the fact that Earth's atmosphere today has 21% oxygen reflects the continuous production of oxygen by photosynthesizers on land and in surface waters. Without life, oxygen cannot be present at more than a few parts per million. Chemically, 21% oxygen provides enough O_2 to form an ozone (O_3) layer in the high atmosphere, helping to shield the surface and its life from UV radiation. Physically, 21% oxygen affects the chemistry of seawater (iron won't dissolve in it, for example), and it affects the reactions by which rocks break down on the surface and turn into sediment. The oxygen is extracted from CO_2, which reduces the concentration of that greenhouse gas in the atmosphere and ocean, and cools the climate. There is nothing magic about a level of 21% oxygen: that level, and Earth's surface temperature, has fluctuated (moderately) for hundreds of millions of years.

In another example, methane is a greenhouse gas, so at times when methanogens were globally important

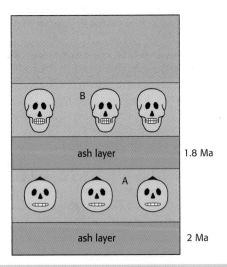

Figure 2.5 The skulls of these two fossil hominids do not contain any radioactive isotopes, but they lie close to two layers of volcanic ash that do. By using relative dating methods, one can say that hominid A is older than hominid B. Using absolute dating methods, the age of hominid A can be fixed closely between 1.8 Ma and 2.0 Ma because there are dated ash layers above and below it. All we know about the age of hominid B from this situation is that it is younger than 1.8 Ma.

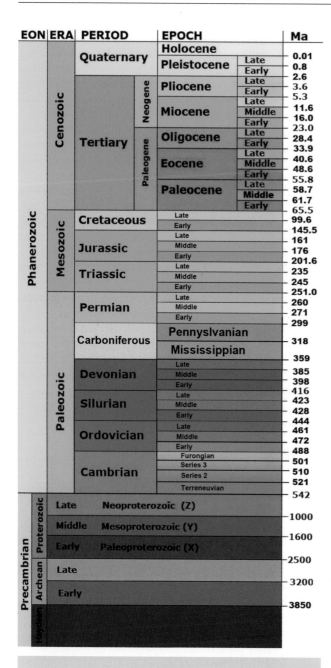

EON	ERA	PERIOD		EPOCH		Ma
Phanerozoic	Cenozoic	Quaternary		Holocene		0.01
				Pleistocene	Late	0.8
					Early	2.6
		Tertiary	Neogene	Pliocene	Late	3.6
					Early	5.3
				Miocene	Late	11.6
					Middle	16.0
					Early	23.0
			Paleogene	Oligocene	Late	28.4
					Early	33.9
				Eocene	Late	40.6
					Middle	48.6
					Early	55.8
				Paleocene	Late	58.7
					Middle	61.7
					Early	65.5
	Mesozoic	Cretaceous		Late		99.6
				Early		145.5
		Jurassic		Late		161
				Middle		176
				Early		201.6
		Triassic		Late		235
				Middle		245
				Early		251.0
	Paleozoic	Permian		Late		260
				Middle		271
				Early		299
		Carboniferous		Pennyslvanian		318
				Mississippian		359
		Devonian		Late		385
				Middle		398
				Early		416
		Silurian		Late		423
				Middle		428
				Early		444
		Ordovician		Late		461
				Middle		472
				Early		488
		Cambrian		Furongian		501
				Series 3		510
				Series 2		521
				Terreneuvian		542
Precambrian	Proterozoic	Late	Neoproterozoic (Z)			1000
		Middle	Mesoproterozoic (Y)			1600
		Early	Paleoproterozoic (X)			2500
	Archean	Late				3200
		Early				3850
	Hadean					

Figure 2.6 The official time scale adopted by the United States Geological Survey. The column is arranged in relative time, with youngest at the top. The absolute time scale is on the right. Obviously, the depths of the divisions are not to scale. USGS diagram.

autotrophs, their methane release may have warmed the Earth.

As we follow the history of life, we shall see that major biological changes led to major environmental changes, which in turn led to further biological events, and so on. And in reverse, major physical changes led to biological changes, and so on. It is the dynamic interplay which is important. Life has such an important effect on a planet

that NASA is working out strategies for detecting the presence of life on extrasolar planets by searching for its chemical signature. We could use the same strategy in trying to work out when life arose on Earth, and what form it took.

Isotope Evidence for Biology

Most chemical elements have two or more isotopes, that is, their atoms may have slightly different masses. Thus, most carbon atoms weigh 12 atomic mass units (the nucleus has 6 protons and 6 neutrons). But a few carbon atoms have an extra neutron, so they weigh 13 units, and are called carbon-13 or 13C.

The extra mass does not affect the chemistry, but it has physical effects. The heavier carbon atom moves a little slower than the lighter. In the molecule CO_2, for example, molecules with 13C move a little slower than molecules with 12C. Photosynthesizers, in air or in water, take in CO_2 and break it up, building the carbon into their tissues. But since they take in 12C molecules more easily than 13C, carbon that has gone through photosynthesis contains more 12C: it has a ratio of 13C to 12C that is different from the ratio in the CO_2 it came from, skewed toward the lighter carbon isotope. The difference is called isotope fractionation, and can be measured in a mass spectrometer (at around $25 per sample). The isotope fractionation is expressed typically in parts per thousand (or "per mil"). Photosynthetic carbon in the ocean has an isotope fractionation, or $\partial 13C$, of about −20 per mil: the negative sign means that it contains more lighter carbon than "normal".

(This has nothing to do with radiocarbon dating, which is based on the radioactive carbon isotope 14C. The isotopes used in the work described here are nonradioactive or stable isotopes.)

Different organisms operating different reactions may cause a different isotope fractionation. So methanogens, which split CO_2 and make methane, produce a $\partial 13C$ of about −60: in other words, methanogenic methane contains very light carbon. Bacteria that are lithotrophs, oxidizing iron and making an energetic profit, fractionate the iron isotopes they process, thus leaving a chemical trace of their activity in the sediments where that iron oxide is deposited.

Nitrogen isotopes are used to help us to assess ancient diets. If a heterotroph eats the tissue of another creature, it will digest, absorb, and perhaps lay down that nitrogen in organic components of bones or teeth. During that digestion, nitrogen isotopes 15N and 14N are fractionated by about +3 per mil. If that heterotroph is eaten by yet another, $\partial 15N$ increases by another +3 per mil. One can assess the ecology of some extinct animals by using N isotopes.

Earth's Oldest Rocks

The first one-third of Earth's history is called the **Archean** (Fig. 2.6), a time when the early Earth was very different

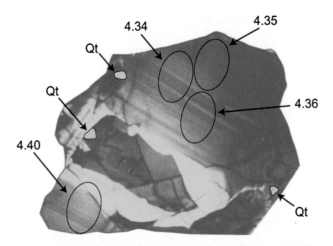

Figure 2.7 "The earliest piece of the Earth." This zircon crystal from Australia has been dated at 4.4 Ga. The crystal is not really blue, but it fluoresces blue in the radiation used to examine it. Image by Professor John Valley of the University of Wisconsin, and used by permission. His zircon Web page is http://www.geology.wisc.edu/zircon/zircon_home.html

from today's planet. There was little or no oxygen in the atmosphere. There was much less life in the seas and none on the land. The Earth was young; its interior was hotter, and its internal energy was greater. Volcanic activity was much greater, but we have no idea whether it was more violent or just more continuous. Very large asteroid impacts smashed into the Earth every 40 million years or so (Kerr 2011), so reconstructing "normal" conditions on the early Earth is difficult.

Rocks older than 3.5 Ga (3500 Ma) are very rare. The oldest minerals on Earth are zircon crystals dated at 4400 Ma (Fig. 2.7), but they have been eroded out of their original rocks and deposited as fragments in younger rocks. These grains do contain evidence that there were patches of continental crust on Earth at or before 4400 Ma (Valley 2005).

This is important because continental crust, dominated by granite, is only found on Earth. Its chemistry is very different from ocean crust, and includes important minerals that release phosphorus and potassium as they are broken down in weathering at the surface. Phosphorus in particular is vital for life (Chapter 1). Continental crust may be yet another unique feature of Earth that encouraged the evolution of life here.

The oldest known rocks on Earth are in northern Canada, but the oldest known *sedimentary* rocks occur in the Isua area of West Greenland, and have been dated by several methods at about 3850 Ma. They have been repeatedly folded, faulted, and reheated, but they can still tell us something about conditions on the early Earth when they were formed. The Isua sediments were laid down in shallow

water along a volcanic shoreline (they include beach-rounded pebbles and weathering products from lava). Temperatures at the time may have been warm, but not extraordinary. So conditions on Earth were hospitable to life by 3800 Ma at the latest, including the fact that there was land as well as ocean (Chapter 1).

The Isua rocks have carbonate rocks in them, and several groups of scientists have examined that carbon. On the face of it, carbon isotope fractionations indicate the activity of life processing that carbon, either photosynthetic or methanogenic. Iron-rich carbonates at Isua have iron isotope fractionations indicating that lithotrophic bacteria processed at least some of the iron. The question is still being debated. However, in younger rocks there would be no argument: the fractionations would be accepted as traces of biological processes, because the alternative hypothesis is more complex.

If these conclusions are correct, then different lineages of cells were flourishing around 3800 Ma when we see the first reasonably well preserved sedimentary rocks on Earth. No cells are preserved: only their chemical traces reveal they were there. At the moment, we have to wait another 300 million years before much better evidence of abundant life is found, including fossil cells.

Stromatolites

Archean rocks are often rich in minerals, and Archean regions have been well explored geologically for economic reasons. The Pilbara region of Northwest Australia is a remote and inhospitable area that originally attracted geological attention because it is rich in the mineral barite. It is now famous for more academic reasons.

Pilbara rocks include the Warrawoona Series, dated to about 3300 to 3550 Ma. The Warrawoona rocks are mainly volcanic lavas erupted in shallow water, or nearby on shore, but there are sedimentary rocks too. The sediments include storm-disturbed mudflakes, wave-washed sands, and minerals formed by evaporation in very shallow pools. The rocks have not been tilted, folded, or heated very much, and the environment can be reconstructed accurately. The rocks formed along shorelines that we can interpret clearly because we can match them to modern environments.

The Warrawoona rocks contain structures called **stromatolites**, which are low mounds or domes of finely laminated sediment (Fig. 2.8). We know what stromatolites are because they are still forming today in a few places. Thus we can study the living forms to try to understand the fossil structures (Fig. 2.9).

Stromatolites are formed by mat-like masses of abundant microbes, usually including photosynthetic cyanobacteria. Stromatolites live today in Shark Bay, Western Australia, in warm salty waters in long shallow inlets along a desert coast (Fig. 2.10). They form from the highest tide level down to subtidal levels, but the higher ones close to

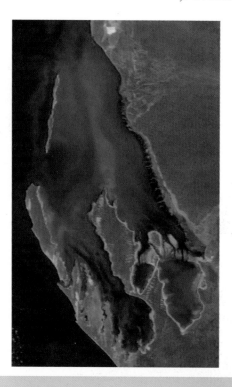

Figure 2.8 The oldest trace fossils on Earth. Cone-shaped stromatolites from the Warrawoona rock sequence in Western Australia. They are found with wave-affected sediments, and therefore formed in very shallow water. Comparing these structures with those forming today (Fig. 2.9), they can be interpreted as having been formed by cyanobacterial mats 3.5 billion years ago. The image shows an eroded surface cutting through the cones. From Macquarie University at http://pilbara.mq.edu.au/wiki

Figure 2.10 Shark Bay World Heritage Site is a series of shallow bays on the coast of Western Australia. Stromatolites form today in the warm salty water close to shore. NASA image taken from space.

Figure 2.9 a) an astonishing panorama of conical stromatolites forming today on the floor of freshwater Lake Untersee in Antarctica. The cyanobacteria grow slowly in light that filters (in summer) through the surface ice. The tallest cones are 50 cm high. Image from Andersen et al. 2011, © 2011 Dale T. Andersen, all rights reserved; used by permission. b) the oldest cyanobacteria on Earth, conical forms in the Warrawoona rocks of Western Australia, 3.5 billion years old. The cones are about 7 cm high. Image courtesy of Dr. Abigail Allwood, used by permission.

shore have been better studied (sea snakes, not sharks, are the problem).

The cyanobacteria that grow and photosynthesize in Shark Bay so luxuriantly thrive in water that is too salty for grazing animals such as snails and sea urchins that would otherwise eat them. Like most bacteria, they secrete slime, and can also move a little in a gliding motion. Sediment thrown up in the waves may stick to the slime and cover up some of the bacteria. But they quickly slide and grow through the sediment back into the light, trapping sediment as they do so. As the cycle repeats itself, sediment is built up under the growing mats. Eventually the mats grow as high as the highest tide, but cannot grow higher without becoming too hot and dry. Some mats harden because the photosynthetic activity of the bacteria helps carbonate to precipitate from seawater, binding the sediment into a rocklike consistency that resists wave action (Fig. 2.11). However, sediment stabilization in stromatolites today works best in the light. Stromatolites placed experimentally in the dark lose stability (Paterson et al. 2008). This implies that all stromatolites, ancient and modern, formed and built rock-like trace fossils through photosynthesis.

Some cyanobacterial mats are so dense that light may penetrate only 1 mm. The topmost layer of cyanobacteria absorbs about 95% of the blue and green light, but just underneath is a zone where light is dimmer but exposure to UV radiation and heat is also less. Green and purple bacteria live here in huge numbers and also contribute to the growth of the mat. Deeper still in the mat, light is too low for photosynthesis, and there heterotrophic bacteria absorb and process the dying and dead remains of the bacteria above them. Oxygen diffuses down into the mat from above, and sulfide diffuses upward from the zone below, creating an extraordinary zone where chemistry can change within minutes and within millimeters.

Night follows day, of course, and photosynthesis stops at night. The oxygen in the top layers of the stromatolite is quickly lost. Sulfide dominates the night-time hours, oxygen dominates the daylight hours, and all the bacteria must be able to adjust quickly to the daily change. The internal chemistry in stromatolites is as complex as the mix of bacteria. There is no reason to suppose that ancient stromatolites were any different.

Ancient Stromatolites

Stromatolites are trace fossils. They are formed by the action of living cells, even if those cells are hardly ever preserved in them as fossils. Because they are large, and because their distinctive structure makes them easy to recognize, stromatolites are the most conspicuous fossils for three billion years of Earth history, from about 3500 Ma to the end of the Proterozoic at about 550 Ma. They are rare in Archean rocks, probably because there were few clear, shallow-water shelf environments suitable for stromatolite growth at the time. The few Archean land masses were volcanically active, generating high rates of sedimentation that probably inhibited mat growth in many shoreline environments. Even so, stromatolites flourished in Australia and South Africa around 3430 Ma (Fig. 2.8), and locally covered miles of shoreline (Fig. 2.12).

Figure 2.11 Stromatolites are forming today in warm salty water along the shore line of Shark Bay, Western Australia. Close study of these modern structures allows us to interpret the stromatolites from Warrawoona (Fig. 2.8) as early trace fossils formed by mats of cyanobacteria. Image from Macquarie University at http://pilbara.mq.edu.au/wiki

Figure 2.12 Abigail Allwood spent three field seasons working in the Pilbara region of Western Australia, studying Archean stromatolites. She mapped a 10-km (6-mile) stretch of them along an ancient shoreline, in a rock formation that is now called "Abby's Reef". This is good evidence that early stromatolites were locally abundant, on the kind of scale we see today at Shark Bay (Allwood et al. 2006). Here Dr. Allwood imagines the Archean landscape from a convenient vantage point: a stromatolite outcrop of "her" reef. Photo credit A. C. Allwood. Image from Macquarie University at http://pilbara.mq.edu.au/wiki

Solar UV radiation was intense in Archean time, with no oxygen (or ozone layer) in the atmosphere. A shield of perhaps 10 m (30+ feet) of water might have been needed to prevent damage to a normal early cell by UV radiation. However, the early evolution of the stromatolitic way of life by cyanobacteria may have been a response to UV radiation. With light (and UV) penetrating only a little way into the mat, bacteria were able to live essentially at the food-rich water surface without damage from UV. Cyanobacteria were not just existing at Warrawoona: they were already modifying their microenvironment for survival and success. Stromatolites were not just the sites of simple microbial mats, but were complex miniature ecosystems teeming with life.

Identifying Fossil Cells in Ancient Rocks

Chert is a rock formed of microscopic silica particles (SiO_2). It does not form easily today because all kinds of organisms, including sponges, take silica from seawater to make their skeletons. But silica-using organisms had not evolved in Archean times, so cherts are often abundant in Archean rocks. As chert forms from a gel-like goo on the seafloor, it may surround cells and impregnate them with silica, preserving them in exquisite detail as the silica hardens into chert. Once it hardens, chert is watertight, so percolating water does not easily contaminate the fossil cells.

Several processes can generate inorganic blobs of chemicals in rocks, and blobs in chert have often been mistaken for fossil cells. But cell-like structures in the Apex Chert, a formation in the Warrawoona rocks dated at 3465 Ma, are genuine Archean cells (Fig. 2.13). This is hardly surprising,

given that locally huge areas of the Warrawoona coastline were covered with stromatolites at about this time (Figs 2.8, 2.12).

These earliest stromatolites formed around 3430 Ma (Fig. 2.8). By 3.1 Ga, there were two distinctly different styles of bacterial mat; by 2.9 Ga bacterial mats were forming on soft silty sea-floors, and by 2800 Ma stromatolites are known from salt-lake environments as well as oceanic shorelines. A diverse set of cells is known from cherts in the Pilbara at 3000 Ma. Bacterial mats were abundant and varied by Late Archean time.

There were important geological changes at the end of the Archean, which is dated at 2500 Ma. The Earth had cooled internally to some extent, and the crust was thicker and stronger. The thicker crust affected tectonic patterns: the way the crust moves, buckles, and cracks under stress. Continents became larger and more stable in Early Proterozoic times, with wide shallow continental shelves that favored the growth and preservation of stromatolites. Most Proterozoic carbonate rocks include stromatolites, some of them enormous in extent. Proterozoic stromatolites evolved new and complex shapes as bacterial communities became richer and expanded into more environments.

Banded Iron Formations: BIF

From the beginning of the Archaean (around 3800 Ma) we find increasing accumulations of a peculiar rock type. **BIF** or **Banded Iron Formations** are sedimentary rocks found mainly in sequences older than 1800 Ma. The bands are alternations of iron oxide and chert (Fig. 2.14), sometimes repeated millions of times in microscopic bands (Fig. 2.15). No iron deposits like this are forming now, but we can

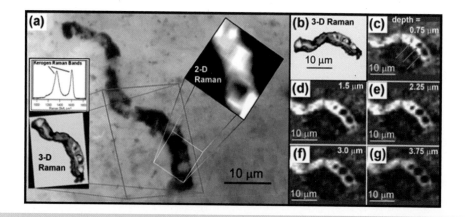

Figure 2.13 This fossil is *Primaevifilum*, from the Apex Chert in the Warrawoona rocks of Western Australia, dated about 3465 Ma. A photograph down a microscope a) shows a rather blurry outline, which caused unnecessary vicious criticism suggesting that it was not a cell and not a fossil. However, new techniques for imaging the fossil show its internal structure and 3D shape, confirming it as the earliest fossil cell so far found. It looks like a cyanobacterium, but that is hard to prove with our current technology. Image from Schopf and Kudryavtsev 2010; courtesy Professor J. W. Schopf.

Figure 2.14 Block of banded iron. Image by André Karwath, and placed into Wikimedia.

5.0 mm

Figure 2.15 Close up of banding in BIF specimen from the Proterozoic of Michigan. Scale bar is 5 mm. Photograph by Mark Wilson of the College of Wooster, and placed into Wikimedia.

make intelligent deductions about the conditions in which BIF were laid down.

The chemistry of seawater on an Earth without oxygen differed greatly from today's situation. Today there is practically no dissolved iron in the ocean, but iron dissolves readily in water without oxygen. Even today, in oxygen-poor water on the floor of the Red Sea, iron is enriched 5000 times above normal levels. So Archean oceans must have contained a great deal of dissolved iron as well as silica.

Silica would have been depositing more or less continuously on an Archean seafloor to form chert beds, especially in areas that did not receive much silt and sand from the land. But iron oxide can only have precipitated out of sea-

water in such massive amounts by a chemical reaction that included oxygen.

Therefore, to form the iron oxide layers in BIF, there must have been occasional or regular oxidation events to produce iron ore, against a background of regular chert formation. Between oxidation events, dissolved iron was replenished from erosion down rivers or from deep-sea volcanic vents. What were these oxidation events, and what started them? The most likely hypothesis to explain BIF formation calls on seasonal changes in sunlight and temperature that in turn affect bacterial action and mineral deposition.

Lake Matano, on the Indonesian island of Sulawesi, gives us an idea of how an Archaean ocean may have worked (Crowe et al. 2008). The lake is small but very deep. The tropical climate means that the surface waters are always warm, so they never sink or mix with the deeper water below. The surface waters are well-lit, and floating cyanobacteria photosynthesize there. But the lake water has few nutrients, so these surface bacteria are not important, except that they keep the surface layers of the lake oxygenated.

Below the surface layer is water with no oxygen, rich in dissolved iron, just as we imagine the Archaean ocean to have been. Sunlight reaches the top layer of this deeper water, but the light is too dim for cyanobacteria to photosynthesise. Instead, huge numbers of green sulfur bacteria, with a variety of chlorophyll that works better in dim light, operate lithotrophy (Chapter 1) about 120 meters down in Lake Matano. They break down water, use the oxygen to oxidise the dissolved iron, and make a metabolic profit from the reaction. Oxidized iron sinks to the lake floor in large quantities.

Laboratory experiments suggest that in this situation, iron oxidation would work fastest in a narrow temperature range, with silica depositing faster at higher or lower temperatures. Using the evidence from Lake Matano, it looks as if sulfur bacteria could also have formed the alternating mineral bands in Archaean BIF. Although they involve oxidation, the reactions occur in environments without oxygen, and they do not produce any. Since sulfur bacteria are very ancient, there is no problem in suggesting that they were involved in producing the first BIF in the Isua rocks at 3800 Ma.

In an extensive Archaean ocean, rather than a small tropical lake like Lake Matano, we would expect that iron oxidation would occur over a large area. Indeed, BIF were often deposited in bands that can be traced for hundreds of kilometers.

Today we probably see only a small fraction of the BIF that once formed on Archean sea floors, because most ocean crust has since been recycled back into the Earth. But even the amounts remaining are staggering. BIF make up thousands of meters of rocks in some areas and they contain by far the greatest deposits of iron ore on Earth. At least 640 billion tonnes of BIF were laid down in the early Proterozoic between 2500 Ma and 2000 Ma (that's an average of half a million tonnes of iron per year). [The

Figure 2.16 An enormous iron mine in BIF of the Pilbara region of Western Australia, abandoned in 2008. Photograph by Philipist, and placed into Wikimedia.

metric tonne that is used internationally is 1000 kilograms, very close to an American ton.] The Hamersley Iron Province in Western Australia alone contains 20 billion tonnes of iron ore, with 55% iron content. At times, iron was dropping out in that basin at 30 million tonnes a year. Most modern steel industries are based on iron ores laid down in BIF during that time (Fig. 2.16).

BIF, Stromatolites and Oxygen

The current best hypothesis for forming BIF requires oxygen-free ocean water with dissolved iron, at least below the surface. However, cyanobacteria in stromatolites were producing free oxygen in the Archaean, starting with the first major stromatolites around 3500 Ma.

Cyanobacteria evolved oxygen dismutase as an antidote to oxygen poisoning (Chapter 1), and that gave them the opportunity to control and then use that oxygen in a new process, **respiration** (biological oxidation). Fermenting sugars leaves byproducts such as lactic acid that still have energy bonded within them. By using oxygen to break those byproducts all the way down to carbon dioxide and water, a cell can release up to 18 times more energy from a sugar molecule by respiration than it can by fermentation.

Cyanobacteria can photosynthesize in light and respire in the dark. To do this, they must be able to store oxygen in a stable, nontoxic state for hours at a time. Most likely, they began to use oxygen in respiration very early: the energy advantages are astounding. The early success of cyanobacteria probably reflects their access to an abundant and reliable energy supply in two different ways: first in photosynthesis, and second, by breaking down food molecules by respiration rather than fermentation.

However, that success poses a problem. If cyanobacteria made such an important biochemical breakthrough, using a process that produces oxygen as a by-product, why did the ocean and atmosphere not become oxygenated quickly? Stomatolites were locally abundant by 3500 Ma, producing "whiffs of oxygen" (as one researcher has written). However, stromatolites lived only along shallow shorelines, and early continents were small. So any effects of free oxygen would at first have been local rather than global.

Nevertheless, even whiffs of oxygen can be detected by skilful geochemists. Molecules called steranes can only be produced in reactions that use free molecules of oxygen: and steranes have been detected in rocks at 2720 Ma (Waldbauer et al. 2011).

Stromatolites increased dramatically at about 2500 Ma, along with the formation of larger continents and more shallow-water habitat. BIF production reached a peak around that time, too, coinciding with massive volcanic eruptions that must have vastly enriched dissolved oxygen supplies in seawater.

By 2500 Ma, the official beginning of the Proterozoic, Earth's surface chemistry had been changed by life for a billion years. Cyanobacteria in stromatolites, and probably in surface waters everywhere, were producing waves of oxygen large enough to oxygenate large areas of shallow ocean waters, especially along continental shores. Green sulfur bacteria were forming huge masses of BIF in anoxic iron-rich waters in the oceans, and Archaea were producing methane. Yet the atmosphere had practically no oxygen until about 2300 Ma.

There are many ways that oxygen can be used up before and after it is produced in photosynthesis. Today, only about 5% of all the oxygen produced in photosynthesis reaches the atmosphere and ocean. The other 95% reacts with iron and sulfur compounds to form iron oxides and sulfates. These and similar reactions must have used up almost every oxygen molecule produced on the early Earth, too, and any surplus oxygen would have used up to oxidize organic molecules in the water. It could easily have taken a billion years before free oxygen began to accumulate in air and water on a global scale.

In addition, photosynthesis (and oxygen production) may have been a lot slower than we might imagine. Photosynthesis doesn't just need light, water, and carbon dioxide: the plants or bacteria that operate it need nutrients as well. Phosphorus is a limiting nutrient in many environments even today (many of our fertilizers contain phosphorus).

Phosphorus comes largely from continental crust, and Archaean continents were small. So phosphorus supplies may have been limited, especially out in the vast expanses of Archaean oceans. In addition, iron minerals absorb some phosphorus as they form, so forming BIF would have used up (and locked up) a lot of phosphorus as well as a lot of iron. Lack of phosphorus would then have slowed cyanobacterial growth, which would have slowed photosynthesis and oxygen production, until renewed weathering and erosion brought new iron and phosphorus supplies to ocean water. Altogether, this would have dramatically slowed the oxygenation of Earth.

In the end, however, the supply of oxygen became large enough that free oxygen began to accumulate in oceans and atmosphere, setting the stage for dramatic changes in Earth's surface chemistry, and its life.

The Great Oxidation Event

BIF production rose to a peak around 2500 Ma, then fell off, with only occasional bursts of activity over the next billion years. Other geological evidence confirms that the ocean surface waters and the atmosphere was oxygenated early in the Proterozoic. The uranium mineral **uraninite** cannot exist for long if it is exposed to oxygen, and it is not found in rocks younger than about 2300 Ma. The sulfur isotopes in rocks suggest that sulfate levels rose in the ocean, lowering methane production by Archaea, while the methane that they continued to produce was quickly broken down by free oxygen. More complex indicators of ancient oxygen levels come from isotope changes in a number of metals, but there is a general agreement on the timing. The great change in surface oxygen levels occurred between 2400 and 2000 Ma, with periods of slow change and periods of rapid change. Oxygen levels rose from perhaps one-millionth of their present atmospheric level, to 1% by 2000 Ma (Kump et al. 2011). In turn, the drop in methane levels in the atmosphere cooled the Earth, and there is evidence of a very large ice age between 2400 and 2200 Ga.

Once oxygen was part of the atmosphere, it would have rusted any iron minerals exposed on the land surface by weathering. Rivers would have run red over the Earth's surface before vegetation invaded the land. On land and in shallow seas, **red beds**, sediments bearing iron oxides, date from about 2300 Ma (Figs. 2.17, 2.18).

Photosynthesis produces oxygen only in surface waters, because that is as far as usable light penetrates water. Surface waters are warmer than deeper layers, so are less dense and tend to stay at the surface. The deep waters of the ocean receive no oxygen directly. In today's oceans, oxygen-rich surface waters can sink, but only if they are unusually dense: if they are very cold, for example, or if they are very salty, or both. (Examples are polar seas—in the North Atlantic and around Antarctica—or hot shallow salty tropical seas such as the Persian Gulf.)

In some seas today, the surface waters do not sink, so there is no oxygen, and little life, below the surface layers. The Black Sea is the best-known example, but the Red Sea also has deep basins that lack oxygen.

Today, enough surface water sinks to carry oxygen to most of the world ocean. But it would have been different in the Proterozoic. Clearly, the surface waters would have become oxygen-rich before the bulk of the ocean did. And the atmosphere, which can exchange gases with the surface waters, would also have become oxygen-bearing before the deep ocean did. We can imagine a Proterozoic world that

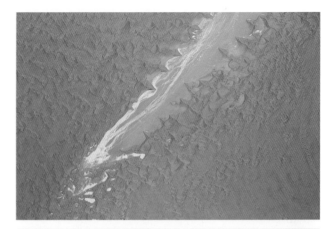

Figure 2.17 Image taken by an alien spacecraft of the Earth's Proterozoic land surface shortly after the Great Oxidation Event. I tell a lie: alien spacecraft don't exist. This is actually an image of red sand dunes in the Namib Sand Sea of Namibia, taken from the International Space Station. NASA Earth Observatory photograph.

Figure 2.18 Red beds have been common rocks on the Earth since the Great Oxidation Event around 2300 Ma. This splendid outcrop of red sandstone in Namibia is called Lion's Head. Photograph by Violet Gottrop, and placed into Wikimedia.

had free oxygen only in surface waters and the atmosphere. The deep ocean would still have been **anoxic**, rich in dissolved iron and silica and sulfide, and inhabited by bacteria and methanogens, while the surface waters had photosynthesizers and oxygen-tolerant microbes.

BIF are rare after 2300 Ma, as oxygen levels in the surface waters extended deeper, driving iron-rich water deeper, and sulfur bacteria so deep that they had no light to form BIF. BIF could form only in rare isolated basins, like the famous ones in Michigan, or at times of crisis when iron-rich waters extended upward into shallow depths.

The term that summarizes all the chemical, geological, and biological changes around 2300 Ma is **The Great Oxidation Event**. (The latest and most precise estimate is 2316 ± 7 Ma) (Konhauser et al. 2011). The surface chemistry of Earth's air, land, and water changed forever, and one indirect result was vital for the further evolution of living things. Solar UV radiation acts on any free oxygen high in the atmosphere to produce ozone, which is O_3 rather than O_2. Even a very thin layer of ozone can block most UV radiation. Earth's surface, land and water, has been protected from massive UV radiation ever since free oxygen entered the atmosphere. It then became possible for organisms to evolve that were more complex than Bacteria or Archaea: the eukaryotes (Chapter 3).

Further Reading

Allwood, A. C. et al. 2006. Stromatolite reef from the Early Archaean era of Australia. *Nature* 441: 714–718. Available at http://spacewardbound.nasa.gov/australia2011/resources/allwood%20etal%202006%20pilbara%20stromatolites.pdf

Andersen, D. T. et al. 2011. Discovery of large conical stromatolites in Lake Untersee, Antarctica. *Geobiology* 9: 280–293. Available at http://150.214.110.65/gtb/sod/usu/$UCOGrepositorio/13101114_13101114.pdf

Arndt, N. T. and E. G. Nisbet. 2012. Processes on the young Earth and the habitats of early life. *Annual Review of Earth & Planetary Sciences* 40: 521–549.

Bosak, T. et al. 2009. Morphological record of oxygenic photosynthesis in conical stromatolites. *PNAS* 106: 10939–10943. [More evidence of local oxygen production before the Great Oxidation Event.] Available at http://www.pnas.org/content/106/27/10939. full

Craddock, P. R. and N. Dauphas. 2011. Iron and carbon isotope evidence for microbial iron respiration throughout the Archean. *Earth and Planetary Science Letters* 303, 121–132. [Microbes were respiring iron at 3.8 Ga.] Available from http://originslab.org/media/articles_library/48_Craddock_Dauphas_EPSL_2011.pdf

Crowe, S. A. et al. 2008. Photoferrotrophs thrive in an Archean Ocean analogue. *PNAS* 105: 15938–15943. [Lake Matano.] Available at http://www.pnas.org/content/105/41/15938.full

Hazen, R. M. 2010. How old is Earth, and how do we know? *Evolution: Education and Outreach.* http://www.springerlink.com/content/uh78u254754ntm17/fulltext.pdf

Javaux, E. et al. 2010. Organic-walled microfossils in 3.2-billion-year-old shallow-marine siliciclastic deposits. *Nature* 463: 934–938. Available at http://web.mac.com/redifiori/Russell_Di_Fiori/Eukaryotes_files/3.2%20billion%20year%20old%20eukaryote.pdf

Kato, Y. et al. 2009. Hematite formation by oxygenated groundwater more than 2.76 billion years ago. *Earth and Planetary Science Letters* 278: 40–49.

Kendall, B. et al. 2010. Pervasive oxygenation along late Archaean ocean margins. *Nature Geoscience* 3: 647–652. [Oxygen oases in the late Archaean.] Available at http://www.geol.umd.edu/~kaufman/pdf/Kendal10_NatGeosci.pdf

Kerr, R. A. 2011. Asteroid model shows early life suffered a billion-year battering. *Science* 332: 302–303. Available at http://lucidthoughts.com.au/wordpress/wp-content/uploads/2011/09/Science-2011-Kerr-302-3.pdf

Konhauser, K. O. et al. 2011. Aerobic bacterial pyrite oxidation and acid rock drainage during the Great Oxidation Event. *Nature* 478: 369–373. Available at http://xa.ylmg.kq/groups/18383638/1745097634/name/nature10511.pdf

Kump, L. R. et al. 2011. Isotopic evidence for massive oxidation of organic matter following the Great Oxidation Event. *Science* 334: 1694–1696.

Paterson D. M. et al. 2008. Light-dependent biostabilisation of sediments by stromatolite assemblages. *PLoS ONE* 3(9): e3176. Available at http://www.plosone.org/article/info%3Adoi%2F10.1371%2Fjournal.pone.0003176

Schopf, J. W. and A. B. Kudryatsev. 2010. A renaissance in studies of ancient life. *Geology Today* 26: 140–145.

Valley, J. W. 2005. A cool early Earth? *Scientific American* 293 (4): 58–65. Available at http://www.geology.wisc.edu/zircon/Valley2005SciAm.pdf

Waldbauer, J. R. et al. 2011. Microaerobic steroid biosynthesis and the molecular fossil record of Archean life. *PNAS* 108: 13409–13414. Available at http://anpron.eu/wp-contentuploads/2011/08/Microaerobic-steroid-biosynthesis-and-the-molecular-fossil-record-of-Archean-life.pdf

Questions for Thought, Study, and Discussion

1. Explain the difficulty of finding the age of a rock bed by the fossils in it, and at the same time finding its age by radiometric methods.

2. If you were able to watch the Earth from a spacecraft during Earth's first two billion years of existence, how would you tell that life had evolved on the planet? This is not necessarily the time when life evolved: it's the time when an observer in space could detect it. The spacecraft has reasonable instruments for remote sensing.

THREE

Sex and Nuclei: Eukaryotes

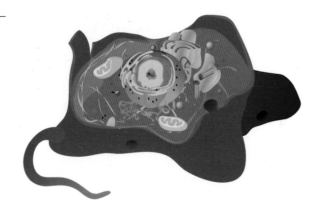

In This Chapter

For the first half of Earth history, life consisted of prokaryotes (archaeans and bacteria), but the evolution of eukaryotes (cells with nuclei) changed the biological world forever. Eukaryotes have complex cells resulting from the combination of one or more prokaryotic cells in an interdependent relationship called symbiosis, and most of them also have sexual reproduction. These are major steps in evolution, and I discuss why and how they happened. Eukaryote complexity evolved into multicellular complex plants and animals, and I describe how we classify these organisms to reflect their evolutionary heritage.

Single-Celled Life

The microbes that were Earth's first life evolved into two different major groups or domains: Archaea and Bacteria. They shared much the same body plan, however, and we group them together as prokaryotes (Fig. 3.1). A third domain of life, the Eukarya, contains all other living organisms, and has a distinctly more complex body structure (Fig. 3.2, Box 3.1). Eukaryotes evolved after the other two, so they must have had ancestors that were some form of prokaryote.

Prokaryotes were and are very successful in an incredible range of habitats, from stinking swamps to the hindgut of termites and from hot springs in the deep sea to the ice desert of Antarctica, and deep in rocks underground. They occur in numbers averaging 500 million per liter in surface ocean waters, 1 billion per liter in fresh water, and about 300 million on the skin of the average human. The diversity of prokaryotes makes it difficult to select a plausible ancestor of eukaryotes among them, and many alternatives have been suggested!

Eukaryotes today are larger and much more complex than prokaryotes. Their DNA is contained in a nucleus with a membrane around it, and the eukaryotic cell also contains organelles, each one wrapped in a membrane, that performs functions in the cell. However, it is clear from

History of Life, Fifth Edition. Richard Cowen.
© 2013 by Richard Cowen. Published 2013 by Blackwell Publishing Ltd.

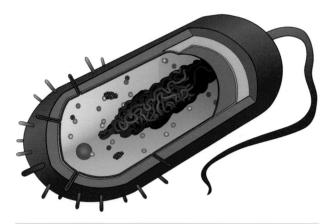

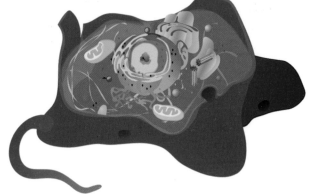

Figure 3.1   A prokaryotic cell. The DNA (red cords) is twisted and folded to fit into the cell and floats free in the cell cytoplasm (blue). This prokaryote is mobile, and propelled by a flagellum. Image by Mariana Ruiz Villareal, Lady of Hats, and placed into Wikimedia.

Figure 3.2   A eukaryotic cell (more specifically, a protist). A cell membrane (red) surrounds the cell's cytoplasm (orange). Among the cell contents are pill-shaped mitochondria (lime green), and a nuclear membrane (dark green) surrounding the nucleus itself, which contains DNA. This motile cell is driven by a flagellum which has a different structure from its prokaryotic equivalent. Image by Mariana Ruiz Villareal, Lady of Hats, and placed into Wikimedia.

Box 3.1   Differences Between Prokaryotes and Eukaryotes

(1)   Eukaryotes have their DNA contained inside a membrane, and under the microscope this package forms a distinct body called the nucleus. Prokaryotes have their strands of DNA loose in the cell cytoplasm.

(2)   Prokaryotes have no internal subdivisions of the cell, but almost all eukaryotes have organelles as well as a nucleus. Organelles are subunits of the cell that are bounded by membranes and they perform some specific function in the cell. Plastids, for example, perform photosynthesis inside the cell, generating food molecules and releasing oxygen. Mitochondria contain the respiratory enzymes of the cell. Food molecules are first fermented in the cytoplasm, then passed to the mitochondria for respiration. Mitochondria generate ATP as they break food molecules down to water and CO2, and they pass energy and waste products to the rest of the cell. They also make steroids, which help to form cell membranes in eukaryotes and give them much more flexibility than prokaryote membranes.

(3)   Eukaryotes can perform sexual reproduction, in which the DNA of two cells is shuffled and redealt into new combinations.

(4)   Prokaryotes have rather inflexible cell walls, so they cannot easily engulf other cells. The flexibility of eukaryotic cell membranes allows them to engulf large particles, to form cell vacuoles, and to move freely. Plant cells, armored by cellulose, are the only eukaryotes that have given up a flexible outer cell wall for most of their lives.

(5)   Eukaryotes have a well-organized system for duplicating their DNA exactly into two copies during cell division. This process, mitosis, is much more complex and precise than the simple splitting found in prokaryotes.

(6)   Eukaryotes are almost always much larger than prokaryotes. A eukaryote is typically ten times larger in diameter, which means that it has about 1000 times the volume of a prokaryote.

(7)   Eukaryotes have perhaps a thousand times as much DNA as prokaryotes. They have multiple copies of their DNA, with much repetition of sequences. The DNA content of prokaryotes is small, and they have only one copy of it. There is little room to store the complex "IF . . . THEN . . ." commands in the genetic program that turn on one gene as opposed to another. Therefore, genetic regulation is not well developed in prokaryotes, which means that they cannot produce the differentiated cells that we and other eukaryotes can. Multicellular colonies of bacteria are all made up of the same cell type, repeated many times in a clone. Therefore, any species of bacterium is very good at one thing but cannot do others; its range of functions is narrow.

their structure and the fact that they contain DNA that these organelles were once cells in their own right, but are now living as part of the eukaryote.

Symbiosis and Endosymbiosis

Symbiosis is a relationship in which two different organisms live together. Often they both get some benefit from the arrangement. Examples range from the symbiosis between humans and dogs to bizarre relationships such as that of acacia plants, which house and feed ant colonies that in turn protect the acacia against herbivorous animals and insects. The ultimate state of symbiosis is endosymbiosis, in which one organism lives inside its partner. Animals as varied as termites, sea turtles, and cattle can live on plant material because they contain bacteria in their digestive system with the enzymes to break down the cellulose that is unaffected by the host's own digestive juices. Many tropical reef organisms have symbiotic partners in the form of photosynthesizing microorganisms. Living inside the tissues of corals or giant clams, these symbiotic partners have a safe place to live. In turn, the host receives a share of their photosynthetic production.

It is now clear that endosymbiosis was a critical step in the evolution of eukaryotes. A prokaryote made a dramatic evolutionary breakthrough: it took in "foreign" prokaryotes that came to live inside it (as endosymbionts). The partnership became permanent, and we now recognize the host as a eukaryotic cell that contains internal partners (called organelles). Mitochondria are organelles that perform respiration inside a eukaryote, oxidizing "food" molecules and releasing the energy to the host cell. Plastids perform photosynthesis inside what we now call "plant" cells: the plastids contain all the chlorophyll in the cell, and turn light energy into "food" for the cell. Flagella may once have been mobile bacteria, but fixed to the host cell, they can move it around.

Once free-living bacteria, these organelles are now so closely integrated into a host cell that they are for practical purposes part of it (Fig. 3.2). At least five major pieces of evidence show that organelles (and therefore eukaryotes) originated by endosymbiosis (Box 3.2).

Mitochondria and their Ancestors

I have chosen to describe one scenario for the origin of eukaryotes, out of the many that have been suggested. It is a simplified version taken from two recent papers (Gross and Bhattacharya 2010, Cotton and McInerney 2010). I prefer it because it fits well with the ecology and environment of stromatolites. It begins with oxygen, and it proposes that two species of closely packed prokaryotes evolved symbiotic relationships to deal with the toxic effects of the oxygen produced by cyanobacteria.

Early cyanobacteria released periodic whiffs and waves of oxygen into the shallow water where they lived, as a by-

Box 3.2   Evidence for Organelle/Eukaryote Symbiosis

(1)   The DNA in mitochondria and plastids is not the same as the DNA in the eukaryotic cell nucleus.

(2)   Mitochondria and plastids are separated from the rest of the eukaryotic cell by membranes; thus they are really "outside" the cell. The cell itself makes the membrane, but inside it is a second membrane made by the organelle.

(3)   Plastids, mitochondria, and prokaryotes make proteins by similar biochemical pathways, which differ from those in the cytoplasm of eukaryotes.

(4)   Mitochondria and plastids are susceptible to antibiotics such as streptomycin and tetracycline, like prokaryotes; eukaryotic cytoplasm is not affected by these drugs.

(5)   Mitochondria and plastids can multiply only by dividing; they cannot be made by the eukaryotic cell. Thus organelles have their own independent reproductive mechanism. A cell that loses its mitochondria or plastids cannot make any more.

product of their photosynthesis. In doing so, they affected the massive populations of other prokaryotes that were living immediately next to them, and feeding from the rich organic glop provided by the protective slime of the cyanobacteria, and their dead and dying cells. Oxygen levels changed from hour to hour, day to night, and from season to season, and would have placed intense chemical stresses on all these prokaryotes. Oxygen tolerance and then oxygen use probably evolved in prokaryotes in these microenvironments. Stromatolites must have been forcing houses of evolution.

Normal oxygen, O_2, is only mildly toxic, but ultraviolet radiation (UV) can turn it into O_3, ozone, or hydrogen peroxide, H_2O_2 which are both highly toxic. The cyanobacteria themselves had evolved the enzyme SOD, superoxide dismutase, to capture oxygen inside the cell and release it safely to the outside.

The prokaryotes living close to cyanobacteria would have been exposed to UV, with the possibility of lethal damage if internal oxygen was turned into a toxic compound. Aerobic bacteria can tolerate oxygen, largely by using it up as they break down organic molecules in the process of respiration. But archaeans can only just tolerate oxygen. It makes sense that such an archaean would gain an advantage living very close to an aerobic bacterium that was absorbing oxygen for its respiration, and as an unintended consequence, made life safer for its less tolerant neighbors. In this scenario, both species routinely succeeded better when they lived side by side in very close contact—in external symbiosis. Eventually, the archaeans took the bacteria inside their tissues (without digesting them), where the bacteria continued to reproduce and became oxygen collectors and internal symbionts for the cell. The bacteria were protected from external stresses, and as they respired food molecules they released some energy to their archaean hosts. Later, the bacteria lost their cell walls and became organelles (mitochondria), and the host was no longer a simple archaean prokaryote but a true eukaryotic cell.

These early eukaryotes now received so much energy from the respiration of their mitochondria that they came to depend entirely on them to provide them with ATP. The number of mitochondria had to be matched closely to the needs of the host cell, so the genes that controlled mitochondrial reproduction were transferred away from the mitochondria and packaged into the host's DNA, inside a nucleus, leaving behind in the mitochondria (as far as we can tell) mainly the genes that control the oxidation they perform for the cell. Now as eukaryotic cells grew and flourished, so did their mitochondria. As a eukaryote divided, each daughter cell took some mitochondria with it.

This scenario produces protists (Fig. 3.2), single-celled eukaryotes, capable of moving, and feeding by engulfing other organisms. Their food is fermented in the cytoplasm and oxidized in mitochondria. The same process occurs in our cells today.

The genetics of all mitochondria are so alike that they likely descended from one single ancestor. In other words, the symbiosis that powers all eukaryotic cells evolved only once.

In another major evolution of cell symbiosis that also happened only once, an early protist took in cyanobacteria as symbiotic partners which became plastids (Fig. 3.3). The cyanobacteria benefited more from nutrients in the host's wastes than they would as independent cells. In time, the protist came to rely so much on the photosynthesis of its partners that it gave up hunting and engulfing other cells, gave up locomotion, grew a strong cellulose cell wall for protection, settled or floated in well-lit waters, and took on the way of life that we now associate with the word plant. Plant photosynthesis is not performed in the cell cytoplasm, but only in the plastids.

This scenario produced the first eukaryotic photosynthesizers (algae) (Figs 3.2, 3.3). Since that event, the plant-animal dichotomy has been one of the most important in the organic world. We rank advanced plants and advanced animals as two different kingdoms. Animals eat plants and one another.

These two symbiotic events can now be traced from the genetics of the symbionts themselves. All chloroplasts in plant cells contain very similar DNA, which is very similar to the DNA of the cyanobacteria that were their ancestors. All mitochondria contain DNA that is very similar to the bacterial lineage they evolved from. And eukaryote nuclei contain an astonishing level of archaean DNA, even though it is now accompanied by DNA that the host cell took from

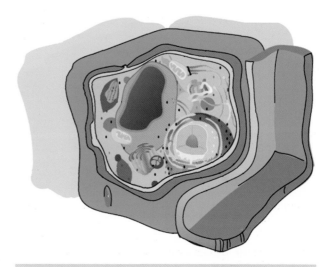

Figure 3.3 A plant cell. Here the cell membrane (green) surrounds the cell cytoplasm (blue), which contains not only the nucleus (red) and mitochondria (pill-shaped, lime-green), but also chloroplasts, organelles that once were free-living cyanobacteria (green). Image by Mariana Ruiz Villareal, Lady of Hats, and placed into Wikimedia.

the organelles. In all eukaryotes, the archaean genes are more important than the bacterial genes.

The word **symbiogenesis** is used to describe the appearance of a dramatically new biological or ecological ability by symbiosis rather than simple mutation. It is a useful term, because we are discovering more and more examples in living ecosystems. For example, many plants are successful because they (must) have symbiotic fungi in or around their roots, which help to break down soil debris and make it available to their plant partners: in turn, they take some nutrition from the plant roots. Nevertheless, although symbiogenesis seems dramatic, it is a normal part of evolution. Species must acquire the mutations that allow them to take part in symbioses. Each species in the symbiosis continues to evolve under natural selection, and individuals that take part in symbioses do so because they reproduce more effectively than those that do not. That is as true for the partners in a eukaryote cell as it is in the later examples of symbiosis. Doctors are increasingly aware that a number of severe human diseases are caused by harmful mutations in the DNA of our mitochondria.

Eukaryotes in the Fossil Record

The endosymbiotic theory for the origin of eukaryotes is based on biological and molecular evidence (Box 3.2). But it is difficult to identify the first fossil eukaryotes. Most fossil cells are small spherical objects with no distinguishing features. Most eukaryotes are much larger than

Figure 3.4   A block of rock showing fossils of *Grypania spiralis*, from the Negaunee Iron Formation of Michigan. *Grypania* is probably the earliest multicellular alga, with fronds about 1 mm across. Photograph courtesy of Dr. James St. John of Ohio State University, Newark.

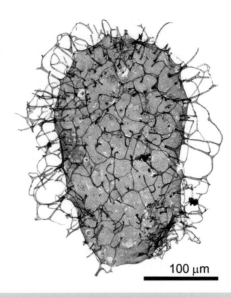

100 μm

Figure 3.5   *Tappania*, an undoubted eukaryote. This specimen dates from about 1400 Ma. It is called an "acritarch" to signify that despite its excellent preservation, we are not sure which group of eukaryotes it belongs to. Image courtesy of Nicholas Butterfield of Cambridge University, who thinks Tappania is a fungal cell (Butterfield 2005).

prokaryotes, but at least one living prokaryote approaches normal eukaryotic size. Experiments to make artificial fossils from rotting prokaryotes have shown that it is almost impossible to distinguish them from eukaryotes after death. After death, the cell contents of prokaryotes can form blobs or dark spots that look like fossilized nuclei or organelles. Rotting colonies of cyanobacteria can look like multicellular eukaryotes, and filamentous bacteria can look like fungal hyphae. And finally, early eukaryotes were probably small and thin-walled, and therefore are most unlikely to be preserved as fossils.

We have to take the geological record and interpret it as best we can. First of all, eukaryotes could not have evolved before oxygen became a permanent component of seawater. But that could have happened in early stromatolites, if "oxygen oases" formed round patches of stromatolites while the rest of the world was anoxic.

The oldest eukaryote is *Grypania spiralis*, a ribbon-like fossil 1–2 mm wide and over 10 cm long. It occurs abundantly in rocks around 1450 Ma in China and Montana, and also from BIF in Michigan dated at 1875 Ma (Fig. 3.4). *Grypania* looks very much like an alga. If so, eukaryotic algae had evolved by 1875 Ma, and simpler eukaryotes could have lived hundreds of millions of years before. It would certainly be plausible that eukaryotes evolved around the time that oxygen production became a serious problem for neighbors of cyanobacteria, but so far that is speculation.

Tappania may not be the earliest eukaryote, but it is among the most beautiful. The specimen in Figure 3.5 is likely a resting stage or cyst, and it dates from about 1.4 Ga.

The Evolution of Sex

Eukaryotes have sexual reproduction, but prokaryotes do not. Prokaryotes sometimes have a limited DNA exchange with other cells, but it is not at all similar to sexual reproduction. Essentially, every prokaryote is its own lineage, either dying, budding off, or splitting into daughter cells that are clones of the parent, in the process of asexual reproduction. Cell division is simple for prokaryotes. There are no mates to find, no organelles to organize. Daughter cells are clones, with the same DNA as the parent cell, so they are already well adapted to the microenvironment. Prokaryotes gamble against a change in the environment: if a change occurs that kills an individual, that change will most likely wipe out all that individual's clones too. Prokaryotes have no way to affect the future of their genes. They can only pass them on unchanged to their offspring.

In eukaryotes, all the DNA of two individuals taking part in sexual reproduction is shuffled and redealt to their offspring. Offspring are therefore similar but are not identical to their parents: in fact, there is an impossibly low chance that any two sexually reproduced individuals are genetically identical, unless they developed from the same egg, as identical twins do.

The offspring of sexual reproduction resemble their parents in all major features, but are unique in their combination of minor characters. Sexually reproducing species have built-in genetic variability that is often lacking in clones of bacteria. Individuals vary in the characters of their bodies, which often means that some individuals are slightly better fitted to the environment than others, so stand a better chance of reproducing. The particular sets of DNA in those individuals are thus differentially represented in future populations.

In organisms that reproduce by cloning, a favorable mutation can spread successfully over many cycles of cloning if it occurred in an individual that divided faster than its competitors. The environment selects or rejects the whole DNA package of the mutant individual, which either divides or dies. This is a one-shot chance, and many potentially successful mutations may be lost because they occur in an individual whose other characters are poorly adapted. On the other hand, a favorable mutation may allow one individual such success that it and its clones outcompete all the others, making the population uniform even though it may contain bad genes along with the good one. Uniform populations of yeast may be desirable to a baker or a brewer, but in nature a uniform population may easily be wiped out by changes in the environment.

In contrast, a mutation in a sexually reproducing individual is shuffled into a different combination in each of its offspring. For example, a mutant oyster might find her mutation being tested in different combinations in each of her 100,000 eggs. Natural selection could then operate on 100,000 prototypes, not just one. Favorable combinations of genes can be passed on effectively. A sexually reproducing population can evolve rapidly and smoothly in changing environments, and in favorable circumstances, evolution can be greatly accelerated by sexual reproduction.

At the same time, sexual reproduction is conservative. Extreme mutations, good or bad, can be diluted out at each generation by recombination with normal genes. The genes may not disappear from the population, but may lurk as recessives, likely to reappear at unpredictable times as recombination shuffles them around.

Eukaryotes are so complex that only approximately similar individuals can shuffle their DNA together with any chance of producing viable offspring. Complex physical, chemical, and behavioral ("instinctive") mechanisms usually ensure that sex is attempted only by individuals that share much the same DNA. Such a set of organisms forms a species, defined as a set of individuals that are potentially or actually interbreeding. The composite total of genes that are found in a species is called the gene pool.

Sexual reproduction has two great flaws. First, a sexual individual passes on only half of its DNA to any one offspring, with the other half coming from the partner. Therefore, to pass on all its genes, a sexual individual has to invest double the effort of an asexual individual. Second, the offspring of sexual parents are not identical. Sets of incompatible genes may be shuffled together into the DNA of an unfortunate individual, which may die early or fail to reproduce. At every generation, then, some reproductive "wastage" occurs.

Many eukaryotes can also reproduce by simple fission, cloning identical copies of themselves. An amoeba is perhaps the most familiar example, but corals, strawberries, Bermuda grass, and aphids often use this method too. But it's likely that sex evolved once, in the earliest eukaryotes, and only a few lineages now resort to asexual reproduction. So many eukaryotes reproduce sexually that there must be very strong counterbalancing advantages of sex. Several books have been devoted to this question, but there are as yet no convincing answers.

The Classification of Eukaryotes

Eukaryotes occur in the natural world in ecological and evolutionary units called species. As we have seen, species are groups of individuals whose genetic material is drawn from the same gene pool but is almost always incompatible with that of another gene pool. Members of the same species, therefore, can potentially interbreed to produce viable offspring. They tend to share more physical, behavioral, and biochemical features (characters) with one another than they do with members of other species. Defining and comparing such characters allows us to distinguish between species of organisms. A species is not an arbitrary group of organisms, but a real, or natural, unit.

Biologists use the Linnean system of naming species, after the Swedish biologist Carl Linné who invented it in the eighteenth century. A species is given a unique name (a specific name) by which we can refer to it unambiguously. Linné gave the specific name *noctua* to the European little owl (Fig. 3.6) because it flies at night. Species that share a large number of characters are gathered together into

Figure 3.6 *Athene noctua*, the little owl of Europe, was named by Carl Linné. Photograph by Trebol-a, and placed into Wikimedia.

groups called genera (the singular is genus) and given unique generic names. Linné gave the little owl the generic name *Athene*. Athena is the Greek goddess of wisdom, and the little owl is the symbol of the city of Athens, stamped on its ancient coins (Fig. 3.7). However, taxonomic names do not have to carry a message, even though a simple and appropriate name is easier to remember. (One must be careful about names, too: *Puffinus puffinus* is not a puffin, but a shearwater, and *Pinguinus* is not a penguin but the extinct Great Auk!) Thus, Linnean names are only a convenience, but a very valuable one. The bird that the British call the tawny owl, the Germans the wood owl, and the Swedes the cat owl, is *Strix aluco* among international scientists.

Genera may be grouped together into higher categories (Fig. 3.8). For example, *Athene* and *Strix* and many other owls are grouped together to form the Family Strigidae, named after *Strix*. Families may be grouped into superfamilies, and then into orders, classes, and phyla. Other subdivisions can be coined for convenience.

Any division or subdivision that is used to group organisms is called a taxon (plural, taxa). Biologists who try to recognize, describe, name, define, and classify organisms are taxonomists or systematists, and the practice is called **taxonomy** or **systematics** or **classification**. Slightly different ranks of categories are used for different kingdoms of organisms, but the basic units of classification recognized by all biologists remain the species and genus. Taxonomy has rather complex rules for applying names to groups of organisms. Botanists, zoologists, and microbiologists have slightly different rules for describing new species.

Describing Evolution

Linné did not believe in evolution. Today we try to organize species, genera and larger taxa so that each contains a group

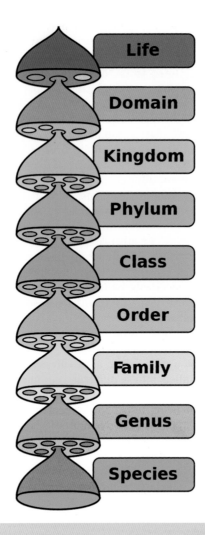

Figure 3.8 A simple diagram of the Linnean hierarchy of taxa as it is used today. Drawn by Peter Halasz, and placed into Wikimedia.

Figure 3.7 Silver coin minted in classical Athens, with the goddess Athene on one side, and the little owl on the other, with the letters Alpha-Theta-Eta indicating Athens. Photograph by PHGCOM, and placed into Wikimedia.

of organisms that evolved from a single ancestral species, and is evolutionarily separate from the rest of the organic world. But Linné's system of naming species and larger set of organisms is so powerful and convenient that we still use it for our modern purposes.

With our new knowledge, recognizing and naming a new species today is a statement about evolution. It reflects the taxonomist's hypothesis that the members of the species share the same gene pool, which is different from the gene pool of any other species because there has been evolutionary divergence over time.

The word **cladistics** describes an approach to classifying species that uses the fact that every species began by branching from another (*klados* is the Greek word for a branch). So every species is a clade, if we have defined it accurately. We then compare species that resemble one another, and try to arrange them into a phylogeny: a branching pattern that reflects the way they evolved.

To use an analogy, a clade is a branch on a tree that represents all life: if all life on Earth is descended from the first living cell by a series of evolutionary branching events, then life as a whole is one clade (the entire tree). Just as trees may branch many times, and branches then branch, and so on, clades of organisms exist in a hierarchy of scales, with the end of each branch representing a single species. Every species is a clade that belongs to a larger clade, which belongs to a larger clade, and so on. Every clade, large or small, began with a single branching event that produced the ancestor of the clade.

Species in a clade share a set of characters that were evolved as new features in a common ancestor, and then passed on to all descendant species. Newly evolved characters represent a change from an ancestral or original state to a novel or derived state.

Three living species, A, B, and C, could be related along three possible evolutionary pathways (Fig. 3.9). Which is correct? Which two of the three species are most closely linked? Two species may look very similar because they share similar characters, but if those are shared ancestral characters that were also present in a common ancestor, they cannot tell us anything about evolution within the group, because they have not changed within that history. The useful characters for solving the problem are the derived characters, because they define how the species have changed since they shared the same ancestral characters in their common ancestor (Fig. 3.9).

Figure 3.9 shows three cladograms: they display the distribution of characters in a visual form. The cladogram that requires the simplest and fewest evolutionary changes is assumed to show the most likely history of the species. A cladogram therefore expresses a hypothesis about the phylogeny of a group. Two species are most closely linked, and form a sister group. In turn, the third species becomes their sister group in a larger clade.

For example, all living mammals have fur, but no other living organisms do. Perhaps fur was inherited from the common ancestor of all living mammals, which evolved a furry skin as a novel character that modified an ancestral

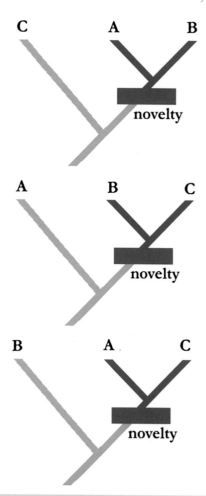

Figure 3.9 Three cladograms that show all the possible relationships between the three species A, B, and C, that had a common ancestor. Two of them may share a newly evolved character that the third does not have: the two species with the derived character are sister species. The cladograms reflect the evolutionary changes that occurred within the group.

one (a scaly skin, say). If that is true, then mammals are a clade. Examining the hypothesis, one finds other shared derived characters of living mammals that strengthen the argument: for example, all living mammals are warmblooded, and suckle their young.

Sometimes problems arise because similar derived characters are found in species outside a clade; those characters have evolved more than once by parallel evolution. For example, bats and birds both have wings, and in each group the wing is a derived character that has been modified from some other structure. But bats and birds share very few other derived characters, and even their wings have a different basic structure. The weight of evidence suggests that birds are a clade, bats are probably a clade, but [bats + birds] is not a clade.

Once the preferred cladogram is drawn to portray the best hypothesis, one can make decisions about the best way to classify the species and to describe its evolutionary history. A cladogram in itself does neither of these things.

One could introduce a formal name for each clade on a cladogram. However, this would lead to a great number of names, not all of which might be needed for everyday discussion around the breakfast table. For most purposes, it is simpler to draw a cladogram and to use a minimum of hierarchical names.

A cladogram is always drawn with all the species under study along one edge (Fig. 3.9). No species in a cladogram is shown as evolving into another. Some cladists claim that one can never know true ancestor–descendant relationships, and in a strict sense this is true because we don't have time machines. But sometimes a fossil is known that could well be an ancestor of a later fossil or of a living organism. At present, for example, it seems more reasonable (to me) to suggest that the earliest bird *Archaeopteryx* is the ancestor of owls and penguins than to suggest that those birds are descended from an ancestor that we haven't found yet. Hypotheses like this are expressed on **phylograms** or phylogenetic trees that include time information: we are allowed to show a suggested ancestor within such a tree (Fig. 3.10). Like cladograms, phylograms are not statements of fact but hypotheses, subject to continuous testing.

Kevin Padian recently introduced a very powerful new kind of diagram to reflect evolution within organisms—the **evogram**. This shows pictorially the organisms, their relationships in time and space, and as much information as can be clearly displayed about the pathway that evolution took. Excellent examples can be found at http:// evolution.berkeley.edu/evolibrary/article/evograms_02

Decisions about the course of evolution are not always obvious, so taxonomic decisions may be revised as new information becomes available. Species are moved around between genera and higher categories as taxonomists refine their classifications to reflect evolutionary history more effectively.

Counterintuitive patterns sometimes emerge in cladistics. We are all used to thinking about living fishes, amphibians, reptiles, birds, and mammals as classes of vertebrates, equal in rank to one another (Fig. 3.11a). But this is not a cladistic classification. Tetrapods are actually a clade within fishes, derived from them by acquiring some novel characters, including feet, and amphibians are a clade within tetrapods. Reptiles, mammals and birds are also clades of derived tetrapods.

There's nothing intimidating about this—it simply takes some time to get used to it. The important feature of a cladistic framework combines a useful evolutionary classification with a cladogram or phylogram to display the classification clearly.

If we classify all living reptiles as one group and draw a cladogram of vertebrates (Fig. 3.11b), we display the well-known fact that living reptiles and birds are more alike than either is to mammals. The cladogram also carries other information. It shows that warm blood, a derived character that living birds and mammals share, must have evolved

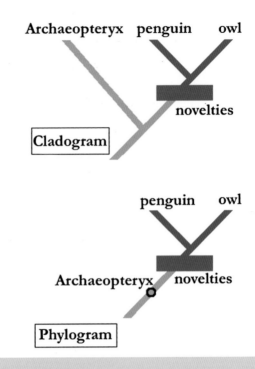

Figure 3.10 A cladogram and a phylogram of three groups of birds. In the cladogram (top), no group is shown as the ancestor of another, because a cladogram seeks only to show the relationship between groups. Penguins and owls have derived, novel, characters that *Archaeopteryx*, the earliest bird, does not have. But suppose I wanted to make the additional hypothesis that *Archaeopteryx* was not only less derived than the other birds, but was actually their ancestor. To do that, I would draw a phylogram, or phylogenetic tree (bottom), to show *Archaeopteryx* in an ancestral position in the body of the tree, below the event that marks the evolutionary branching between owls and penguins.

independently at least twice, unless living reptiles have lost warm blood.

As we consider smaller subgroups of living and fossil reptiles, we find that this neat picture of reptile classification breaks down, so we must revise our ideas about tetrapod evolution. Figure 3.11c shows that "living reptiles" is not a clade. We could define a clade called "reptiles," but we would have to include mammals and birds in it. Turtles, mammals, birds, crocodiles, and snakes are all clades that diverged from a common ancestor, although some are more derived than others in the sense that they have evolved more novel characters that the common ancestor did not have. In the same way, humans are derived fishes, derived amniotes, and derived primates, all at the same time. Once one becomes used to cladistic thinking, evolution becomes much more real, and we can see, for example, that humans, tapeworms, and the bacterial scums of Shark Bay

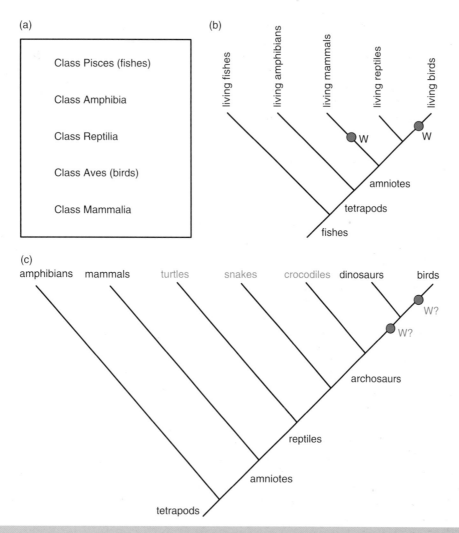

Figure 3.11 a) a traditional classification of the five classes of living vertebrates, showing them as equal rank. b) a cladogram of those five classes, showing that cladistically they are not equal in rank: for example, living mammals are the sister group of [living reptiles + living birds]. Warm blood (W) was independently evolved as a derived character in both mammals and birds, according to the hypothesis expressed in this cladogram. c) when we look more deeply into "living reptiles" (shown in blue), and we add the extinct dinosaurs to the cladogram, we have to change our assessment of vertebrate evolution. Reptiles are not a clade unless they also include birds, and "living reptiles" are not a clade either. Dinosaurs are the sister group of birds, which brings up the question of where warm blood (W) evolved in that lineage. Is it a derived character of birds, or is it a derived character shared by both birds and dinosaurs? Drawing a cladogram forces us to look at such evolutionary questions!

are all prokaryotes—though some have accumulated more derived characters than others.

Further Reading

Cotton, J. A. and J. O. McInerney. 2010. Eukaryotic genes of archaebacterial origin are more important than the more numerous eubacterial genes, irrespective of function. *PNAS* 107: 17252–17255. Available at http://www.pnas.org/content/107/40/17252.long

Gross, J. and D. Bhattacharya. 2010. Uniting sex and eukaryote origins in an emerging oxygenic world. *Biology Direct* 5:53. Available at http://www.biology-direct.com/content/5/1/53

Longsdon, J. R. 2010. Eukaryotic evolution: the importance of being archaebacterial. *Current Biology* 20: 1078–1079 [Comment on Cotton and McInerny.]

Question for Thought, Study, and Discussion

If sexual reproduction is so inefficient and wasteful, why is it that we don't simply evolve virgin birth? Women would give birth to babies identical to themselves, obviously completely capable of living successful lives. Men would be redundant, of course. Without testosterone, the world would be peaceful and civilized.

Yet no mammal or bird has ever evolved that ability (though a few reptiles have).

FOUR
The Evolution of Metazoans

In This Chapter

In this chapter I follow the evolution of single-celled eukaryotes into multicellular organisms, and then into multicellular plants and animals with complex structures that include different organ systems. I start with a new discovery about the evolution of multicellular life in the laboratory. But different ways of life evolve in specific environments on the Earth, so we have to discuss how climate change on Earth affected major breakthroughs in evolution, during a cold period called Snowball or Slushball Earth. Then as climate changed again, we see a dramatic explosion of animal groups in Earth's seas about 540 million years ago. Complex animals evolved, each with organ systems that evolved to give each group the ability to take on a specific ecological role: as floating, swimming, crawling, and burrowing animals, some eating mud, others filtering food from the water, and others as predators that ate other animals.

Proterozoic Microbes

The Proterozoic ocean was anoxic, rich in dissolved iron and silica and sulfide, and for more than a billion years after the Great Oxidation Event, we see little change in ocean chemistry. The deep waters must have been largely inhabited by bacteria and methanogens, because eukaryotes require some oxygen to run their mitochondria. Only a zone of surface waters had photosynthesizers, and even then the oxygen levels were low and probably varied from time to time and from place to place. Oxygen-tolerant microbes and single-celled eukaryotes were confined to this surface zone.

Microfossils occur in these Proterozoic rocks, but they are not abundant or diverse. Beginning about 1800 Ma (Lamb et al. 2009), we find **acritarchs**, spherical microfossils with thick and complex organic walls (Fig. 3.5). They were organisms, most likely eukaryotes of various kinds (Fig. 4.1), that grew thick organic walls (cysts) in a resting stage of their life cycle, but spent the rest of their lives

History of Life, Fifth Edition. Richard Cowen.
© 2013 by Richard Cowen. Published 2013 by Blackwell Publishing Ltd.

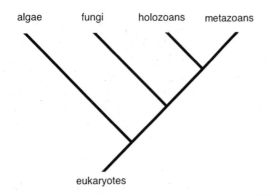

algae   fungi   holozoans   metazoans

eukaryotes

Figure 4.1 Cladogram of eukaryotes that diverged in the Proterozoic. For paleontologists, it is important to note that all these organisms, including the first micrometazoans, were capable of forming resting stages or cysts that would look much alike when fossilized as acritarchs.

floating in the plankton, the organisms that live in the surface waters of oceans and lakes. For the next 800 million years, we see microfossils that are difficult to interpret, but probably reflect a slow diversification of eukaryote protists.

In the late Proterozoic, starting about 800 Ma, increased oxygen levels seem to have gradually extended the oxygen-bearing zone deeper into the ocean, cutting down methane production in deep waters. More of the shallow seafloor would have become inhabitable by protists as well as bacteria and Archaea. For the first time in global history, we can envisage seafloors with successful populations of protists.

No automatic barrier forces protists to be single-celled. A protist that divided with its daughter cells remaining together could form a multi-celled colony in only a few cycles of division. A process like this in an algal cell could have produced the multicellular seaweed frond *Grypania* (Chapter 3). However, a major breakthrough on this question came very recently. A team led by Will Ratcliff (Ratcliff et al. 2012) applied very simply artificial selection to a unicellular strain of brewer's yeast. They selected out all yeast that sank slowly in water, allowing the fastest 5% to keep on reproducing. This encouraged the yeast to form small "colonies" of cells that stayed together after dividing, because larger cell clusters sink faster than single cells. After 60 days, Ratcliff et al. had yeast cultures that were entirely multicellular, little spherical colonies that looked like snowflakes. The snowflakes reproduced like multicellular organisms, breaking apart the parent snowflake into two. And when the sinking selection was removed, none of the cultures reverted back to a single-celled existence. They had become biologically and functionally, multicellular organisms (and their DNA had changed, so the switch to multicellularity was an evolutionary process). What this means for paleontologists is that we can stop worrying about how

multicellular organisms could evolve: it was (comparatively) easy!

The Variety of Eukaryotes

There are thousands of species of single-celled eukaryotes. Several separate lineages of photosynthetic eukaryotes are loosely grouped as "algae". Several heterotrophic eukaryote lineages are grouped as "holozoans", others as "fungi". Figure 4.1 is a simplified cladogram that attempts to relate these groups as they diverged some time in the Proterozoic.

For a long time, holozoans have received little attention from paleontologists, despite their position in Figure 4.1. That's because there is little evidence to separate holozoans from other poorly preserved eukaryote fossils. That has changed.

Snowball or Slushball Earth

The Earth went through a series of dramatic cold periods late in the Proterozoic, in a time period that is informally called the **Cryogenian**, roughly between 750 Ma and 620 Ma. Many deposits from this period contain glacial debris, and many of them occur in regions that are reliably reconstructed near the Equator at the time. These deposits imply episodes of massive and widespread glaciation, much more extensive than any glaciations that have occurred since. One scenario that attempts to explain their wide distribution is "Snowball Earth".

Immediately after the Cryogenian we see a radiation of new fossil animals, so it seems likely that the dramatic physical events of "Snowball Earth" are somehow linked with the metazoan radiation.

The Snowball Earth model (Hoffman et al. 1998) proposes that the ocean surface was frozen all the way to the Equator (except immediately around volcanoes). Surface temperatures dropped to about −40° C. As the ice spread, photosynthesis was choked off, and most life in the oceans died off. The only surviving life would have been around seafloor hot vents, and (perhaps) in surface ice. Solar radiation would be reflected back into space, so you would think Earth would be locked permanently into a snowball state.

However, volcanoes continued to erupt, putting carbon dioxide back into the atmosphere, until there was once again enough carbon dioxide to trap solar heat and melt the ice. But calculations suggest it would have taken an enormous amount of carbon dioxide to break the grip of Snowball Earth. The ice cover did not melt until volcanoes had erupted around 350 times more carbon dioxide than there is in our present atmosphere.

The ice then melted quickly, but the enormous reservoir of carbon dioxide in the atmosphere rocketed the whole Earth directly into a "greenhouse" hot period, with temperatures averaging around 50°C (over 120°F). Tremendous (acid) rains then acted on the sterilized continents, the ocean was flooded with carbonate, and thick limestones formed very quickly on top of the glacial deposits. Finally, weathering and photosynthesis brought down carbon dioxide levels, and the world recovered biologically. However, the geographic set-up that had begun the Snowball Earth cycle was still present, so the cycle then repeated itself, perhaps as many as three times.

Snowball Earth therefore calls for a catastrophe to trigger the metazoan revolution. That simply doesn't happen: major crises actually cause major extinctions. Repeatedly wiping the ocean free of oxygen is not the way to foster the evolution of metazoans. Any eukaryotic survivors of a Snowball Earth would have been few indeed.

However, there is good evidence that the Snowball Earth scenario is too extreme (Fig. 4.2). The glacial sediments include dropstones, rocks which fall from floating icebergs into soft sea-floor sediment. Those icebergs must have been floating freely, in open water. A "Slushball Earth" concept, which calls for less than a completely frozen Earth, is supported by computer models that suggest a stable climate, with low-latitude continental ice-sheets and seasonal floating sea-ice over much of the world's oceans. These models project open tropical waters at cool to mild temperatures (up to 10°C at the Equator). The model atmosphere has only 2.5 times today's carbon dioxide, and the stability of the model implies that any small rise in those levels will easily revert conditions to Earth normal, without the extreme greenhouse called for by the Snowball Earth idea.

Most important in terms of evolution and paleontology, I suggest that Slushball Earth could have encouraged major evolution among eukaryotes. Ocean water below the polar Slushball ice sheets would quickly have become anoxic all the way to the ice at the surface. Thus polar waters and deep waters would have contained an "Archaean biology" of anaerobic bacteria and Archaea.

But conditions in the open surface waters of the tropics would have been radically different (Fig. 4.3). There would have been little or no seasonal fluctuation in climate. Active erosion by mountain glaciers on the equatorial continents provided a steady year-round supply of nutrients (especially phosphate) to ice shelves along the coastline and, via icebergs, to the surrounding waters. Iron enrichment from wind-blown dust gave an important supplement to normal nutrient supply. Solar radiation in the tropical areas would have been uniform and intense, no matter what the surface temperature was. Within tropical ocean waters, there would have been active surface mixing and upwelling of nutrient-rich water in the open equatorial areas, like the perennial

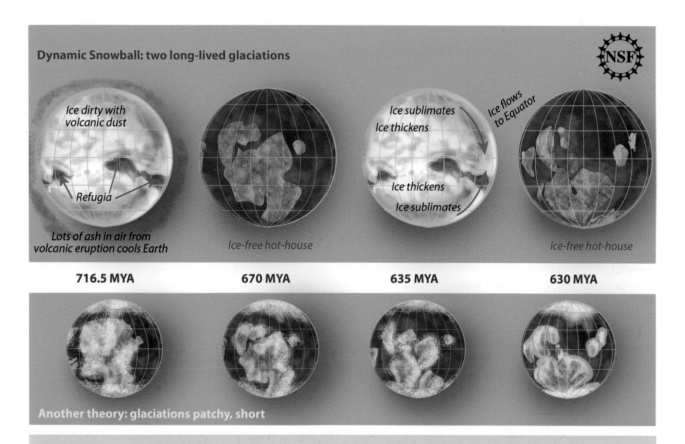

Figure 4.2 Snowball Earth and Slushball Earth: two alternative hypotheses with very different implications for the physical and biological Earth. Graphic by Zena Deretsky: courtesy National Science Foundation.

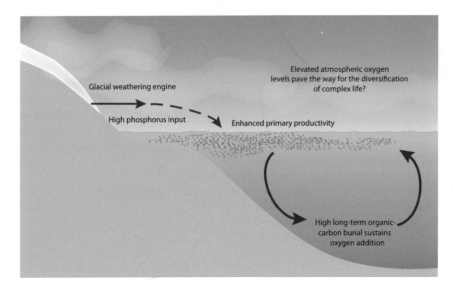

Figure 4.3 Diagram of likely conditions on the Equator during Slushball Earth. Glacial grinding delivers nutrients to well-lit shallow ocean waters, where plankton (including eukaryotes) flourish. Diagram kindly provided by Noah Planavsky, after Filippelli 2010.

upwelling in the Southern Convergence today, or along the Equator.

The nutrient-rich surface equatorial waters of Slushball Earth would have supported dramatic year-round productivity of cyanobacteria and algae, uninterrupted by the seasonal darkness that cuts down today's polar productivity in the winter months. Fall-out of dead organic matter from the surface productivity would have driven the deeper layers to anoxia, no matter how oxygen-rich the surface was. This fits with evidence that iron-rich sediments were deposited on the sea floor during the glacial periods.

The surface tropical waters of Slushball Earth would have been a paradise for plankton. The extraordinary, and permanent, productivity along the Slushball Equator would have provided a lot of oxygen in a shallow surface zone, an ideal setting for the evolution of many different tiny eukaryotic predators on surface plankton (Fig. 4.3). Much of that oxygen would have diffused off into the atmosphere, so the oxygen levels in surface waters would have still been much lower than today's.

By the time Slushball Earth and the great glaciations ended, phosphate levels in the ocean had reached the highest levels ever recorded in Earth history. With a milder climate and more open water, eukaryotes evolved rapidly. Oxygen levels in surface waters and the atmosphere increased also, likely allowing increased size and energy output in metazoans (Fig. 4.3).

At first the eukaryotes would all have been planktonic, feeding on bacteria and protists. They would have reproduced at these very small sizes, perhaps with a great deal of cloning (as echinoderm larvae do today). This ecological reconstruction makes sense in terms of the oxygen levels in a "normal" Late Proterozoic ocean: most likely highest in the surface layers where photosynthesis occurred, and likely to have been low in the sediments of the ocean floor (Fig. 4.3).

All this is reasonable speculation: but does it fit the evidence from Cryogenian rocks? There are convincing reports of eukaryotic protists (specifically, Foraminifera) from glacial sediments dating back to 700 Ma or so (Bosak et al. 2011), which implies that a good number of other protists would have evolved by that time too.

Astonishingly, fossils that look very much like tiny sponges have been recently discovered in Cryogenian rocks from Namibia, in southern Africa (Brain et al. 2012). Sponges are the most basal metazoans, but 1000 specimens have been found, some of them in rocks that are certainly older than 710 Ma and may be closer to 760 Ma. The species *Otavia antiqua* would be cheerfully accepted as a sponge if it had been found in more recent rocks, so this discovery of the earliest and simplest metazoans looks convincing (Fig. 4.4).

The Ediacaran Period

Soon after 570 Ma, in rocks found worldwide from Canada to Russia to Australia, we find soft-bodied animals that make up the Ediacaran fauna, named after rocks found in the Ediacara Gorge in the Flinders Ranges near Adelaide. These fossils form the basis for recognizing a new period of geological time, the Ediacaran Period, between the end of the Cryogenian at 575 Ma and the base of the Cambrian at 543 Ma.

The Doushantuo Formation

Early in the Ediacaran, the Doushantuo Formation was laid down as a set of rocks in South China, very soon after the last major glaciation. It contains some exquisite fossils that are the key to understanding the life of this time. The Doushantuo rocks are so rich in phosphate that they are mined for fertilizer, so they have been well studied. One layer, dated about 570 Ma, contains tiny fossils that were preserved so soon after death that phosphate replaced the individual cells, preserving them in 3D. The simplest

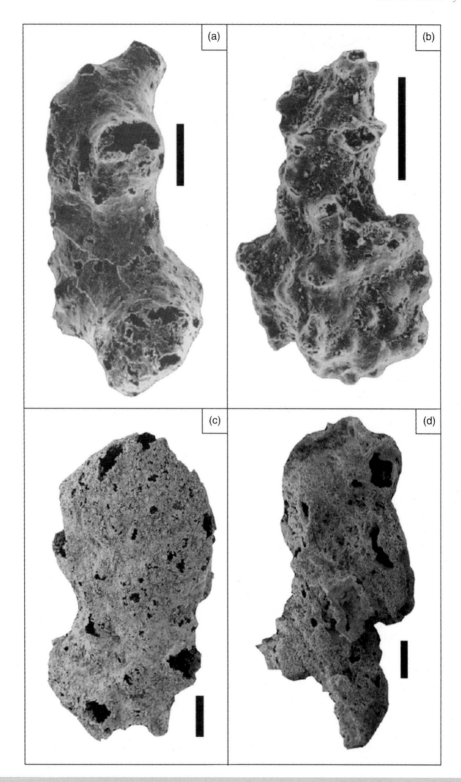

Figure 4.4   The earliest fossil sponge, *Otavia antiqua*, from the Cryogenian of Namibia. Scale bars, 100 μm. The species ranges in age from more than 710 Ma to about 550 Ma. From Brain et al. (2012). Courtesy of D. Herd, Department of Earth Science, University of St. Andrews, Scotland.

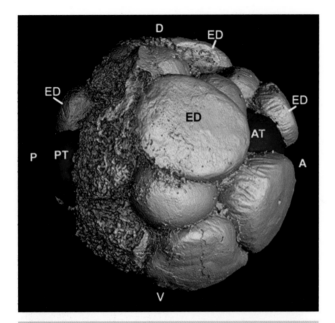

Figure 4.5 A fossil from the Doushantuo Formation, showing beautifully preservation in phosphate, about 0.5 mm across. Its anatomy has been interpreted as showing the features of a metazoan embryo (A, anterior, P, posterior, V, ventral, D, dorsal). ED marks ectodermal (outer) cells and AT and PT mark endodermal (inner) cells. From Figure 4 of Chen et al. (2009). © Dr. Jun-Yuan Chen, used by permission. But this fossil may be a holozoan, not a metazoan (see text).

Figure 4.6 A large acritarch from Ediacaran rocks of Russia, about 0.5 mm across, easily interpreted as a cyst: but made by what organism? Image from Cohen et al. (2009). © Phoebe A. Cohen, used by permission.

interpretation of the fossils is that they fell out of a rich planktonic biota flourishing in productive surface waters above an organic seafloor.

Thousands of these Doushantuo microfossils have been studied. Figure 4.5 gives an idea of the quality of the preservation. Enthusiastic accounts of their astounding variety identified protists and algae, tiny metazoans (sponges and cnidarians), and eggs and embryos of yet more metazoans.

However, many small creatures build cysts (Fig. 3.5) as resting stages while conditions are bad, and cysts are more likely to be preserved as fossils than other stages of the life cycle. The difficulty of dealing with fossil cysts is that it is very difficult to see interior details (Fig. 4.6). The cyst stages of living holozoans (Fig. 4.1) have much the same structures as the Doushantuo microfossils (Huldtgren et al. 2011). They reproduce by cloning dozens, or hundreds, or thousands of identical individuals before releasing them. Metazoan embryos don't do that. Each cell, programmed differently, is destined to become a separate individual cell in the metazoan body. This is a controversial topic, but the weight of evidence now suggests that the Doushantuo fossils are simple holozoans, not true metazoans.

This statement could change overnight, of course. There may be very early metazoans in the Doushantuo. A group

of big acritarchs in the early Ediacaran is not found in the later Ediacaran (Fig. 4.6). They seem large to be single-celled, so they may be micrometazoan cysts (Cohen et al. 2009).

We shall look first at the variety of living metazoans, then in Chapter 5 look at the Ediacaran fossil record. There are only three kinds of metazoans: **sponges** and their relatives; **cnidarians** and their relatives; and **bilaterians** (3D animals with distinct bilateral symmetry). All of them solved the problems of evolving to greater size and complexity, but in different ways.

Making a Metazoan

A flagellate protist is a single cell with a lashing filament, a **flagellum** (plural, flagella), that moves it through the water. Some flagellates called **choanoflagellates** (*choana* is the Greek word for collar) build a conical collar around the flagellum (Fig. 4.7a) that has pores through it, rather like a coffee filter. As the flagellum beats, it pulls water through the collar, which collects tiny food particles from the water such as bacteria. A choanoflagellate can also anchor to the seafloor and feed using this system. Instead of dividing into two independent daughter cells, some choanoflagellates form **colonies**: they bud off new individuals which then stay together to form a group (Fig. 4.8a). The flagella of all the members of the colony now beat together to generate a powerful water current that makes a very efficient filtering system.

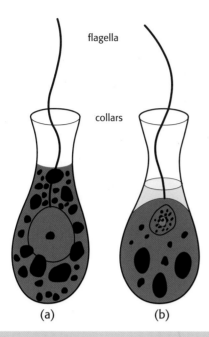

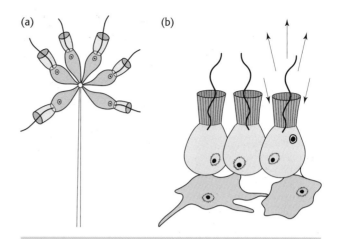

Figure 4.7 From eukaryote to metazoan. a, a choano-flagellate that collects food as the flagellum pulls water through the collar. b, a collar cell or chanocyte from a sponge. Here the collar cell is firmly anchored in the body of the sponge. Adapted from Barnes et al., *The Invertebrates: A Synthesis*. 3rd edition © Blackwell Science 2001.

Figure 4.8 a) a colonial choanoflagellate. It has divided several times, but the daughter cells have stayed together, and generate a powerful feeding current for the colony as a whole. b) a group of choanocyte cells from a sponge. They are embedded in other tissues in this metazoan animal. Both images from Barnes et al., *The Invertebrates: A Synthesis*. 3rd edition © Blackwell Science 2001.

The process of sticking cells together to form a multicel-lular animal is promoted by two kinds of proteins coded in the DNA of choanoflagellates. Cadherins help cells to stick together, and integrins allow cells to react to the presence of a neighbor cell. No single-celled organisms except cho-anoflagellates have these proteins and the genes that code for them. This shows what zoologists had suspected for a long time, that choanoflagellates were ancestors or very close relatives of the first metazoan animals.

Metazoans most likely evolved only once. They all origi-nally had one cilium or flagellum per cell, for example. They also share the same kind of early development. They quickly form into folded balls of internal cells which are often free to move, and are covered by outer sheets of cells that form an external skin-like coating for the young animal. Sponges probably branched off first from the ancestral metazoan, by extending the choanoflagellate way of life to large size and sophisticated packaging.

Metazoans are not just multicellular. They have different kinds of cells that perform different functions. **Sponges** are the simplest metazoans living today. They contain many flagellated cells called **choanocytes** (Fig. 4.7b), which are arranged so that they generate efficient feeding currents (Fig. 4.8b). In turn efficient groups of choanocytes pump water (and the oxygen and bacteria they capture from it) through the sponge, in internal filtering modules (Fig. 4.9).

Sponges surely evolved from choanoflagellates (Figs 4.7, 4.8), but they are much more advanced because they also have other specialized cells. One breakthrough was to link cells firmly together to form a body wall, using a gene complex that is also found in all later metazoans. The sponge body wall can (very slowly) contract the sponge as a defense mechanism, even though it has no muscle cells. Other cells digest and distribute the food that the choano-cytes collect, and yet others construct a stiffening frame-work, often made of mineral, that allows sponges to become large without collapsing into a heap of jelly (Fig. 4.10).

Cnidarians (or coelenterates), including sea anemones, jellyfish, and corals, are built mostly of *sheets* of cells, and they exploit the large surface area of the sheets in sophisti-cated ways to make a living. The cnidarian sheet of tissue has cells on each surface and a layer of jellylike substance in the middle. The sheet is shaped into a baglike form to define an outer and an inner surface (Fig. 4.11). A cnidar-ian thus contains a lot of seawater in a largely enclosed cavity lined by the inner surface of the sheet. The neck of the bag forms a mouth, which can be closed by muscles that act like a drawstring. A network of nerve cells runs through the tissue sheet to coordinate the actions of the animal.

In most cnidarians the outer surface of the sheet is simply a protective skin. The inner surface is mainly diges-tive, and absorbs food molecules from the water in the enclosed cavity. Because cnidarians are built only of thin sheets of tissue, they weigh very little, and can exist on small amounts of food. They can absorb all the oxygen they need from the water that surrounds them.

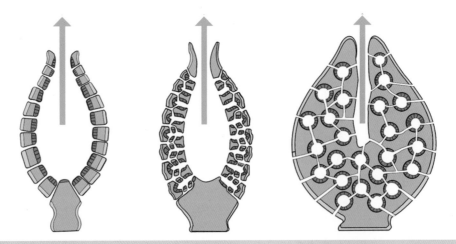

Figure 4.9 Sponges build modular filtering units that can reach high complexity. In these three examples of sponge structure, sets of choanocytes are in red, the outer skin of the sponge is yellow, and the outgoing water current of water that has been filtered is blue-green. Diagram created by Philcha and placed into Wikimedia.

Figure 4.10 The skeleton of a deep-sea glass sponge, made of silica spicules. Photograph by Randolph Femmer for the United States Geological Survey.

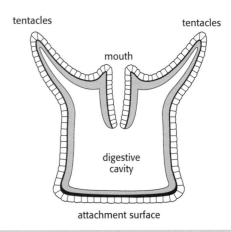

Figure 4.11 Basic structure of a cnidarian. A two-sided sheet of tissue defines the inside and outside surfaces of a bag-shaped digestive cavity. From Boardman et al., *Fossil Invertebrates*. © Blackwell Scientific 1987.

Cnidarians have **nematocysts** or stinging cells set into the outer skin surface. The toxins of some nematocysts are powerful enough to kill fish, and people have died after being stung by swarms of jellyfish. Nematocysts are usually concentrated on the surfaces and the ends of tentacles, which form a ring around the mouth. They provide an effective defense for the cnidarian, but they are also powerful weapons for catching and killing prey, which the tenta-cles then push through the mouth into the digestive cavity (Fig. 4.12). The tissues of the prey are then broken down by powerful enzymes, and the food molecules are absorbed through the cells of the inner lining of the cavity. A cnidarian can thus eat prey without jaws or a real gut.

Hardly any sponges can tackle food particles larger than a bacterium, though there are a few exceptions. Yet living cnidarians routinely trap, kill, and digest creatures that outweigh them many times by using their nematocysts. However, there is no guarantee that the first cnidarians had nematocysts. They may simply have absorbed dissolved organic nutrients from seawater.

The third and most complex metazoan group contains all the other metazoans, including vertebrates. These are the Bilateria or **bilaterians**, metazoans with a distinct bilat-

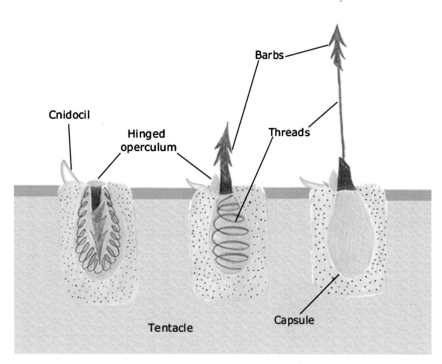

Figure 4.12 How a cnidarian fires a stinging netamocyst. Public domain image from a NOAA document.

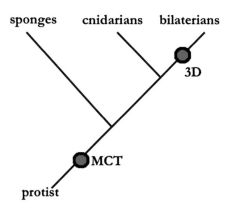

Figure 4.13 Simple phylogram of metazoans. MCT indicates the evolutionary point at which multicellular tissues evolved. 3D indicates a 3D structure of the body.

eral symmetry that influences their biology enormously. They consist basically of a double sheet of tissue that is folded around with the inner surfaces largely joining to form a three-dimensional animal. In contrast to sponges and cnidarians, they have complex organ systems made from specialized cells, and those organ systems are built as the animal grows by special regulatory mechanisms coded in the genes (Fig. 4.13). Worms are simple bilaterians.

All sponges and most cnidarians are attached to the seafloor as adults, and depend on trapping food from the water. But many bilaterians were and are mostly free-living animals, making a living as mobile scavengers and predators. The bilateral symmetry is undoubtedly linked with mobility: any other shape would give an animal that could not move efficiently.

The first bilaterians would have been worm-like. Worms creep along the seafloor on their ventral (lower) surface, which may be different from the dorsal (upper) surface. They prefer to move in one direction, and a head at the (front) end contains major nerve centers associated with sensing the environment. A well-developed nervous system coordinates muscles so that a worm can react quickly and efficiently to external stimuli. The mobility of early bilaterians on the seafloor probably led to the differentiation of the body into anterior and posterior (head and tail) and into dorsal and ventral surfaces, as the various parts of the animal encountered different stimuli and had to be able to react to them.

The front end of bilaterians usually features the food intake, a mouth through which food is passed into and along a specialized one-way internal **digestive tract** instead of being digested in a simple seawater cavity. No sponge cell or cnidarian cell is very far away from a food-absorbing (digestive) cell, so these creatures have no specialized internal transport system. But the digestive system of bilaterians needs an oxygen supply, and the nutrients absorbed there have to be transported to the rest of the body. Bilaterians therefore have a **circulation system**, and the larger and more three-dimensional they are, the better the circulation system must be.

All but the simplest bilaterians have an internal fluid-filled cavity called a **coelom**, which may be highly modified in living forms. In humans, for example, the coelom is the sac containing all the internal organs. The coelom may have evolved as a useful hydraulic device. Liquid is incompressible, and a bilaterian with a coelom (a coelomate) can squeeze this internal reservoir by body muscles. Such squeezing pokes out the body wall at its weakest point, which is usually an end (Fig. 4.14). Such a hydraulic

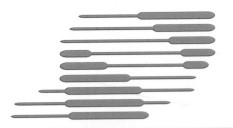

Figure 4.14 This worm-like bilaterian uses its coelom to burrow from right to left. It squeezes fluid forward to push out the front end, then makes it into a bulb. The back can then be pulled forward, and the cycle repeats.

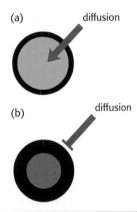

Figure 4.15 Animals with thin tissues (a) can rely on diffusion to supply the entire body with oxygen. But diffusion alone cannot supply oxygen to the interior of animals with thick tissues (b). Special respiratory systems are needed to make sure the interior organs are not starved of oxygen.

extension of the body can be used as a power drill for burrowing into the sediment, to find food, or safety.

The coelom could have provided another great advantage for bilaterians. Oxygen must reach all the cells in the body for respiration and metabolism. Single celled organisms can usually get all the oxygen they need because it simply diffuses through the cell wall into their tiny bodies. Sponges pump water throughout their bodies as they feed, and cnidarians and flatworms are at most two sheets of tissue thick. But larger animals with thicker tissues cannot supply all the oxygen they need by diffusion (Fig. 4.15). Oxygen supply to the innermost tissues becomes a genuine problem with any increase in body thickness or complexity. If the animal evolved some exchange system so that its coelomic fluid was oxygenated, the coelom could then become a large store of reserve oxygen. Eventually the animal could evolve pumps and branches and circuits connected with the coelom to form an efficient circulatory system.

Many advanced bilaterians have **segments**: their bodies are divided by septa that separate the coelom into separate chambers connected by valves. This arrangement is more efficient for burrowing than a simple, single coelomic cavity (Fig. 4.14). The segmentation of many animals, including earthworms, may be derived from this invention on the Precambrian seafloor.

Respiration problems probably prevented early coelomates from burrowing for food in rich organic sediments, which are very low in oxygen. But a coelomate burrowing for protection might have evolved some special organs to obtain oxygen from the overlying seawater at one end while the main body remained safely below the surface. Many coelomates that live in shallow burrows have various kinds of tentacles, filaments, and gills that they extend into the water as respiratory organs. It is a very short step from here to the point where a coelomate collects food as well as oxygen from the water by filter feeding (Fig. 4.16), as in all bryozoans and brachiopods, in some molluscs, worms, and echinoderms, and in simple chordates.

Evolution and Development

We can now read the entire genetic code of a large number of organisms: all it takes is time and money. We can begin to recognize certain strings of DNA as genes, and understand what many of them do within the living organism. For example, the entire genome (genetic code) of the human parasite *Mycoplasma genitalium*, the smallest genome so far discovered, contains only 583,000 nucleobases (humans have 3 billion). Sorting, slicing, and dicing this genome, geneticists have concluded that *M. genitalium* has 470 genes that code for proteins, and 37 that code for RNA. We understand the emphasis on proteins, because they perform so many cell functions: building lipids for the cell membrane, transporting phosphate, breaking down glucose, and so on.

But it is more complex to grow a viable metazoan than a single-celled protist. The genome must contain the information to build many kinds of cells rather than just one, and the information to grow them at the right time, to place them accurately in the body, and to develop the control mechanisms, sensory systems, transport systems, and whole-body biochemical reactions that operate in a metazoan.

The genetic programming that builds a metazoan from a single cell need not specify individual cells one by one. Like a well-written computer program, there can be tricks that promote efficiency. For example, one could program a computer to draw a flower, specifying the size, shape, and position of each petal. But the petals of any given flower are typically much alike, so one can use the same shape and size for each petal, and simply tell the computer to move the pen to the right place and draw the same petal each time (Fig. 4.17).

In the same way, metazoans have **structural genes** to build each piece of the animal, and **regulatory genes** that make sure the piece is built in the right place at the right time. For example, a set of regulatory genes could be used in combination with a set of "segment" genes to build all

Figure 4.16 These two worms build tubes to make their own burrows. With the body safe inside the tube, they extend tentacles to collect food and oxygen from the water. a) a feather duster worm, *Eudistylia*, from the California Channel Islands. Photograph by Chris Gotschalk for NOAA. B) the Christmas tree worm *Spirobranchus* from the tropical Pacific. Photograph by Nick Hobgood, and placed into Wikimedia.

Figure 4.17 One can construct a complex object by careful placing of identical simple units.

the segments along a growing worm. The same sort of regulatory genes could easily be used to build legs on, say, a millipede or a crab, by calling on a "leg" gene the appropriate number of times instead of a "segment" gene. By calling on slight modifications of the leg gene as growth developed, regulatory genes could build an animal whose legs were different along its length (as in insects), or build a vertebrate with different bones along the length of a backbone. For example, embryonic snakes have genetically programmed limb buds that show us where once there were legs on ancestral snakes. Today those buds do not develop into legs because the regulatory genes do not send a growth instruction to them.

Developmental geneticists can identify regulatory genes by checking what goes wrong when a particular gene is damaged. As a result, we know that there are regulatory genes that control which way up an animal is formed, which is front and back, and how the animal varies along its length or around its edges. The most thrilling discovery is that much the same master genes occur throughout metazoans. These "homeobox" genes, or **Hox genes** for short, are so similar that they must have evolved from a common ancestor. They are sets of regulatory genes that sit close to one another in the DNA. Although they use much the same master "program", they can build an astonishing variety of metazoan bodies by calling on a variety of structural genes in a variety of patterns at different times and at different places in the body. In other words, one might well discover that two metazoan groups have Hox genes that are fairly similar, though the bodies those Hox genes code for look very different.

How could we do classification in such cases? In living animals, we have to look at evidence of body morphology and evidence from genetics together, to suggest the simplest hypothesis that would connect these two metazoan groups with others. However, we cannot do genetic analyses on extinct animals: especially for early metazoans, we have to do the best we can with the morphological evidence from fossils that are close to the time when the metazoan groups actually originated. As with any data, discrepancies are going to occur and arguments will rage. However, evolution took one pathway, and we are all trying to find what that pathway was. "**Evo-devo**", the study of evolutionary development among living animals, is giving us dramatic new insights. This new approach has helped to clarify fundamental aspects of early metazoan evolution, in many cases overturning ideas that had been accepted for 100 years or more.

Hox Genes

Sponges have one set of Hox genes (and have simple structures), whereas mammals have 38 sets in four clusters, and goldfish have 48 sets in seven clusters. Hox genes control the growth of nerve nets, segments, and limbs throughout metazoans, and their evolution and divergence must have accompanied the divergence in anatomy and physiology and ecology and behavior that we can interpret from the fossil record. So Hox genes provide separate but complementary evidence to help us read the evolutionary history of the metazoans.

Protists don't need Hox genes to form a multicellular adult. But in early metazoans, Hox genes provided the genetic tool kit to guide the construction of viable complex animals. Hox genes control the lay-out of a sponge that gives efficiency of water currents passing through the body. In the simplest worms, Hox genes lay out the nerve nets that allow the worm to sense the environment all along the body. The earliest metazoans, wherever, whenever, and however they evolved, could quickly have radiated into a great variety of body shapes and structures, with natural selection acting equally quickly to weed out the shapes that were poor adaptations, and leaving a scrapbook of successful prototypes that proliferated.

The Variety of Metazoans

When one animal group is radically different from another, and is also considered to be a clade that evolved from some single ancestral species (Chapter 3), it is a **phylum**, defined by its own particular body structure, ecology, and evolutionary history. Mollusca and Arthropoda are familiar phyla. They must once have had a common bilaterian metazoan ancestor, but that ancestor wasn't a mollusc or an arthropod (by definition as well as common sense). There are arguments about the number of phyla among living metazoans, mostly because there is a bewildering variety of worm-like organisms, but most people would count about 30 phyla. Because only creatures with hard parts are easy to recognize as fossils, only nine or ten phyla are or have ever been important in paleontology (Box 4.1).

It is stunning to realize that all these phyla are known from Cambrian rocks, but only two of them are known for

Box 4.1   The Major Phyla of Fossil Invertebrates

(† indicates an extinct group)
- Porifera or sponges (includes †Archaeocyatha)
- Cnidaria
- Bryozoa
- Brachiopoda
- Mollusca
- Arthropoda
- Echinodermata
- Hemichordata (with †graptolites)
- Chordata (including vertebrates)

certain in rocks older than Cambrian. That means two things, on the face of it: first, that there was an "explosive" burst of evolution at the beginning of the Cambrian, and second, that we have no fossil record of the metazoan evolution that gave rise to the phyla that we recognize today.

Without fossil evidence of the metazoan radiation, we are forced to look at evidence from their living descendants half a billion years later, and hope it tells some semblance of the truth! One would hope that the results would be compatible with the rich fossil record that begins with Cambrian rocks.

As we have seen, the two most simple metazoan phyla are sponges and cnidarians. The bilaterians pose more of a problem: they are all complex and three-dimensional, and they all have Hox genes controlling the placement of structures along their axis of symmetry. It is difficult to find compelling reasons for ranking them in terms of order of evolutionary branching, using standard arguments from anatomy or ecology. But molecular and genetic evidence has helped us attack the problem.

Advanced bilaterians form three major clusters of phyla: Ecdysozoa, Lophotrochozoa, and Deuterostomia.

Ecdysozoa are animals that molt off their outer skins as they grow. This can be an important way of getting rid of unwanted external parasites. Molting is characteristic of the Arthropoda, for example, and for many of them it is a major evolutionary burden as well as an advantage. Crabs and lobsters must molt many times over the years of their (natural) lives, and each time they do so they are very vulnerable to predators and must spend a considerable time hiding while their new shell hardens. Some ecdysozoans have found a way to avoid this evolutionary constraint. For example, insects do not molt as adults. However, insects can only do this by having a very short adult life, with all their growth taking place in earlier life stages (as larvae). (A short adult life is an extreme but successful way of avoiding a major evolutionary constraint!) Arthropoda often have hard shells, and are the dominant members of the Ecdysozoa in the fossil record.

Lophotrochozoa are animals with a cute fuzzy little floating larva and a way of life that originally involved filter-feeding from the water. The Mollusca are the best known phylum of Lophotrochozoa, well fossilized, well understood, and very varied in their anatomy and ecology. Brachiopoda and Bryozoa are important fossil groups. Annelids (worms) have a poor fossil record.

Deuterostomia also seem to have been originally filter-feeders with floating larvae, but their larvae are so different from those of lophotrochozoans that they cannot belong to the same clade. Deuterostomes include Echinodermata (sea-stars and relatives), and Chordata (including ourselves) and minor related groups.

Current *speculation* includes the suggestion of a period during which ancestral bilaterians diverged from the metazoans, evolving maybe as many as seven sets of Hox genes as they did so. These first bilaterians may have looked like little flatworms, or perhaps like the planktonic larvae of flatworms. Then, shortly before the Cambrian, around 555 Ma, bilaterians become large enough to leave traces on seafloor sediment. The three bilaterian groups diverged, but in ways that left no significant record of body fossils. Finally, the groups split into the phyla, many with hard parts, that did leave a rich fossil record, beginning at the base of the Cambrian.

But there is an ecological twist to the evolution of bilaterians. They all build embryos as they begin cell division from the egg. They develop into free-living larvae that feed in the plankton. The larvae are bilateral; they are made of perhaps 2000 cells with only a few cell types. At the end of the larval period, cells that have simply been riding along in the larva, without specific function, begin to divide and are organized, positioned, and differentiated under the direction of the Hox genes into a complex adult metazoan animal that usually bears no resemblance to its larva, and usually has a completely different habitat and ecology.

The earliest bilaterians may have been tiny animals that looked and functioned like the simple larval stages of many of their modern-day descendants, floating and feeding in the plankton. These micrometazoans would have been small and soft-bodied, and unlikely to be fossilized, even as they diverged from one another.

At some point, conditions changed so that metazoans could flourish on the sea floor, and grow bigger. Larger size may require hard parts to work efficiently, so many early metazoans seem to have evolved skeletons of one sort and another, and began to show up in numbers and variety in the fossil record. More complex development sequences, with planktonic larvae changing into large seafloor adults, led to "explosive" evolution of many different phyla in the Cambrian. But what triggered all this?

Further Reading

Bosak, T. et al. 2011. Possible early foraminiferans in post-Sturtian (716–635 Ma) cap carbonates. *Geology* 40: 67–70. Abstract at http://geology.gsapubs.orgcontent/40/1/67.abstract

Brain C. K. et al. 2012. The first animals: ca. 760-million-year-old sponge-like fossils from Namibia. *South African Journal of Science* 108 (1/2), published online. [Otavia.] Available at http://www.sajs.co.za/index.php/SAJS/article/view/658/994

Butterfield, N. J. 2001. Probable Proterozoic fungi. *Paleobiology* 31: 165–182.

Butterfield, N. J. 2011. Terminal developments in Ediacaran embryology. *Science* 334: 1655–1656.

Cohen, P. A. et al. 2009. Large spinose microfossils in Ediacaran rocks as resting stages of early animals. *PNAS* 106: 6519–6524. Available at http://www.pnas.org/content/106/16/6519.full

Filippelli, G. M. 2010. Phosphorus and the gust of fresh air. *Nature* 467: 1052–1053.

Hoffman, P. F. et al. 1998. A neoproterozoic Snowball Earth. *Science* 281: 1342–1346. Available at http://www.ndsu.nodak.edu/pubweb/~ashworth/webpages/g440/Hoffman%26Shrag_Science.pdf

Huldtgren, T. et al. 2011. Fossilized nuclei and germination structures identify Ediacaran "animal embryos" as encysting protists. *Science* 334: 1696–1699. Available at http://eis.bris.ac.uk/~jc1224/page2/assets/Huldtgren%20et%20al%202011.pdf

Lamb, D. M. et al. 2009. Evidence for eukaryotic diversification in the 1800 million-year-old Changzhougou Formation, North China. *Precambrian Research* 173: 93–104. [Age of very early eukaryotes from China.]

Planavsky, N. J. et al. 2010. The evolution of the marine phosphate reservoir. *Nature* 467: 1088–1090.

Ratcliff, W. C. et al., 2012. Experimental evolution of multicellularity. *PNAS* 109: 1595–1600. Available at http://www.pnas.org/content/109/5/1595.full.pdf+html

Question for Thought, Study, and Discussion

Describe the science behind this limerick by Elizabeth Wenk:

The paleontologist's view
Puts worms together with you
This is based on the claim
That instead of a plane
A worm's three-dimensional too.

FIVE

The Metazoan Radiation

In This Chapter

We begin the chapter with the Ediacaran animals that populated the sea floors right after Snowball/Slushball Earth. Rangeomorphs are totally extinct, and may have fed by simply absorbing organic molecules through intricate collecting organs. The Ediacaran animals also included ancestors or relatives of living animals. Bilaterians were bilaterally symmetrical, with front, back, and sides, clearly related to moving on or in the seafloor. As they did so, they left characteristic marks in the sediments, trace fossils of their activity. Body motion for animals requires oxygen, and we feel confident that oxygen levels in the ocean were rising by the end of Ediacaran time.

The other great animal characteristic that evolved at the end of the Ediacaran was hard parts, mostly for protection. Cambrian animals are more abundant in the fossil record because many of them had hard parts. But a few special deposits have preserved animals with soft parts, showing that the Cambrian increase in fossil diversity was not just related to hard parts. It was a feature of the ocean-floor biology and ecology in general. That demands an explanation, and I discuss that issue to end the chapter.

After Snowball/Slushball Earth

The end of the great glaciations of the Cryogenian begins a major change in Earth's physical and biological evolution. There was a biological revolution as metazoans became major players in the oceans, Earth's atmosphere became more oxygen-rich, and Earth's climate was moderated so that there have never again been glaciations of Cryogenian magnitude. All these changes are connected, and if we had to use one word for the link that joins them, it would be

History of Life, Fifth Edition. Richard Cowen.
© 2013 by Richard Cowen. Published 2013 by Blackwell Publishing Ltd.

oxygen. Several new studies have revealed how the physical and biological world changed during the Ediacaran, though there is no convenient summary yet.

Large Ediacaran Animals

Many Ediacaran fossils belong to an extinct group called **rangeomorphs**, but there are Ediacaran sponges, cnidarians and bilaterians, too. Rangeomorphs became extinct at the end of the Ediacaran, at or before 543 Ma, but the others were the ancestors of the Cambrian animals that followed.

All Ediacaran animals were soft-bodied. It is only when their corpses were colonized after death by layers of bacteria that we see them at all, typically as "ghost" outlines where biofilms of bacteria compacted the sediment. We also see a few tracks and traces where mobile Ediacaran bilaterians moved on or just under the surface sediment. These Ediacaran animals colonized the seafloor, from shallow water to well below the well-lit surface zone.

In the Mistaken Point Formation in Newfoundland, Canada, we find the earliest large organized Ediacaran animals, from about 565 Ma. Here masses of rangeomorphs (and a few other animals) were killed and buried where they lived by very fine-grained volcanic ash falling through the water. The animals are preserved in great detail in three dimensions, giving us a unique opportunity to interpret their mode of life.

Rangeomorphs are animals built from small blade-shaped units ("frondlets") about 1 cm long. Young forms have only a few frondlets, but larger ones have multiple branching supports, each one bearing multiple frondlets, and growing up to a meter long. The animal is fixed to the seafloor by a circular disk or holdfast (Fig. 5.1).

There are no openings in the rangeomorph body wall, and the simplest hypothesis for their biology is **osmotrophy**: taking up dissolved nutrients from the water directly through the skin by osmosis. Each frondlet thus obtains its own nutrition, but clearly there must be some nutrient transport through the body to grow the non-feeding holdfast and the supporting tissues. The fractal arrangement of branches and frondlets approaches a mathematical optimum for an array of osmotrophic collectors (Fig. 5.2).

Many marine invertebrates get some nutrition this way, through skin, gills, or tentacles: jellyfish are just one example. Even a vertebrate, the ghastly hagfish, can burrow inside a whale carcass and absorb dissolved nutrients from the rotting flesh, through its gills and its skin. However, nutrients are not concentrated enough in most environments today to feed larger animals entirely by osmotrophy. Probably Ediacaran seafloors had more dissolved nutrients because there were few organisms eating plankton at the surface, or intercepting and eating dead and dying plankton before they decayed to release nutrients. Rangeomorphs may have died out as larger metazoans radiated at the base of the Cambrian and depleted their nutrient supplies.

Figure 5.1 The rangeomorph *Avalofructus* from the Mistaken Point Formation in Newfoundland. Reconstruction of a large specimen showing branches, frondlets, and the basal holdfast. Image from Narbonne et al. (2009). © Guy Narbonne and The Paleontological Society, used by permission.

Other Ediacaran animals include distinct bilaterians. *Dickinsonia* (Fig. 5.3) is flat and large, and also seems to have fed by osmotrophy through its lower side as it moved across the sea floor. *Kimberella* (Fig. 5.4) may be evolving toward a slug-like early mollusc, and may have grazed on algal mats. Some Ediacaran fossils resist interpretation.

How could these large metazoans survive if Ediacaran environments had low-oxygen conditions, as would certainly have been the case for the rangeomorphs at Mistaken Point? Osmotrophic animals today have very low metabolic rates, so the Ediacaran rangeomorphs, and *Dickinsonia*, probably had the same low oxygen requirements.

However, some Ediacaran animals left trace fossils of their burrowing activity in and on the surface, and these presumably were bilaterians using a coelom to move through the sediment: a relatively high-energy way to move about.

A large coral reef complex lies off the north coast of Venezuela, around the Las Roques islands. Some shallow lagoons are warm and very salty, so that normal marine animals do not live there. Cyanobacterial mats flourish in very shallow water, and produce oxygen by day under the tropical sun. The water immediately around the mats can

(a)   SELF-SIMILAR ORGANIZATION

A   B   C   D

FIRST ORDER   SECOND ORDER   THIRD ORDER   FOURTH ORDER
'PRIMARY'   'SECONDARY'   'TERTIARY'   'QUATERNARY'

Primary Branch = B   Secondary Branch = C   Tertiary Branch = D   Quaternary Branch

(b)   BRANCH PIVOTING

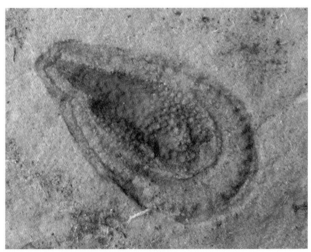

Figure 5.2 a) diagram to show how rangeomorph frondlets are organized, with serial levels of complex branching. b) diagram showing how rangeomorphs were able to pivot branches, presumably at all levels, to best intercept nutrient-laden water currents. Both images from Narbonne et al. 2009. © Guy Narbonne and The Paleontological Society, used by permission.

Figure 5.3 *Dickinsonia*, a bilaterian from the Ediacaran. Scale in cm. Image by Merikanto, and placed into Wikimedia.

Figure 5.4 *Kimberella*, a bilaterian from the Ediacaran of Russia About 1 cm long. Image by Aleksey Nagovitsyn, and placed into Wikimedia.

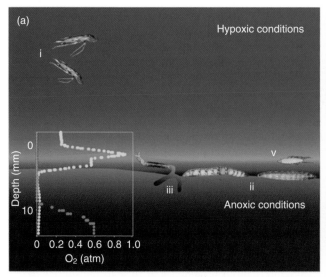
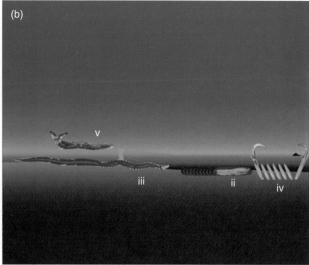

Figure 5.5 Oxygen miners. Left, results found today at Las Roques, Venezuela. An oxygen-rich algal mat covers the surface under oxygen-poor lagoon water and above sea floor sediment with no oxygen. (Daytime oxygen levels are shown as yellow dots, sulfur levels as red dots.) Small swimming metazoans make forays into the water, but spend a lot of time on the bottom. Others crawl on or burrow into or under the algal mats, mining them for oxygen and food. Right, analogous conditions inferred for some Ediacaran sea floors, with small metazoans living and feeding close to oxygen-rich mats. From Gingras et al. (2011), used by permission.

contain up to four times normal oxygen levels. At night photosynthesis stops, and oxygen levels drop sharply toward zero, except for oxygen bubbles trapped in and under the mats. Thus the shallow water of the lagoon has generally low oxygen levels, but the mats form "oxygen oases" in an "oxygen desert". A few metazoans flourish around the mats. Some of them graze the mats, others eat organic sediment. All of them live close to the mats, burrowing into them or under them, "mining" oxygen to support their metazoan metabolism (Fig. 5.5, left).

These modern studies (Gingras et al. 2011) are important because metazoans left fossil burrows and trails in Ediacaran rocks. Many Ediacaran burrows are very similar to the burrows in the mats at Las Roques (Fig. 5.6), so Gingras et al. confidently infer that Ediacaran metazoans survived low oxygen conditions by oxygen mining in and around bacterial mats (Fig. 5.5, right). In this sense, Ediacaran bacterial mats were forcing-houses of metazoan evolution, in the same way that stromatolites were forcing-houses of eukaryote evolution. In each case, biological events in just a tiny portion of the global ecosystem took on an importance much greater than the area involved.

Oxygen and Metazoans

Atmospheric oxygen levels seem to have risen slowly in the Ediacaran, as slightly more oxygen was generated than was consumed. Usually oxygen produced in photosynthesis is used up in the water by plants and animals respiring it to

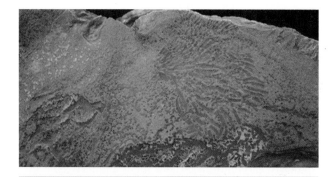

Figure 5.6 Trace fossil of an Ediacaran oxygen miner from Australia. This trace was made by a burrowing metazoan in a horizontal plane immediately under a fossil bacterial mat, probably mining it for oxygen and food. From Gingras et al. (2011), used by permission.

oxidize their food, and oxygen is also used up when organic tissue decays. However, if oxygen bubbled or diffuses off into the atmosphere, organic matter may fall unoxidized to the sea floor. If it is then buried quickly it is taken out of the chemical cycle, and oxygen levels in the ocean and atmosphere increase.

During the later Ediacaran, this process featured a positive feedback between metazoans and ocean chemistry, in one of those revolutionary, co-evolutionary changes that have radically affected Earth history. This change is comparable in significance with the "Great Oxidation Event" of the early Proterozoic (Chapter 3). It set the stage for the Phanerozoic Era that followed the Ediacaran. The idea is not new, but it has been summarized and strengthened recently by Nicholas Butterfield (Butterfield 2011).

The argument goes like this. Metazoans, especially bilaterians, have guts (Chapter 4). They process their food, and produce carbon-rich waste in compact fecal pellets. Those fecal pellets drop quickly through the water and if they are buried quickly, that will increase oxygen levels in the sea and in the atmosphere. So the rise of metazoans big enough to produce quantities of fecal pellets led to a rise in oxygen. In turn, higher oxygen levels permit larger, more active metazoans to evolve, and so on.

As oxygen levels increase and reach deeper and deeper water, metazoans could then settle on and exploit sea floor sediments for the first time. The sea floor had been accumulating rich organic sediment for many millions of years. New bottom-dwelling metazoans could have evolved adaptations for crawling and deposit-feeding, by modifying the Hox gene complexes that now coded for larger and more complex animals, comparable in size and power to the "adult" metazoans we see today. Sponges too could have adapted to seafloor life by specializing for capturing bacteria by filtering water. Cnidarians, perhaps already large feeders in surface waters, may have evolved the sessile polyp configuration at this time.

All of this happened during the later Ediacaran, when metazoans were still soft-bodied. But it laid the anatomical framework for the Cambrian period that followed, when many metazoans evolved hard parts.

The Ediacaran evolution of larger metazoans may have prevented the recurrence of the extreme glaciations of Snowball or Slushball Earth. Surface productivity could no longer draw down carbon dioxide to critically low levels because primary producers were eaten back by new planktonic predators (small or larval metazoans). Stronger metazoan burrowers (arthropods and segmented worms) dug up buried carbon from seafloor muds and recycled it into carbon dioxide. And complex populations in nearshore waters intercepted nutrients before they reached the oceanic sea surface.

All this is another example of the continuous interplay between life and Earth's physical environment. The evolution of metazoans was made possible by the increase in oxygen levels that resulted from increased photosynthesis, which in turn resulted from increased nutrients released by major glaciations. But the evolution of metazoans also acted to moderate the dramatic shifts in Earth's climate.

Geneticists have been arguing that metazoan roots are deep in the Precambrian, and paleontologists have been arguing that if so, there is no fossil evidence of them. This controversy is resolved by the fossils from Doushantuo: diverse and abundant, tiny, but rarely preserved. As oxygen levels increased and encouraged the evolution of larger metazoans, this set the stage for the demise of the osmotrophic Ediacaran animals on the sea floor and the success of benthic metazoans. Simon Conway Morris wrote that

"the Cambrian explosion was largely ecological", and it is clear that he is right.

The Evolution of Skeletons

One of the most important events in the history of life was the evolution of mineralized hard parts in animals. Beginning rather suddenly at the beginning of the Cambrian, the fossil record contains skeletons: shells and other pieces of mineral that were formed biochemically by animals. Humans have one kind of skeleton, an internal skeleton or endoskeleton, where the mineralization is internal and the soft tissues lie outside. Most animals have the reverse arrangement, with a mineralized exoskeleton on the outside and soft tissues inside, as in most molluscs and in arthropods (Fig. 5.7). The shell or test of an echinoderm is technically internal but usually lies so close to the surface that it is external for all practical purposes. The hard parts laid down by corals are external, but underneath the body, so that the soft parts lie on top of the hard parts and seem comparatively unprotected by them. Sponge skeletons are simply networks of tiny spicules that form a largely internal framework. There is incredible variety in the type, function, arrangement, chemistry, and formation of animal skeletons: biomineralization is a whole science in itself.

With the evolution of hard parts, the fossil record became much richer, because hard parts resist the destructive agents that affect the soft parts of bodies. The evolution of hard parts defines the beginning of a new eon in Earth

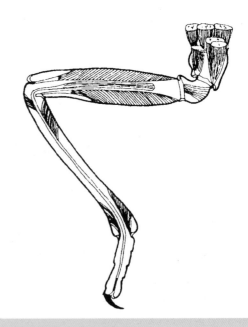

Figure 5.7 An arthropod leg. Arthropods have jointed exoskeletons operated from inside by muscles and ligaments. From Barnes et al., *The Invertebrates: A Synthesis*, 3rd edn., © 2001 Blackwell Science.

history, the **Phanerozoic**, a new era, the **Paleozoic**, and its oldest subdivision, the **Cambrian** Period. Many different ways of life, using many different hard parts in many different body plans, seem to have been explored as soon as animals evolved the biochemical pathways for making hard parts. Thus the Cambrian "explosion" was genetic as well as ecological, and dramatic indeed to paleontologists collecting their fossils. The new animals evolved the features that allow us to identify most of them as members of the metazoan phyla that survive today.

Late Ediacaran and Early Cambrian: Small Shelly Fossils

In Siberia and China, rocks in the very latest Ediacaran contain some small shells in the form of tiny cones and tubes that we don't understand properly (Fig. 5.8). Some survive into the earliest Cambrian, where they are joined by sponges and tiny molluscs. Soon archaeocyathid sponges were forming large reef patches (Fig. 5.9).

Small shelly fossils are now known worldwide, but for perhaps 20 m.y., there were no animals larger than a few millimeters long except for the archaeocyathid sponges. The next stage of the Cambrian saw the appearance of more abundant and more complex creatures, worldwide, in a few million years after 520 Ma. Dominant among these animals were trilobites, brachiopods, and echinoderms.

Larger Cambrian Animals

Trilobites are arthropods, complex creatures with thick jointed armor covering them from head to tail (Fig. 5.10). They had antennae and large eyes, they were mobile on the seafloor using long jointed legs, and they were something like crustaceans and horseshoe crabs in structure. They did not have the complex mouth parts of living crustaceans, so

their diet may have been restricted to sediment or very small or soft prey. They burrowed actively, leaving traces of their activities in the sediment, and they are by far the most numerous fossils in Cambrian rocks. The number of fossils they left behind was increased by the fact that they molted

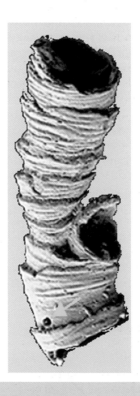

Figure 5.8 The small shelly fossil *Cloudina*, from the late Ediacaran of China. The animal, whatever it was, grew a tube-like shell about 0.5 mm in diameter. It presumably collected food from the water. But notice the trace fossil: a hole bored through the shell, presumably by an unknown predator. Image by Philcha, and placed into Wikimedia.

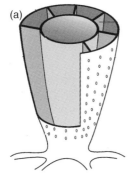 

Figure 5.9 a) diagram of a Cambrian archaeocyathid sponge. Water was taken in through the side of the colony, filtered through a compartmented body, and expelled into a central exit cone. Image by Muriel Gottrop, and placed into Wikimedia. b) an Early Cambrian archaeocyathid reef exposed by erosion on the coast of southern Labrador. Undergraduate researcher Hannah Clemente of Smith College for scale. Image © Dr. Sara Pruss of Smith College, used by permission.

their armor as they grew, like living crustaceans. Thus, a large adult trilobite could have contributed twenty or more suits of armor to the fossil record before its final death. Even allowing for this bias of the fossil record, it is clear that Cambrian seafloors were dominated by trilobites. Other large arthropods are also known from Early Cambrian rocks, although they are much less common.

Brachiopods are relatively abundant Cambrian fossils, creatures that had two shells protecting a small body and a large water-filled cavity where food was filtered from sea-

water pumped in and out of the shell (Fig. 2.1). Cambrian brachiopods lived on the sediment surface or burrowed just under it.

These animals are large, and they are easily assigned to living phyla. For the first time, the seafloor would have looked reasonably familiar to a marine ecologist. Trilobites probably ate mud, and brachiopods gathered food from seawater.

Soft-Bodied Cambrian Animals

I have so far discussed the "Cambrian explosion" as if it related entirely to the evolution of skeletons. While this is basically true in terms of fossil abundance, there was also dramatic evolution at the same time among animal groups with little or no skeleton. Trace fossils—tracks, trails, and burrows—increase in abundance at the beginning of the Cambrian, and soft-bodied animals appeared with some amazingly sophisticated body plans.

Many soft-bodied animals were preserved by quirks of the environment in Early Cambrian rocks in South China (the Chengjiang Fauna), and in Middle Cambrian rocks in the Canadian Rockies (in the Burgess Shale). Similar fossils are now known from Cambrian rocks in several other places. I shall call them all the "Burgess Fauna".

More than half the Burgess animals burrowed in or lived freely on the seafloor, and most of these were deposit feeders. Arthropods (such as *Marrella*, Fig. 5.11a) and worms dominate the Burgess Fauna. Only about 30% of the species were fixed to the seafloor or lived stationary lives on it, and these were probably filter-feeders, mainly sponges and worms. Thus, the dominance of most Cambrian fossil collections by bottom-dwelling, deposit-feeding arthropods is not a bias of the preservation of hard parts: it occurs among soft-bodied communities too. Trilobites are fair representatives of Cambrian animals and Cambrian ecology.

The main delights of the Burgess Fauna are the unusual animals, which have provided fun and headaches for pale-

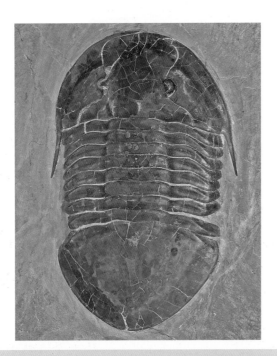

Figure 5.10 A trilobite, *Megalaspides* from the Ordovician of Ohio. Two prominent eyes are set on the headshield, with lines of weakness running past them to make molting easier. Image by Llez (H. Zell), and placed into Wikimedia.

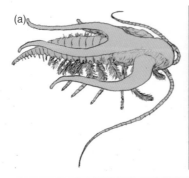
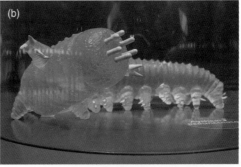

Figure 5.11 a) a complex arthropod, *Marrella*, from the Burgess Shale. Drawing by Ghedoghedo and placed into Wikimedia. b) a compelling glass-fiber model of the lobopod *Aysheaia*. Image by Eduard Solà Vázquez, and placed into Wikimedia.

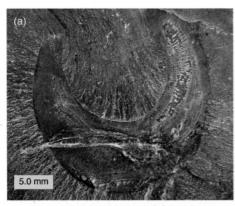
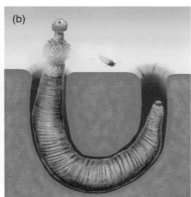

Figure 5.12 The priapulid worm *Ottoia* from the Burgess Shale. A typical *Ottoia* is about 30 mm. long (stretched out!). a) photograph by Dr. Mark Wilson of the College of Wooster, and placed in Wikimedia. b) reconstruction of an *Ottoia* in its burrow. Image by Smokeybjb and placed in Wikimedia.

ontologists. *Aysheaia* is a lobopod: it looks like a caterpillar, with thick soft legs (Fig. 5.11b). It has stubby little appendages near its head (Fig. 5.11c) that may be slime glands for entangling prey. *Hallucigenia*, named for its bizarre appearance, is a lobopod with spines.

There are predators in the Burgess fauna. Priapulid worms today live in shallow burrows and capture soft-bodied prey by plunging a hooked proboscis into them as they crawl by. The Burgess priapulid *Ottoia* (Fig. 5.12) probably did the same.

Anomalocarids are the most spectacular Cambrian predators. They are an extinct group of animals related to arthropods: pieces of Burgess animals suggest that they could have been a meter long! Anomalocarids have been difficult to reconstruct because they are usually found as pieces that have to be fitted together. They were swimmers, and all of them had very large grasping appendages, and a mouth with scraping or piercing saw-like edges. Although they have a very lightly built outer skeleton, they would have been powerful predators, especially on equally thin-skinned prey. *Anomalocaris* itself was the largest, at a meter long, and *Peytoia* is the best known (Fig. 5.13). *Opabinia* was highly evolved, long and slim, with a vertical tail fin. It had five eyes and only one large grasping claw on the front of its head (Fig. 5.14).

The eyes of *Anomalocaris* are astounding. Eyes discovered in Early Cambrian rocks of South Australia are preserved in such detail that one can estimate that each eye had 16,000 little lenses (Paterson et al. 2011). This would have given *Anomalocaris* a finely detailed image of its surroundings and its potential prey. The discovery adds to the picture of a highly effective predator in the Early Cambrian. Don't forget, too, that an eye with 16,000 lenses must have had a complex system of visual receptors and nerve networks to transmit the images from each eye to the brain. And since arthropods molt their outer covering in order to grow, every anomalocarid would have molted off its eye lenses with each growth stage, and then would have had to

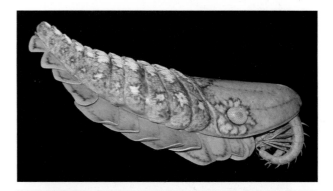

Figure 5.13 A model reconstruction of the anomalocarid *Peytoia* from the Burgess Shale. The grasping appendages are curled around the mouth, and the animal is posed in swimming position. *Peytoia* was typically about 60 cm (2 feet) long. (The blue eyes are appealing, but the color is speculation!) Model by Espen Horn. Photograph by Llez (H. Zell), and placed by him in Wikimedia.

grow a new lens system, precisely coordinating with the exposed visual system under it. A meter-long anomalocarid could easily have molted twenty times during its life, growing a new and larger lens system each time.

Wiwaxia (Fig. 5.15) is a flat creature that crept along the seafloor under a cover of tiny scales that were interspersed with tall strong spines. Halkieriids, best known from the Burgess fauna of Greenland, look like flattened worms, with perhaps 2000 spines forming a protective coating embedded into the dorsal surface. Yet two distinct subcircular shells are embedded in the upper surface close to each end. We do not yet know what the nearest living relatives of these creatures are.

The Burgess animals also include wormlike creatures that are identified as early chordates and vertebrates: in

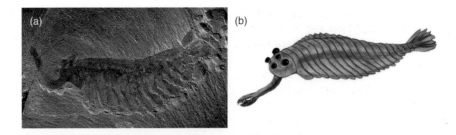

Figure 5.14 *Opabinia* from the Burgess Shale, about 6 cm long. It is clearly an anomalocarid, but has only one large appendage, and multiple eyes. a) photograph by Jstuby, and placed into Wikimedia. b) reconstruction by Nobu Tamura, and placed into Wikimedia.

Figure 5.15 *Wiwaxia* from the Burgess Shale, about 5 cm long. Photograph by Jstuby, and placed into Wikimedia.

other words, the remote ancestors of ourselves and all other vertebrates (Chapter 7).

Altogether, the Burgess faunas give us a good idea of the exciting but extinct soft-bodied creatures that may always have lived alongside the trilobites but were hardly ever preserved. They show that the variety of Cambrian life on and near the seafloor was greater than the typical fossil collection would suggest.

The Cambrian Explosion

The waves of evolutionary novelty that appeared in the seas during the Early Cambrian have few parallels in the history of life. Many groups of fossils appeared quite sud-

denly in the fossil record, thanks to their evolution of skeletons, sometimes at comparatively large body size. Given the Ediacaran legacy of metazoans and relatively high oxygen levels, however, it is most likely that the Cambrian explosion simply records the invention and exploitation of skeletons for many good reasons associated with locomotion (walking, digging and swimming), size, support, defense, and other functions, made even more complex by the fact that animals interact ecologically with other species as they evolve.

A skeleton may support soft tissue, from the inside or from the outside, and simply allow an animal to grow larger. Therefore, sponges could grow larger and higher after they evolved supporting structures of protein or mineral (Fig. 4.10), and they could reach further into the water to take advantage of currents and to gather food. Large size also protects animals from predators large and small. A large animal is less likely to be totally consumed, and in an animal like a sponge that has little organization, damage can eventually be repaired if even a part of the animal survives attack. As skeletons evolved, even for other reasons, they helped animals to survive because of their defensive value.

Early echinoderms had lightly plated skeletons just under their surfaces, and the most reasonable explanation of their first function is support, accompanied or followed by defense.

For other animals, skeletons provided a box that gave organs a controlled environment in which to work. Filters were less exposed to currents, so perhaps they would not clog so easily from silt and mud (Fig. 2.1). A boxlike skeleton would also have given an advantage against predation. Molluscs and brachiopods may have evolved skeletons for these reasons.

In yet other animals, hard parts may have performed more specific functions. We have already seen that worms tend to burrow head-first in sediment. But after penetrating the sediment they squirm through it (Fig. 4.14). A worm that evolved a hardened head covering could use a different and perhaps better technique, shoveling sediment aside like a bulldozer. Richard Fortey suggested that the large head shield of trilobites was evolved and used in this fashion.

But arthropods, and especially trilobites, are strongly armored all over their dorsal surfaces, not just in the head region. Most likely their armor served for the attachment of strong muscles. Muscles pull and cannot push. Worms move by using internal hydraulic systems, as we have seen. On the other hand, walking and digging demands that limbs push on the sediment, and that is very unrewarding if the other end of the leg is unbraced. Arthropods evolved a large, strong dorsal skeleton against which their jointed legs were firmly braced, allowing them to move much more efficiently than worms or lobopods do.

But despite all the discussion of skeletons, the Burgess animals show that dramatic evolution took place also in animals that did not have strong skeletons. Many of these animals had outer coverings that were tough, but lightly mineralized: the Burgess arthropods are particularly good examples.

The common factor along successful groups of Cambrian animals is larger body size. This suggests that in some way the world had become hospitable to large animals, and in turn that tells us that the Cambrian event was driven by worldwide ecological factors. Those factors could have been related to a change in food supply in the sea, which in turn depends on upwelling, which in turn depends on global climatic and geographic patterns. They could have been related to oxygen levels (large animals need more oxygen than small ones), but oxygen levels depend on productivity and burial of organic matter. We don't yet know enough about Cambrian geography, climate, and geochemistry to say anything sensible about these factors, but this global level is being investigated as our knowledge increases.

Some specific mechanisms have been suggested to explain the Cambrian explosion.

Predation. The predation theory has two aspects. The first is a general ecological argument. The ecologist Robert Paine removed the top predator (a starfish) from rocky shore communities on the Washington coast and found that diversity dropped. In the absence of the starfish, mussels took over all available rocky surfaces and smothered all their competitors. Paine suggested that a major ecological principle was at work: effective predators maintain diversity in a community. If a prey species becomes dominant and numerous, the top predator eats it back, maintaining diversity by keeping space available for other species.

Steven Stanley used Paine's work to suggest that the evolution of predation triggered the Cambrian radiation. Stanley made an intellectual jump to suggest that predators can cause additional diversity in their prey. He argued that if predators first appeared in the Early Cambrian, they may have caused the increase in diversity at that time. Perhaps predators also encouraged the evolution of many different types of skeletonized animals.

Geerat Vermeij supported Stanley's idea, suggesting how new predators might indeed cause diversification among prey (at any time). In response to new predators, prey creatures might evolve large size, or hard coverings made from any available biochemical substance, or powerful toxins, or changes in life style or behavior (such as deeper burrowing), or any combination of these, all to become more predator-proof. And as the new predators in turn evolve more sophisticated ways of attacking prey, the responses and counter-responses might well add up to a significant burst of evolutionary change.

The rules of the predator/prey game probably changed radically as large metazoans evolved. Many Early Cambrian fossils have hard parts that look defensive. Some sharp little conical shells called sclerites may have been spines that were carried pointing outwards on the dorsal and lateral sides of animals, to fend off predators. There are armored and spined Early Cambrian animals, and some Early Cambrian trilobites have healed injuries that may indicate damage by a predator. Defensive structures made of hard parts could therefore have contributed to the increase in the number of fossils in Early Cambrian rocks.

So predation played an important part in generating the Cambrian event. The only major predators we have discovered are the anomalocarids, but they are certainly impressive animals. However, predation alone does not explain the timing of the Cambrian explosion: why not 100 m.y. earlier, or later? And predation alone cannot account for all the variety of skeletons that we see.

Oxygen Levels. The evolution of large bodies and skeletons was made possible, or encouraged, by high oxygen concentrations. Shells and thick tissues prevent the free diffusion of oxygen into a body (Fig. 4.15), so they could not have evolved unless there was a high enough oxygen level to push oxygen into the body through the few remaining areas of exposed tissue: through gills, for example. This also cannot be the whole story, because sponges and cnidarians could have evolved their skeletons (which do not inhibit respiration) in low oxygen conditions. Again, oxygen levels could explain much of the Cambrian explosion.

Exploiting the Sea Floor. Cambrian animals, whether they had skeletons or not, lived in and close to the sea floor. There was likely a rich supply of organic matter there. Ediacaran sea floors and Cambrian sea floors were ecologically different, so that people sometimes refer to the "Cambrian substrate revolution" (Fig. 5.16), marked by larger animals that dug and burrowed deeper and more powerfully than before, leaving trace fossils and sediment disturbance to document their life style.

Jack Sepkoski's "Cambrian Fauna" (Chapter 6) is dominated by trilobites, and the dominant disturbers of Cambrian seafloor sediment were trilobites. Their multiple limbs were effective at walking, digging, and stirring up the organic-rich Cambrian seafloor mud. They had small mouths and rather ineffective appendages round the mouth, so they probably ate mud: a lot of it. These "ecosystem engineers" were also "evolutionary engineers", because their activity helped to underpin the "Cambrian substrate revolution". This vivid phraseology helps to

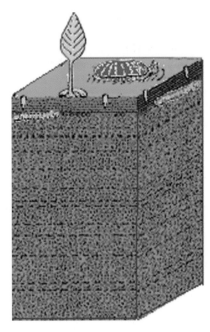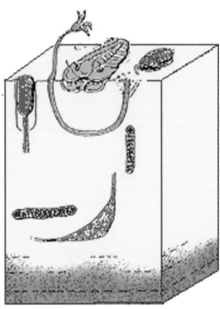

Figure 5.16 The Cambrian Substrate Revolution. Left, an Ediacaran sea floor, with oxygen miners concentrated immediately around a surface bacterial mat. Right, a Cambrian sea floor, with powerful burrowers and diggers. Image by Pilcha, and placed into the public domain.

describe one of the fundamental changes in the Earth system at the beginning of the Cambrian. Butterfield (2011) gives a concise and convincing summary.

After all these specific suggestions for the Cambrian explosion, we come back to the general overall idea of the synergistic relationship between metazoans and oxygen levels as the root cause of the dramatic change in the physical and biological worlds (giving the timing), with predation as a major accelerant for the dramatic changes in body plans, especially in hard parts of the skeleton.

After the dramatic changes in the early Cambrian, the continued increase in numbers and diversity of fossils later in the period seems anticlimactic. Cambrian fossil collections are not complex ecologically; they are dominated by trilobites, most of which lived on the seafloor and were deposit feeders. Filtering organisms are very much secondary, and large carnivores are represented only by anomalocarids.

The Cambrian explosion is spectacular, but it is not unique; the spectacular diversification of the diapsid reptiles, especially archosaurs, in the Triassic is an analogous case (Chapter 11), as is the diversification of the mammals in the Paleocene (Chapter 17). These radiations stand out from "normal" evolutionary events just as "mass extinctions" stand out from the rest (Chapter 6). On a real planet inhabited by real organisms, evolutionary rates are likely to vary in time and space, and evolutionary events are likely to vary in magnitude, duration, and frequency. We should not expect that ideal rules we might propose for an ideal planet would be followed by the natural world; instead, we have to find out from that natural world what the rules actually were.

Given a healthy fossil record, we can explore from the Cambrian onward how life varied through time, along with the physical changes on Earth. So the next chapter deals with changing life on a changing planet, with much more evidence to help us.

Further Reading

Budd, G. E. 2008. The earliest fossil record of the animals and its significance. *Philosophical Transactions of the Royal Society B* 363, 1425–1434.

Butterfield, N.J. 2011. Animals and the invention of the Phanerozoic Earth system. *Trends in Ecology & Evolution* 26: 81–87.

Gingras, M. et al. 2011. Possible evolution of mobile animals in association with microbial mats. *Nature Geoscience* 4: 372–375.

Laflamme, M. et al. 2009. Osmotrophy in modular Ediacara organisms. *PNAS* 106: 14438–14443. Available at http://www.pnas.org/content/106/34/14438.full

Laflamme, M. et al. 2012. Ecological tiering and the evolution of a stem: the oldest stemmed frond from the Ediacaran of Newfoundland, Canada. *Journal of Paleontology* 86: 193–200.

Narbonne, G. M. 2004. Modular construction of early Ediacaran complex life forms. *Science* 305: 1141–1144. Available at http://geol.queensu.ca/people/narbonne/NarbonneScienceRangeomorphs2004.pdf

Narbonne, G. M. 2005. The Ediacara biota: Neoproterozoic origin of animals and their ecosystems. *Annual Review of Earth and Planetary Sciences* 33: 421–442. Available at http://geol.queensu.ca/people/narbonne/NarbonneAREPS2005Final.pdf

Narbonne, G. M. et al. 2009. Reconstructing a lost world: Ediacaran rangeomorphs from Spaniard's Bay, Newfoundland. *Journal of Paleontology* 83: 503–523.

Narbonne, G. M. 2010. Ocean chemistry and early animals. *Science* 328: 53–54.

Paterson, J. R. et al. 2011. Acute vision in the giant Cambrian predator Anomalocaris and the origin of compound eyes. *Nature* 480: 237–240.

Peterson, K. J. et al. 2008. The Ediacaran emergence of bilaterians: congruence between the genetic and the geological fossil records. *Philosophical Transactions of the Royal Society B* 363, 1435–1443.

Sperling, E. A. and J. Vinther. 2010. A placozoan affinity for Dickinsonia and the evolution of late Proterozoic metazoan feeding modes. *Evolution & Development* 12: 201–209.

Sperling, E. A. et al. 2011. Rangeomorphs, Thectardis (Porifera?) and dissolved organic carbon in the Ediacaran oceans. *Geobiology* 9, 24–33.

Xiao, S. and M. Laflamme. 2009. On the eve of animal radiation: phylogeny, ecology and evolution of the Ediacara biota. *Trends in Ecology and Evolution* 24: 31–40.

Questions for Thought, Study, and Discussion

1. Briefly describe the rangeomorphs and their unusual body construction. What is the best explanation (at the moment) of their way of life? Why do so few animals today live like this?

2. Choose one of the stranger Burgess animals (that is, not a crustacean or a sponge or other familiar animal). Decribe its body and explain how it might have lived.

3. How can predators encourage animal diversity to increase? After all, they are eating them!

SIX
Changing Life in a Changing World

In This Chapter

The great diversity of Earth's life today depends heavily on the energy flow through ecosystems. This flow, mostly food supply, varies across the Earth with geography and climate, so today we see geographic provinces, each with its own ecosystems. That pattern is seen for the Earth's past 500 million years, preserved in the fossil record. So counts of the diversity of Earth's fossils through time and space have been invaluable in piecing together broad patterns in the diversity of life. Because Earth's geography is altered by the motion of plates of crust (oceans and continents), so its

ecology is affected too. Even more significant, three major Faunas can be defined, with the Modern Fauna succeeding the Paleozoic Fauna, which in turn had replaced the Cambrian Fauna. There have been episodes of dramatic extinction, called "mass extinctions", but it turns out that each mass extinction is unique, with its own set of potential causes. I have room to discuss the biggest one, the PermoTriassic Extinction at length, and the others only in outline. Finally, there are times of rapid evolutionary innovation, each again with its own causal factors.

Today's World

Life has evolved on a planet that has changing geology, geography, and climate. Life did not evolve in random patterns, either; to understand the fossil record, we have to look at the relationships between the physical and biological world on which it lives. With a marine fossil record that has left us a reasonable picture of more than 500 million years of evolution, we can look at the patterns of life through Phanerozoic time: the Paleozoic, Mesozoic, and

History of Life, Fifth Edition. Richard Cowen.
© 2013 by Richard Cowen. Published 2013 by Blackwell Publishing Ltd.

Cenozoic eras. But let us look at the modern world first.

Energy

Today's world teems with life in the oceans and on land. Land plants reach roots into the soil to extract nutrients and water. They bind surface sediment and make it resistant to water erosion. Their old leaves and their dead tissues enrich the surface with organic matter. The fungi and bacteria that live in and around their roots work on rocks and sediment to break them down chemically, again deepening and enriching the soil. Plant production runs off down streams and rivers to the coast and into the ocean, carrying nutrients into the marine world.

In the ocean, complex sets of organisms operate photosynthesis in the surface waters, enriched by land-derived nutrients and by upwelling of water from below. A cascade of creatures feeds on this surface nutrition, and in the end, organic matter is cycled and re-cycled, but some of it ends up in seafloor sediment, where it powers benthic communities and then may be buried.

The global ecosystem is thus powered by energy flows between one part of the system and another. Although energy seems to flow mainly downward in the scheme I have just described, physical mechanisms such as upwelling in the ocean reverse that to some extent. But we also have to recognize that the system is largely solar-powered by photosynthesis, so the downward flow is based on a real factor.

As we look backwards in geological time, we recognize immediately that today's global ecosystem has evolved dramatically. It is only a little over 100 million years ago that flowering plants evolved. They were better able than their predecessors to colonize difficult environments on land, and their success added greatly to the supply of energy reaching the oceans. It is only about 400 million years ago that the first forests evolved on Earth, and they too added a lot of energy to the land-based photosynthesis.

Before 500 million years ago, with no important life on land, there would have been very little soil. Every rain would have caused a flash flood, leaving mostly bare, unweathered rock on the continents. Chemical and organic nutrient flow to the ocean would have been small, so marine productivity must have been some unknown but small fraction of today's.

There were other "feedback loops", too. Lower productivity means less carbon being caught up into organic tissues and sediments. With less carbon being buried, oxygen levels in the oceans and atmosphere would not only have been lower than today's, but would have been more vulnerable to severe swings. With more frequent and more dangerous changes in the global environment, ecosystems would have been less stable, more prone to disruption or even destruction. In addition, with low oxygen levels, individual organisms would have been less energetic, smaller and weaker, and unable to evolve some of the spectacular ways of life that we admire and respect in today's animals.

It is clear that in the overall history of life on Earth, one great trend has been an increase in energy available, and energy used to carve out specific ways of life. For example, it is probably not an accident that the first great powerful predators evolved in the ocean in the Devonian, as land plants spread quickly. (Placoderms, discussed in Chapter 7). Flight in insects evolved as the first great global forests spread from Equator to the poles (Chapter 7).

But other ways of assessing changes in life depend more on the geography of Earth at a particular time, which changes but does not show a directional trend.

Provinces

We all know that most creatures live only in a certain ecology in a certain part of the world: for example, kiwis eat insects at night in New Zealand, and sloths eat leaves in South American rain forests. In the sea, organisms also occur in characteristic sets of species called **communities**, living together in certain types of habitat—rocky shore communities, mudflat communities, and so on. As an example, the northwest coast of North America, bathed by cool water, has a characteristic rocky-shore community of plants and animals that looks much the same from British Columbia to Central California. In turn, the coastal communities of the world can be arranged into geographically separate **provinces** (Fig. 6.1), with each province containing its own set of communities, such as the Oregonian and Californian Provinces of western North America.

Provinces are real phenomena, not artifacts of a human tendency to classify things. There are natural ecological breaks on the Earth's surface, usually at places where geographic or climatic gradients are sharp, so that one may pass from one environmental regime to another in a short distance. A classic example is at Point Conception on the California coast. Here, the ocean circulation patterns cause a sharp gradient in water temperature. In human terms, Point Conception marks the northern limit of west coast beaches for surfing without a wet suit, but marine creatures surely feel that difference too. The communities on each side of Point Conception are very different, so a provincial boundary is drawn here, with the Oregonian province grading very sharply into the Californian province (Fig. 6.1).

As provinces are identified around the coasts of the world, it seems that the number of species in common between neighboring provinces is usually 20% or less. About 30 provinces have been defined along the world's coasts, mostly on the basis of molluscs, which are obvious, abundant, and easily identified members of coastal communities. Some provinces are very large because they inhabit long coastlines that lie in the same climatic belt (the Indo-Pacific, Antarctic, and Arctic Provinces); some are small, like the Zealandian Province, which includes only

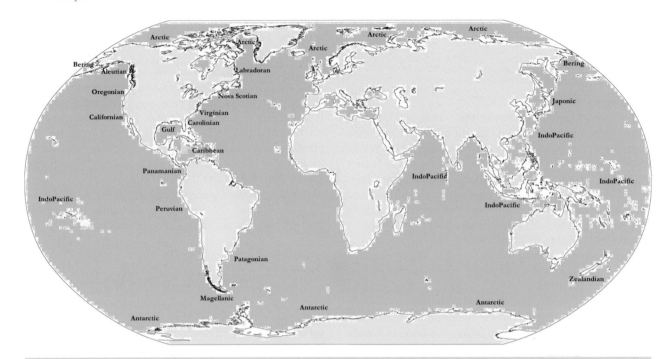

Figure 6.1   James Valentine pointed out forty years ago that today's marine biosphere includes 30 or so biological provinces along the coasts of the world. This figure does not show all of them, but I have included all the provinces around the Americas, to stress the differences between east and west coasts and the strong latitudinal gradient that sets many provinces along north-south coastlines. In contrast, the Arctic province is very large, because organisms migrate easily east-west around the Arctic Ocean, and the Indo-Pacific province is enormous, because marine organisms migrate easily along east-west coastlines and island chains. The Zealandian province is small because it can occupy only a restricted area of shallow shelf. Land masses, yellow; ocean, blue; low-lying land and shallow coastal waters, white.

the communities around the coasts of New Zealand (Fig. 6.1).

Each province contains its own communities and therefore carries unique sets of animals that fill various ecological niches. For example, the intertidal rocky-shore community in New Zealand has its ecological equivalent in British Columbia, even though the families and genera of animals are quite different in the two communities.

The total diversity of the world's shallow marine fauna directly reflects the number of provinces, which in turn reflects climate and geography. But if tectonic movements were to change Earth's geography enough, they would also alter the number of provinces of organisms, and that in turn would increase or decrease world diversity. Other things being equal, a world with widely split continents would have greater lengths of shoreline, scattered around the world, giving many marine provinces and high diversity of life.

Poles and Tropics

The Equator has a fairly uniform climate, and the same applies to the broad tropical zone on Earth, which lies between 23.5° North and South. The sun is always strong in the tropics, and the temperature variation between seasons is small. The general result is that food supply is stable, available at about the same level all year round. Therefore a species can rely on one or two particular food sources that are always available. As each species comes to depend on a narrow range of food sources, it adapts so well to harvesting them that it cannot easily switch to alternatives. Thus, a great variety of specialized species may evolve, competing only marginally with one another, at least for food. For example, on the Serengeti plains of East Africa, several species of vultures are all scavengers on carcasses. But the lappet-faced vulture has a head and beak capable of tearing through the tough hide of a fresh carcass, the white-backed vulture can eat the soft insides from an opened carcass, and the slim-beaked hooded vulture is adept at cleaning bones and gleaning scraps (Fig. 6.2). In the sea, the tremendous diversity of life in and around coral reefs is a major contributor to the overall diversity of the tropics.

In high latitudes, on the other hand, food supplies may vary greatly from season to season and from year to year. The total amount of food supply may be high. Tundra vegetation blooms in spectacular fashion in the spring. There are rich plankton blooms in polar waters in spring and summer, and millions of seabirds and thousands of

Figure 6.2 Serengeti vultures. Left, the lappet faced vulture, with its huge beak, can tear through tough hides to scavenge a carcass. Image by Lip Kee Yap. Center, the white-backed vulture can feed on a carcass once it is opened. Image by Frank Wouters. Right, the hooded vulture, with its slim weak beak, is well adapted to gleaning scraps and cleaning bones. Image by Atamari. All three images placed into Wikimedia.

whales migrate there to share in the abundant food that is produced. Antarctic waters teem with millions of tons of tiny crustaceans (krill) that eat plankton and in turn are fed on by fish, seabirds, penguins, whales, and seals. Yet for organisms that live all year in polar regions, spring abundance contrasts with winter famine. Plants will not grow in winter darkness. Food variability is a major problem. The Arctic tern migrates almost from pole to pole, timing its stay at each end of the world to coincide with abundant food supply.

Where food supplies vary, animals cannot be specialists on only one food source; they must be versatile generalists. Generalists share some food sources, and probably compete more than specialists do. If so, fewer generalists than specialists can coexist on the same food resources. In seasonal or variable environments, where organisms must be generalists, diversity is lower. So there is a rather dramatic global diversity gradient, with high diversity at the equator and low diversity at the pole.

Tectonic movements can move continents around the globe. When many continents are in the tropics, global diversity may be higher than when many continents are in high latitudes.

Islands, Continents, and Supercontinents

Island groups tend to have milder climates—"maritime" or "oceanic" climates— compared with nearby continents, no matter whether they are tropical or at high latitudes. Thus the British Isles and Japan have milder climates than Siberia; the West Indies have milder climates than Mexico; and Indonesia has a milder climate than Indochina.

Large continental areas have especially severe climates for their latitudes. Asia, for example, is so large that extreme heat builds up in its interior in the northern summer, forming an intense low-pressure area. Eventually the low

pressure draws in a giant inflow of air from the ocean, the **summer monsoon**, that brings a wet season to areas all along the south and east edges of the continent, from China to Pakistan (Fig. 6.3). In winter the interior of Asia becomes very cold, a high-pressure system is set up, and an outflow of air, the **winter monsoon**, brings very chilly weather to India, Pakistan, China, and Korea (Fig. 6.3). Land organisms respond to the great seasonality of the monsoon climate, and organisms in the shallow coastal waters are affected strongly too. As nutrient-poor water is blown in from the surface of the open ocean in the summer monsoon, food becomes scarce; as water is blown offshore in the winter monsoon, deeper water is sucked to the surface and brings nutrients and high food levels. As a result, the diversity of marine creatures along the coasts of India is far less than it is in the Philippines and Indonesia, which are far enough away from the center of Asia that they feel the effects of the monsoons much less strongly. Reefs are scarce and poor in diversity along the Asian mainland coast; but they are rich and diverse in a great arc from the Philippines to the Australian Barrier Reef.

Thus the effects of continental geography as opposed to oceanic geography have an important effect on global diversity, though their effects are still directly linked to variation in food supply.

We now turn to the fossil record to explore evidence of diversity through time rather than looking just at the modern Earth.

Diversity Patterns in the Fossil Record

Jack Sepkoski spent over 20 years compiling data on the fossil record of Earth through the Phanerozoic, concentrating most on marine fossils. At first he simply counted the number of families of marine fossils that had been defined by paleontologists from Ediacaran to Recent times; later he compiled genera. These data on global (marine) diversity

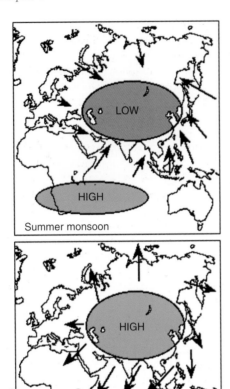

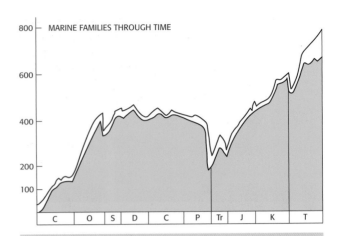

Figure 6.4   Jack Sepkoski's compilation of all marine fossil families for the Phanerozoic. The periods are indicated along the x-axis. Major extinctions occurred at the ends of the Paleozoic and Mesozoic eras (vertical lines). Based on Sepkoski (1981, 1984).

Figure 6.3   The monsoons of Asia. In summer, heat builds up over the continent and generates low pressure that draws in moist air from the surrounding oceans. In winter, high pressure over the continent generates cold winds that blow offshore. As a result, southern Asia is more seasonal than most regions that lie on or near the Equator.

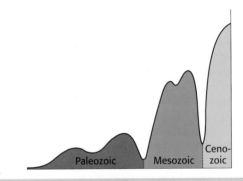

Figure 6.5   John Phillips published this graph in 1860, showing the diversity of life through time as he knew it then, and showing that it helped to define the Paleozoic, Mesozoic, and Cenozoic Eras.

show clear and reasonably simple trends (Fig. 6.4). Few families of marine animals existed in Ediacaran times, but the beginning of the Cambrian saw a dramatic increase that followed a steep curve to a Late Cambrian level. A new, dramatic rise at the beginning of the Ordovician raised the total to a high level that remained comparatively stable through the rest of the Paleozoic. In the Late Permian there was a dramatic diversity drop in a very large extinction that marks the end of the Paleozoic Era. A steady rise that began in the Triassic has continued to the present, with a small and short-lived reversal (extinction) at the end of the Triassic, and a steeper and deeper extinction at the end of the Cretaceous, which also marks the end of the Mesozoic Era.

This general pattern had been known since 1860 (Fig. 6.5). The pattern is also familiar to any invertebrate paleontologist who has spent time rummaging broadly through the collections of a major museum. Sepkoski's contribution was to put the pattern in quantitative terms, and to lay out

the data for anyone to analyze. (People seem to like to deal with numbers rather than reality!)

It's easy to think of possible problems with Sepkoski's approach. For example, only some parts of the world have been thoroughly searched for fossils; some parts of the geological record have been searched more carefully than others; older rocks have been preferentially destroyed or covered over by normal processes such as erosion and deposition. Different researchers mean different things as they define species, genera, and families in the group they study. Lengthy recent discussions have not helped us much to see past these biases. It is fair to say that Sepkoski's data is more likely to reflect events in the sea than on land.

I suspect that the trends that Sepkoski identified are real, even if the numbers attached to them may change with new research. When he counted genera instead of families, the trends were the same, but with sharper peaks and valleys.

Paleontologists have been searching the world for fossils for 200 years. The best-sampled fossil communities are shelly faunas that lived on shallow marine shelves, and our estimate of their diversity through time is likely to be a fair sample of the diversity of all life through time. Larger groups of animals are harder to miss than smaller groups, so we have probably discovered all the phyla of shallow marine animals with hard skeletons. Perhaps we have only found a few percent of the species in the fossil record, but we've probably discovered many of the families. In any case, if the search for fossils has been roughly random (and there's no reason to doubt it) the shallow marine fossil record as we now know it is a fair sample of the fossil record as a whole. So we can now ask what influenced the patterns that Sepkoski documented. Were they affected by the changing geography of Earth, and if so, what were the causal connections?

Global Tectonics and Global Diversity

We have known for over 40 years that the Earth's crust is made up of great rigid *plates* that move about under the influence of the convection of the Earth's hot interior. As they move, the plates affect one another along their edges, with results that alter the geography of the Earth's surface in major ways. Two plates can separate to split continents apart, to form new oceans, or to enlarge existing oceans by forming new crust in giant rifts in the ocean floor. Two plates can slide past one another, forming long *transform faults* such as the San Andreas Fault of California. Plates can converge and collide, forming chains of volcanic islands and deep trenches in the ocean, volcanic mountain belts along coasts, or giant belts of folded mountains between continental masses. At times the Earth has had widely separated continents; at other times the continental crust has largely been gathered into just one or two "supercontinents." These movements and their physical consequences are studied in the branch of geology called **plate tectonics.**

Plate tectonic movements affect the geography of continents and oceans, which can in turn affect food supply, climate, and the diversity of life. In other words, the tectonic history of Earth should have been a first-order influence on the diversity of the fossil record. Do we see a correlation? The brief answer is yes.

In the Early Cambrian, most continental pieces were more or less close together in a great belt across the South Pole. They split progressively during the Cambrian and Ordovician, to form a number of small continents that were generally more scattered and in lower latitudes (Fig. 6.6). This dispersion of continents coincides with the great diversity rise of the Cambrian and Ordovician. There were several continental collisions from the Middle Paleozoic through the Permian, and larger land masses were formed (Fig. 6.7a). The great extinction at the end of the Permian coincides with the final merger of the continents into a giant global supercontinent, **Pangea** (Fig. 6.7b), composed of a large northern land mass, **Laurasia**, and a southern land mass, **Gondwana**.

The rise in diversity that began in the Triassic and continued into the Cenozoic coincides very well with the progressive breakup of Pangea. The breakup was under way by the Jurassic, and reached a climax in the Creta-

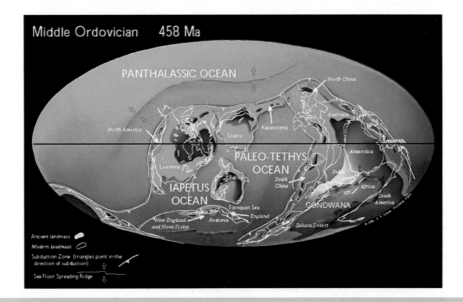

Figure 6.6 Ordovician paleogeography. There are several separate continents in the tropical regions, and the southern continents make the supercontinent Gondwana. Paleogeographic map by C.R. Scotese © 2012, PALEOMAP Project (www.scotese.com).

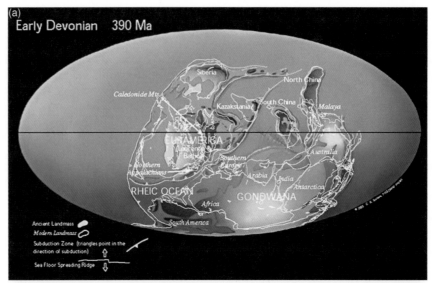

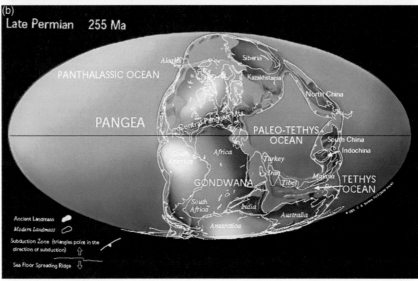

Figure 6.7   Late Paleozoic paleogeography. a) by the Early Devonian, 390 Ma, Gondwana is drifting north into warmer latitudes and other continents are converging together. b) by the Late Permian (260 Ma), a complete Laurasia has united with Gondwana to form the global supercontinent Pangea. Paleogeographic maps by C.R. Scotese © 2012, PALEOMAP Project (www.scotese.com).

ceous (Fig. 6.8). The continental fragments have continued to drift, and today the continents are perhaps as well separated as one could ever expect, even in a random world, with a diversity at an all-time high level.

Thus the tectonic events that affected Earth over the past 550 million years are reflected in the diversity curve. What are the connecting factors?

In an oceanic world, with continents small and widely separated, so that there are many provinces, each community in a province tends to have stable food supplies and high diversity. Therefore, the more the continents are fragmented into smaller units, the more oceanic the world's climate becomes, and the more diverse its total biota.

The other extreme occurs when all the world's continents are together in a supercontinent, such as Pangea: not only are there fewer provinces, but each province has low-diversity communities. Supercontinents had super monsoons.

The overall pattern of diversity data through time does receive a first-order explanation from plate-tectonic effects. But that cannot be the whole story, for several reasons.

1. **Changing faunas through time.** If plate tectonics were the only control on diversity, much the same groups of animals should rise and fall with the changes in global

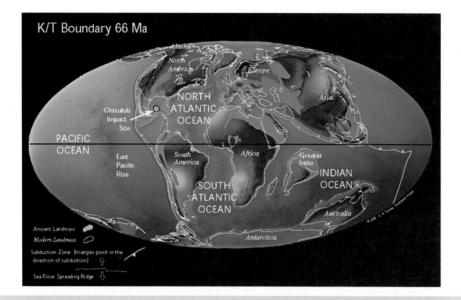

Figure 6.8   Cretaceous paleogeography. Gondwana and Laurasia are split into pieces, with Australia just leaving Antarctica. Paleogeographic map by C.R. Scotese © 2012, PALEOMAP Project (www.scotese.com).

geography. Instead, we see dramatic changes in different animal groups that succeed one another in time.

2. **Increase in global diversity.** The overall increase in global diversity from Ediacaran to Recent times is not predicted on plate tectonic grounds.

3. **Mass extinctions.** Major extinctions are much more dramatic than major radiations. For example, the Permian extinction did not occur gradually over the 150 m.y. of the later Paleozoic, as the continents collided and assembled piece by piece. Most likely, the continental assembly set up the world for extinction, then an "extinction trigger" was pulled. There are too many sudden "mass extinctions" in the fossil record for a plate tectonic argument to be completely satisfactory. Even if plate tectonic factors set the world up for an extinction, we seem to need some separate theory to explain the extinctions themselves.

Changing Faunas Through Time

Three Great Faunas

Jack Sepkoski sorted his data on marine families through time to see if there were subsets of organisms that shared similar patterns of diversity. He distinguished three great divisions of marine life through time, which accommodate about 90% of the data (Figs 6.9 and 6.10). Sepkoski called them the Cambrian Fauna, the Paleozoic Fauna, and the Modern Fauna. The faunas overlap in time, and the names are only for convenience. But they do reflect the fact that different sets of organisms have had very different histories.

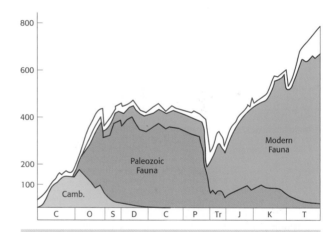

Figure 6.9   The three great faunas defined by Jack Sepkoski in his analysis of the marine fossil record through time. They are subsets of the data shown in Figure 6.4. Scale in numbers of families (data from Sepkoski 1981, 1984).

The Cambrian Fauna contains the groups of organisms, particularly trilobites, that were largely responsible for the Cambrian increase in diversity. But after a Late Cambrian diversity peak, the Cambrian Fauna declined in diversity in the Ordovician and afterward, even though other marine groups increased at that time.

The success of the **Paleozoic Fauna** was almost entirely responsible for the great rise in diversity in the Ordovician, and slowly declined afterward. The Paleozoic Fauna suffered severely in the Late Permian extinction, and its

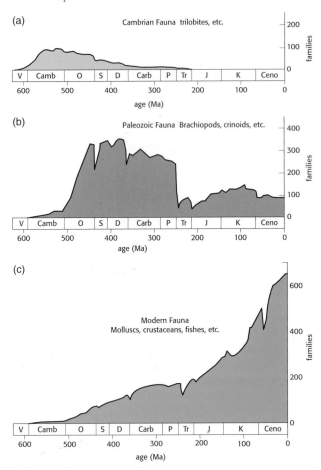

Figure 6.10 The individual histories through time of the three great Faunas defined by Jack Sepkoski. In particular, note the difference between them at the end of the Permian: the Paleozoic Fauna suffers a tremendous extinction, while the others are hit less hard. Based on Sepkoski 1981, 1984.

recovery afterward was insignificant compared with the dramatic diversification of the Modern Fauna.

Figure 6.10 also shows the major animals making up the three faunas. Their definition is approximate, because Sepkoski tried to make his analysis simple by using animal groups at the level of classes or subphyla. In hindsight, one could subdivide the marine animals into groups that would give even sharper divisions between the three faunas: for example, one could separate Paleozoic corals from later ones. There is no zoological affinity connecting the members of the three faunas, but they do have ecological meaning.

Explaining the Three Great Faunas

The diversity patterns imply that ecological opportunities in the world's oceans somehow changed through time to favor one particular ecological mixture and then to allow the diversification of others. Obviously there can be many different explanations of the facts, and I have room to discuss only a few suggestions. The diversity patterns have been known in outline for some time, so some of the explanations predate Sepkoski's analysis.

In the 1970s, James Valentine pointed to the different ways of life that are encouraged under different types of food supply. In the Cambrian, he argued, the continents were not widely separated, food supplies were variable, and the most favored way of life would have been deposit feeding: there is always some nutrition in seafloor mud. Thus Cambrian animals, wrote Valentine, are "plain, even grubby." The Burgess fossils may not be plain, but many of them were certainly mud grubbers. Even among soft-bodied animals, arthropods dominate Cambrian faunas in numbers and diversity, and most of them were deposit feeders.

The Paleozoic Fauna lived in more tightly defined communities, with a more complex ecological structure. The continents were more separated in the Ordovician, so one might expect much more reliable food supply in the plankton, which would have favored the addition of filter feeders to marine communities. One would also expect that a larger food supply in the form of stationary benthic filter feeders would have allowed slow-moving carnivores to become more diverse. Indeed, filter feeders reached higher in the water and fed at different levels, and there was more burrowing in the sediment. The overall trend was to add new ways of life, or **guilds**, to marine faunas. Altogether, Paleozoic animals seem to have subdivided their ways of life more finely over time.

If the Permian extinction was induced by the continental collisions that formed Pangea, one would expect that Paleozoic filter feeders and the predators that depended on them would have suffered a greater crisis than did other groups, because the food supply in the world's oceans would have become much more variable. In general, this prediction is correct: corals, brachiopods, cephalopods, bryozoans, and crinoids felt the Permian extinction most acutely. Again, predictably, the Permian communities that suffered the most in the Permian extinction were the reef faunas.

But it is more difficult to explain the rise of the Modern Fauna. Other things being equal, one would predict that as continents split again in the Mesozoic, Paleozoic-style predators and filter feeders would again have been favored. They had not become completely extinct, and could surely have been expected to recover. In fact, they did, but in a very subdued fashion. Most of the Mesozoic diversification was achieved by other groups that stand out in Sepkoski's analysis as the Modern Fauna. These new groups included new guilds, especially more mobile animals, more infaunal burrowing animals, and new predators.

Steven Stanley and Geerat Vermeij suggested that predation was a major factor in the rise of the Modern Fauna, in what Vermeij called the Mesozoic Marine Revolution. New predators that appeared in the Middle Cretaceous seem to have been more effective than their predecessors at attacking animals on the seafloor. Modern gastropods evolved, capable of attacking shells with drilling radulae backed with acid secretions and poisons. Advanced shell-crushing crustaceans became abundant, and so did bony fishes with effective shell-crushing teeth. The filter feeders of the Paleozoic Fauna, which were largely fixed to the open surface of the seafloor, perhaps became too vulnerable to predation. They were replaced by animals that build and live in burrows, and pump water down into their safe havens, where they filter food from the sea water. Burrowing bivalves with siphons and burrowing echinoids make up very important components of the Modern Fauna, together with effective, wide-roaming predators such as gastropods and fishes.

Increase in Global Diversity

Why is there more diversity in the oceans today than there was during the diversity peak of the Paleozoic? It is not just the success of molluscs and crustaceans, because there were molluscs and crustaceans in the Paleozoic. Some kind of overall change in world ecology must have favored greater diversity since the late Mesozoic. Richard Bambach proposed a **seafood hypothesis** for an energy-related idea originally outlined by Geerat Vermeij. The additional energy pumped into marine ecosystems by runoff from the land, as it was covered first by advanced gymnosperms and then by angiosperm floras, could and did support more complex animals and ecosystems in very high diversity.

Mass Extinctions

Extinction happens all the time. Martha, the last passenger pigeon left in the world, died of old age in the Cincinnati Zoo in 1914 (Fig. 6.11). This officially made her species, *Ectopistes migratorius*, extinct, though of course the species had been ecologically doomed as soon as the last breeding birds were gone.

Extinction occurs on all scales, from local to global, and it occurs at different rates at different times and in different regions. Some species have small populations that depend on a particularly narrow range of food, or habitat, and are vulnerable to even small scale ecological disturbance. So there must be a steady leakage of species, through extinction, out of the global biosphere. Occasionally, by bad luck, perhaps, one of these species will be the least of its family, and the loss of that family would become visible in a compilation like Sepkoski's.

Sooner or later, every species has some chance of becoming extinct: extinction is the expected fate of species, not a rarity. In a world with fairly steady diversity through time, existing species (and families) would become extinct about as often as new species (and families) evolved. We might expect that, other things being equal, global diversity might typically be fairly steady, or fluctuate gently up or down. This does seem to be the case for long stretches of the Phanerozoic. However, there have been times of extremely

Figure 6.11 An icon of extinction: Martha the passenger pigeon, last of her species, who died in 1914 at the Cincinnati Zoo. Photograph in the public domain.

rapid extinction (Fig. 6.4), and these events need explanations. Was some special extinction mechanism at work? If so, we have to try to identify it. Were large extinctions just extreme examples of normal (ordinary) extinction processes, or were they catastrophic events that were truly extraordinary?

Extinction events vary greatly in size. David Raup and Jack Sepkoski sifted through Sepkoski's data and identified extinction events that were large enough and sudden enough to be called **mass extinctions** (Raup and Sepkoski 1982). Thirty years after Sepkoki's early estimates, the "Big Five" mass extinctions are, in geological order:

- At the end of the Ordovician
- At the end of the Frasnian stage of the Late Devonian (F–F)
- At the end of the Permian (Permo-Triassic or P–T)
- At the end of the Triassic
- At the end of the Cretaceous (Cretaceous-Tertiary or K–T)

Of course, these mass extinctions are identified by the number of families or genera becoming extinct. But all families and all genera are not the same. A genus may be a worldwide taxon with hundreds of species and millions of specimens, or it may be like the California condor *Gymnogyps*, with one species and only a few dozen individuals.

Over the last decade, some paleoecologists have been sorting Sepkoski's data in a different way. Instead of counting taxa lost in mass extinctions, they have been trying to recognize major ecological systems that were destroyed or seriously damaged in an extinction event (McGhee et al. 2004). For example, the disappearance of all the large dinosaurs at the end of the Cretaceous, together with almost all the large marine reptiles and the pterosaurs, most likely destroyed or damaged huge parts of ecosystems on land and sea as collateral damage. This would have made the end-Cretaceous mass extinction much more of a global ecological disaster than you would expect from simply counting extinct reptiles. In fact that extinction, as an ecological disaster, is second only to the end-Permian event (McGhee et al. 2004).

This approach is not quantitative: it requires careful and knowledgeable analysis on a very broad scale. But it does give us a chance to examine mass extinctions as the global ecological events that they were.

It is already clear that mass extinctions did not all have the same cause. In fact, each one may have had a unique cause or combination of causes. That becomes important if we are to assess the ongoing human-induced mass extinction currently affecting the globe: it looks as if it will come into the category of a sixth mass extinction (Chapter 21).

Explaining Mass Extinction

Mass extinctions were global phenomena, so they have to be explained by global processes. The first that comes to mind is plate tectonics. However, tectonic changes are relatively slow in geological terms, so if tectonic extinctions were to happen, they would be slow. But mass extinctions are relatively sudden, so we would have to suggest something else *in addition to* tectonic movements to make a case for a tectonic extinction.

Some plausible agents for global extinctions are:

- A failure of normal ocean circulation affects ocean chemistry enough to cause global changes in climate and atmosphere.
- A failure of normal ocean circulation affects ocean chemistry enough to cause global changes in climate and atmosphere.
- A rapid change in sea level affects global ecology and climate.
- An enormous volcanic eruption affects global ecology and climate.
- An extra-terrestrial impact by an asteroid affects global ecology and climate.

Of these possible agents, enormous volcanic eruptions leave behind enormous masses of volcanic rocks, so they are relatively easy to detect in the rock record. However, their indirect effects caused by releases of gases and dust are more difficult to judge. Global changes in sea level will change the distribution of sediments laid down on the Earth's surface: as long as the sea level change lasts long enough to leave behind this kind of evidence, we will be able to find it. But a failure of ocean circulation typically would be expected to leave behind only subtle chemical evidence, and it is likely to be short-lived, so evidence might be difficult to find and interpret. And an extra-terrestrial impact is by definition an instantaneous event that might leave behind only a thin layer of evidence. Unless we find a crater, or some unique piece of evidence that only an impact can produce, it may be very difficult to identify an impact, especially in more ancient rock. Three major indicators are:

1. A defined layer or spike of the element iridium (Ir), which occurs in greater abundance in meteorites than in Earth's crust.
2. Tektites are tiny glass blobs (spherules), which are formed as a meteorite or asteroid splashes molten drops of rock at high speed into the atmosphere.
3. Shocked quartz: quartz crystals with characteristic damage that can only be caused by intense shock waves.

At present, the leading hypotheses for the causes of the largest six extinctions are:

- End-Ordovician Climate (a short-lived ice age)
- Late Devonian (F–F) Oceanic crisis
- End-Permian (P–Tr) Giant eruptions, plus tectonic events
- End-Triassic Very large eruptions, methane spike
- End-Cretaceous (K–T) Giant eruptions, plus a huge asteroid impact

I will discuss the K–T event in Chapter 16. Here I will discuss the four others, but I will concentrate on the largest of them all, at the Permo-Triassic boundary.

The Ordovician Mass Extinction

The mass extinction at or near the end of the Ordovician seems to be closely linked with a major climatic change. A first pulse of extinction happened as a big ice age began, and the second occurred as it ended. This "mass extinction" included the loss of a lot of shelly fossils, but ecologically it was a comparatively minor event (McGhee et al. 2004). There was minimal life on land at the time, so the Ordovician extinction is purely a marine event.

The Late Devonian (F–F) Mass Extinction

A mass extinction took place, possibly in several separate events, at the boundary between the last two stages of the Devonian, the Frasnian and Famennian (the F–F boundary). There was a major worldwide extinction of coral reefs and their associated faunas, and many other groups of animals and plants were severely affected too. The land plants in wet lowland areas and the first amphibians that lived at the water's edge do not seem to have been affected by this extinction.

Evidence suggesting an asteroid impact has been reported from China and Western Europe at or near the F–F boundary. But there are also indications of climatic changes, and major changes in sea-level and ocean chemistry, at the same time. Carbon isotope shifts indicate that global organic productivity changed rapidly before the boundary.

George McGhee (McGhee 1996) favored an impact scenario. There are two cautions, however. First, the geological evidence suggests that there were several closely-spaced but medium sized impacts over perhaps two or three million years, rather than the one tremendous impact that seems to have occurred at the K–T boundary; second, our understanding of the timing of events and of world geography at the time is inexact. There is no "magic marker" of impact phenomena at the extinction event, as there is at the K–T boundary, and that makes the F–F boundary difficult to work with.

The Permo-Triassic (P–Tr) Extinction

The extinction at 250 Ma, at the end of the Permian, is the largest of all time, numerically and ecologically. It was recognized and used by John Phillips 150 years ago to help define the end of the Paleozoic Era and the beginning of the Mesozoic (Fig. 6.5). By Sepkoski's count, an estimated 57% of all families and 95% of all species of marine animals became extinct (Fig. 6.9). The Paleozoic Fauna was very hard hit (Fig. 6.10), losing very many suspension feeders and carnivores, and almost all the reef dwellers.

The P–Tr extinction was rapid, probably taking place in much less than a million years. It was much more severe in the ocean, but it affected terrestrial ecosystems too. Overall, the P–Tr extinction is a major watershed in the history of life on Earth, especially for life in the ocean; the K–T extinction is small in comparison (Fig. 6.9).

The Permian extinction coincides with the largest known volcanic eruption in Earth history: one of a few giant **plume eruptions**. Occasionally, an event at the boundary between the Earth's core and mantle sets a giant pulse of heat rising toward the surface as a plume (Fig. 6.12). As it approaches the surface, the plume melts or distorts the crust to develop a flat head of molten magma that can be 1000 km across and 100 km thick. Melting the crust, the plume generates enormous volcanic eruptions that pour millions of cubic kilometers of basalt—**flood basalts**—out on to the surface. If a plume erupts through a continent, it blasts material into the atmosphere as well. After the head of the plume has erupted, the much narrower tail may continue to erupt for 100 m.y. or more, but now its effects are more local, affecting only 100 km or so of terrain as it forms a long-lasting **hot spot** of volcanic activity.

Plume events are rare: there have been only eight enormous plume eruptions in the last 250 m.y. The most recent is the Yellowstone plume: at about 17 Ma it burned through the crust to form enormous lava fields that are now known as the Columbia Plateau basalts of Oregon and Washington, best seen in the Columbia River gorge. North America

Figure 6.12 A plume of hot material rises toward the Earth's crust. This image is a computer model of a Rayleigh-Taylor instability, the idea that the plume concept is based on. The model was calculated for homogeneous substances, and the Earth's interior is certainly not homogeneous. However, it is acceptable for a first estimate of what a plume looks like, especially the way it separates into a broad "head" and a narrow "tail". Image from Los Alamos National Laboratory, in the public domain.

drifted westward over this "hot spot," which continued to erupt to form the volcanic rocks of the Snake River plain in Idaho (Valley of the Moon and so on), and it now sits under Yellowstone National Park. The hot spot is in a quiet period now, with geyser activity rather than active eruption, but it produced enormous volcanic explosions about 500,000 years ago that blasted ash over most of the mountain states and into Canada. Even so, it was not large enough to cause a mass extinction.

At 252 Ma, a massive plume burned through the continental crust in what is now western Siberia to form the Siberian Traps, gigantic flood basalts that about 5 million sq km in area (Fig. 6.13) and are perhaps 3 million cu km in volume. The eruptions lasted at full intensity for only about half a million years: these are the largest known, most intense eruptions in the history of the Earth (Saunders and Reichow 2009), and they are dated exactly at the P–Tr boundary, 252 Ma (Shen et al. 2011). Can this be a coincidence?

Various scenarios have been proposed to link the Siberian Traps eruption and the P–Tr extinction. Sobolev et al. (2009) envisage the plume reaching the base of the thick cold continental crust and slowly melting it over a few hundred thousand years. When the plume breaks through to the surface it is loaded with carbon dioxide and hydrochloric acid gases that are released into the atmosphere in gigantic amounts, along with sulfur gases and other toxic

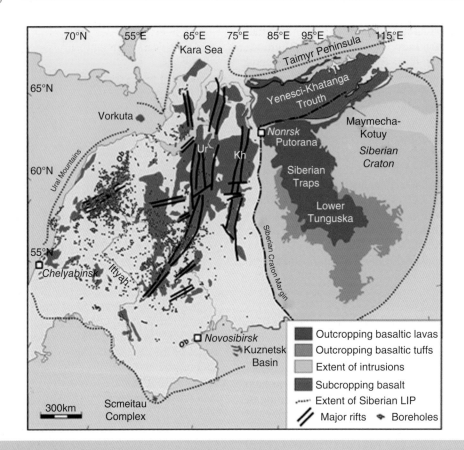

Figure 6.13 A map of the total extent of the Siberian Traps. Dark and medium green marks surface outcrops of lava. Light green: intrusions of lava into the continental sediments. Red: lava discovered underground by drilling the crust. The total extent of the igneous rock is shown by dotted lines. Image from the Siberian Traps Web site http://www.le.ac.uk/gl/ads/SiberianTraps/Index.html maintained by Andy Saunders and Marc Reichow at the University of Leicester, and used by permission of Professor Saunders.

substances including fluorine gas. Sobolev et al. feel that the gases and accompanying dust and aerosols would alone be enough to cause the mass extinctions. The tremendous amount of aerosols would cool the global climate and destroy the ozone layer (temporarily). The gases would upset biological cycles of carbon and sulfur (and we see that in the isotope record).

But there is plenty of evidence that the plume caused much more dramatic environmental damage (Svensen et al. 2009). As the magma from the plume broke through the lower crust, it encountered (by chance) very large oil and gas fields in Ediacaran and Cambrian sediments. As the oil and gas were heated, they formed even more gases to add to those already in the magma. Then (by chance) the rising magma reached and heated a giant salt-bearing field in the Cambrian rocks, which added chlorine gases like hydrochloric acid aerosols to the mix. Even higher in the crust were thick coal-bearing strata, so huge quantities of carbon were added, as carbon dust, or as carbon dioxide, or as methane gas. The cumulative pressure in the trapped gases was great enough to punch hundreds of giant "pipes"

through the crust, many of them hundreds of meters across, and gigantic fountains of toxic gases were blasted high into the atmosphere along with volcanic ash. Ash produced by burning Siberian coal has been identified in end-Permian sediments in Arctic Canada, far away from the eruption region (Grasby et al. 2011).

Finally, Siberia at the time was close to the North Pole, just as it is now. Today, methane hydrate (methane trapped in a gel-like form) builds up in sediments under the Arctic tundra. If there were methane hydrates in the Permian Arctic, the Siberian Trap eruptions could have triggered their release. Vermeij and Dorritie (1996) suggested that methane release formed in this way would have added to the effects of the eruptions.

The Siberian Traps eruptions probably set off the greatest chemical insult to the atmosphere in the last billion years, and it could have had catastrophic effects as the gases altered climate. Rain-out of the ash and gas could have devastated primary production on a world-wide scale, with a cascading effect up the food chain. As we have seen, the extinction in the sea killed off roughly 95% of the species.

Extinctions on land do not seem to have been so lethal, but at least regionally, they were impressive. There are extreme abundances of fossil fungal cells in sediments at the P–T boundary in widespread regions. The fungal "layer" may record a single, world-wide crisis, with the fungi breaking down massive amounts of vegetation that had been catastrophically killed, or even killing the trees themselves (Visscher et al. 2011). (Some people think the "fungi" are algae, which spoils this story).

The best evidence we have from paleobotany suggests that there were indeed major extinctions among gymnosperms, in Europe and Asia and among the coal generating floras of the Southern Hemisphere. Early Triassic vegetation in Europe looks "weedy," that is, invasive of open habitats.

Most eruptions do not cause extinctions. For example, the eruption of Krakatau in 1883 destroyed all life on the island and severely damaged ecosystems for hundreds of miles around. But those ecosystems recovered completely in 100 years, a geologically insignificant time. There is no biological trace of the much larger eruption of Toba, in Sumatra, 75,000 years ago. No North American extinctions coincided with explosive eruptions from Long Valley caldera, California, from Crater Lake, Oregon, or from Yellowstone, all of which blew ash as far as Canada within the last two million years.

However, there may be a threshold effect: if an eruption is not big enough it will do nothing, but if it is big enough it will do everything. Three giant eruptions in the last 500 m.y. occurred at the same time as mass extinctions, at the P–Tr boundary, at the end of the Triassic, and at the K–T boundary (Chapter 16). Three out of five seems compelling, though clearly we should also look at other explanations—or scenarios that would make a volcanic extinction worse!

Evidence has been accumulating that the ocean went through a chemical crisis at the P–Tr boundary. Many of the marine organisms that went extinct made skeletons of calcium carbonate. They would have been particularly susceptible to very high levels of acid in the sea water because that would inhibit the reactions they used to make their skeletons (Knoll et al. 2007). (The most likely candidate for acidifying the ocean is carbon dioxide: in fact, human activity that involves burning fossil fuel is acidifying the oceans slightly today.) In addition, limestone sediments exposed to ocean water in the late Permian often show signs of being etched by acidic sea water. Knoll et al. (1996) suggested that the extinction was caused by a catastrophic overturn of an ocean that was supersaturated in carbon dioxide. Others have suggested hydrogen sulfide and methane as components in the ocean. If a mass of anoxic water in the deep oceans, loaded with dissolved carbon dioxide, methane, and hydrogen sulfide, were brought suddenly to the surface, it would degas violently on a global scale and would likely trigger greenhouse heating and a major and sudden climatic warming. Some people have suggested that a crisis on this scale would be enough to account for the P-Tr extinction, no matter what the Sibe-

rian Traps eruptions did. Others feel that the Siberian Trap eruptions alone were powerful enough to account for the extinctions.

The idea of a crisis that resulted from suddenly overturning an anoxic ocean has not been well defined, though it is clear that such a crisis would devastate the oceans, and damage ecosystems along the shores. In particular, the Permian ocean has long been understood to have been one giant ocean, **Panthalassa**, as a counterpart to the giant continent Pangea. But it is very difficult to generate a crisis in a world ocean. Global wind patterns circulate it efficiently, at least in the surface waters. Today, water circulates vertically in the ocean when surface waters become dense and sink to the bottom, stirring up deep waters that upwell to the surface.

Thus, around the poles and especially round Antarctica, very cold salty water sinks to the bottom and flows almost all over the ocean floors, carrying surface oxygen with it that mingles with and refreshes the world ocean. Thus it is very difficult to imagine conditions in which the modern ocean could become anoxic and filled with dissolved toxic gases: and it is very difficult to imagine how Panthalassa could do so either.

A relatively recent idea offers a new scenario for an anoxic crisis. Celal Sengör and Saniye Atayman of Istanbul Technical University noticed that in the Late Permian, Panthalassa contained an anomalous region called "Paleo-Tethys" where a very large oceanic expanse was close to being landlocked (Fig. 6.7, right, and Fig. 6.14), and they explored the consequences of this geography (Sengör and Atayman 2009).

As plate tectonic movements formed a narrow isthmus between Siberia and Australia (Fig. 6.14), they gradually choked off the earlier global circulation. With no downward currents to keep it ventilated, Paleo-Tethys became anoxic from the bottom up during the Permian. Finally, late in the Permian, the anoxic water reached the edge of the continental shelf and began to kill the benthic animals living there. This creeping extinction affected an enormous length of shallow-water habitat around the perimeter of Paleo-Tethys, and late in the Permian, extinction over this huge region begins to show in the fossil record. Sengör and Alayman propose that at the very end of the Permian, conditions in Paleo-Tethys became so bad that toxic gases reached the sea surface and began to diffuse off in great clouds, killing much of the life on the shallow lowlands around Paleo-Tethys as well as the last vulnerable animals along the shores. It is clear that they do not believe the Siberian Trap eruptions played much of a role in the extinctions. They see the extinctions in and around Paleo-Tethys as the dominant feature in the catastrophe, with remnant faunas along the other coasts of Pangea living in a kind of refuge or asylum, away from the crisis zone.

It is difficult to reconcile these volcanic and oceanic models for extinction, but we have to try. As I noted, tectonic crises are slow-developing, and Sengör and Alayman's scenario is true to that. Their argument would be that the crisis doesn't become a mass extinction until toxic sea water

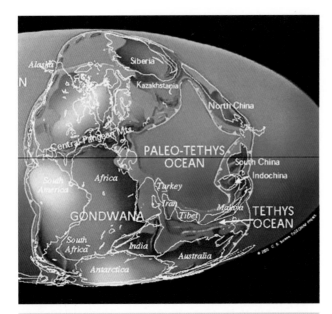

Figure 6.14 Paleo-Tethys: the vital part of Figure 6.7 right. At the end of the Permian, a large part of the world ocean Panthalassa was close to being landlocked. Only surface water could exchange between Paleo-Tethys and Panthalassa: the deep water of Paleo-Tethys was cut off from the deep water of Panthalassa. Paleo-geographic map by C.R. Scotese © 2012, PALEOMAP Project (www.scotese.com).

Figure 6.15 The Black Sea is almost landlocked, except where it connects with the Mediterranean through the Bosphorus Strait in its southwestern corner. Photosynthetic plankton can be seen in this satellite image as swirling clouds in the surface waters, producing oxygen that keeps the surface from become anoxic. Image by NASA. The Visible Earth, http://visibleearth.nasa.gov

upwells to sea level. But why did that happen right at the end of the Permian? Furthermore, why did Paleo-Tethys become anoxic all the way to the surface? Today's Black Sea is a small analog for Paleo-Tethys (Fig. 6.15). It is almost land-locked (a small amount of water is exchanged at the Bosphorus). It is 4 km (7000 feet) deep, and all but a tiny fraction is anoxic. But planktonic bacteria and algae photosynthesize and provide oxygen to the surface waters, which never become anoxic— at least they haven't in recorded history! The only reason I can envisage for the Paleo-Tethys to become anoxic is that something killed off all the surface plankton. And the toxic emissions from the Siberian Traps would do that, right at the end of the Permian, setting off all the rest of the biological devastation from the toxic gases emitted from Paleo-Tethys as well as the eruptions.

The great anoxic event in PaleoTethys can be timed precisely. It turns out that uranium is preferentially absorbed into sediments that are laid down in anoxic conditions. And exactly at the extinction horizon in China, the P–Tr boundary is marked by a uranium spike (Brennecka et al. 2011).

Therefore, I think we need both scenarios for the greatest extinction ever recorded. The odds of two such catastrophes happening at the same time are very unlikely. But so was the double catastrophe of the Cretaceous-Tertiary extinction, which we will look at in Chapter 16. Perhaps

catastrophes of unlikely size are caused by agents that are very unlikely to occur.

Early in the Triassic, plate tectonic movements broke open the narrow isthmus that had confined Paleo-Tethys, re-uniting it with Panthalassa (Fig. 6.16). And, of course, the Siberian Traps eruptions lasted only half a million years. Thus the survivors of the P–Tr extinction rebounded quickly: the closer we look, the faster the recovery was. The survivors that filled the Earth with their descendants, which were a small and biased sample of the Permian faunas. It is not an accident that the rise of the Modern Fauna begins with the Triassic (Fig. 6.9).

The end-Triassic Extinction

Fifty million years after the P–Tr extinction, at the end of the Triassic about 201 Ma, there were more huge eruptions as Pangaea began to split apart and the Atlantic Ocean began to form. Eleven million sq km (4 million sq miles) of basalt lava were erupted, most of it within only half a million years, to form the so-called CAMP (Central Atlantic Magmatic Province) that stretches from France to Brazil (Fig. 6.17). Carbon isotopes suggest major changes in ocean productivity, and there were important extinctions in the sea (Fig. 6.8, and the Paleozoic Fauna in Fig. 6.9). Extinctions on land are not so well defined.

Here, the eruptions are the prime suspect for the extinction. That amount of lava would have released enormous quantities of carbon dioxide, sulfur gases, and aerosols that could have affected ocean chemistry directly, and probably affected productivity in the surface waters and on several continents. The amount of carbon involved in the carbon

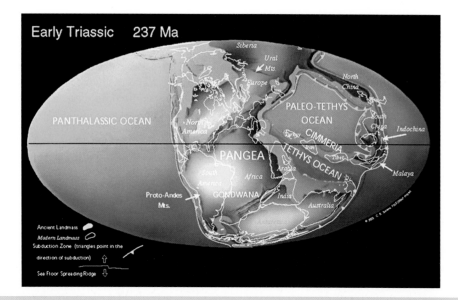

Figure 6.16 Lower Triassic geography: a broad gap between Malaya and Australia now connects Panthalassa and Paleo-Tethys through a deep-water strait. Paleogeographic map by C.R. Scotese © 2012, PALEOMAP Project (www.scotese.com)

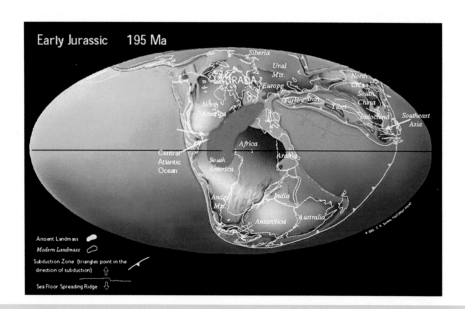

Figure 6.17 The Earth in the Early Jurassic, just a few million years after the end-Triassic extinction. The CAMP basalts are shown in red. They look bigger than the Siberian Traps, but that is because of the map projection used. Paleogeographic map by C.R. Scotese © 2012, PALEOMAP Project (www.scotese.com).

isotope changes suggests that very large quantities of methane were released as well (Ruhl et al. 2011).

But this was not another P–Tr event. The Triassic event is much smaller. As we work out the details, it will be very interesting to see which of the special circumstances of the P–Tr might also have occurred during the end-Triassic event. One thing is sure: at all scales, over all geological time, the geology of the Earth at any time is unique. To outbid the Greek philosopher Heraclitus, you can never look at the same Earth twice.

Evolutionary Radiations

New species appear all the time, just as species become extinct all the time. Occasionally we can look back into the record and see that a particular new species happened to be the first of a very successful group that we define as a family. The appearance of that species would thus be an event that would show up in a Sepkoski compilation of global diversity of families. As described above, a "normal" period in Earth history would have new families more or less balancing older ones that became extinct.

However, just as with extinctions, there are times when new families appeared much more often than old ones became extinct, so that we see a steep diversity rise. These events are called **evolutionary radiations**, and the name has meaning because one can often identify clades that entered a new way of life and evolved into several or even many families. Because of this, it is easier to understand the specifics of individual radiations than the specifics of individual extinctions. Radiations are likely to be evolutionary, whereas extinctions are likely to be disasters.

One can perceive one general theme about radiations. A radiation is a response to an opportunity. So what kind of opportunity would set off a radiation so large that it would show up in a compilation of global diversity? I can think of three.

1.   **Mass Extinctions.** By their very nature, mass extinctions remove many organisms from the biosphere. If the mass extinction was a one-time massive physical disaster (plume eruption; asteroid impact), the physical world would probably recover quickly to "normal", yet have a biology that was missing major components. This situation provides a major opportunity for surviving organisms to evolve to fill those ecological gaps. The newcomers will not have the same anatomy, and will not re-evolve the same characters as their extinct predecessors, so we are likely to see a wave of evolutionary novelty wash across the world.

Obvious examples include the radiation of the Modern Fauna after the P–T extinction; the radiation of land mammals after the extinction of most dinosaurs at the K–T extinction (Chapter 17); and the radiations of bats and whales after the extinction of most flying and swimming reptiles, again at the K–T extinction.

Studying such recoveries is not going to give us any fundamental principles we don't already know. Recoveries from mass extinctions are consequences of the extinctions that provided the necessary opportunities for the survivors.

One can say that mass extinctions remove the **Incumbent Effect**. This powerful metaphor is easily understood by Americans, who live with a political system in which an elected representative to Congress (let's say) is very difficult to remove from office, once elected, even though the election process is entirely open and democratic. The reason is that the incumbent has name recognition, and has a lot of power and access to money, while any prospective challenger typically does not.

The incumbent effect works in biology too. Any species is well adjusted to its normal environment; it evolved in that environment, and its adaptations have been honed by natural selection for success there. Any invading species is likely to be less well fitted to that environment. As the incumbent effect pervades communities and provinces as well, ecosystems typically are stable over long time periods. Yet just as hurricanes can smash a local area of forest and allow weeds to flourish, or clean off a low-lying island that must be re-colonized, so disasters such as mass extinctions can remove incumbents and allow survivors their place in the sun. Frightful as mass extinctions may be, in global terms they give surviving creatures an opportunity for major evolutionary innovation.

2.   **Invading a New Habitat.** Evolution works by natural selection, which implies the continual testing of new mutations against the environment. Some organisms are always "pushing the envelope", and occasionally a lineage will evolve a body plan that allows it to invade a new habitat that may have been available for a long time, but had been unexploited. If successful, that lineage may expand into a radiation as sub-clades explore the different ways of life that are possible in that new habitat. Obvious global examples include the first land plants and the first land animals (Chapters 8 and 9), and the first flying animals (Chapter 13).

This kind of opportunity probably exists at all scales. Land animals reaching a biologically "empty" isolated continent or island may radiate there: obvious examples are the marsupials of Australia and the mammals of South America during the Cenozoic (Chapter 18), not to mention the reptiles and birds of the Galápagos that influenced Darwin so much.

3.   **New Biological Inventions.** Occasionally a lineage will evolve a body plan that allows it to do things that no organism has done before. If successful (if the timing and the ecology are just right), that lineage may expand into a radiation as new sub-clades explore the different ways of exploiting that new invention. Obvious examples include the first eukaryotes (Chapter 3), the early metazoans (Chapters 4 and 5), and (again) the various groups that evolved the apparatus for flight (Chapter 13). Both dinosaurs (Chapter 12) and (eventually) mammals evolved warm blood and the erect limbs that allowed them a very active life style on land. Bats and whales evolved sonar, hominids invented the capacity to make tools… I could go on for pages, and do so in the chapters listed above.

Further Reading

Diversity through Time

Bambach, R. K. 1993. Seafood through time: changes in biomass, energetics, and productivity in the marine ecosystem. *Paleobiology* 19: 372–397.

Schiermeier, Q. 2003. Setting the record straight. *Nature* 424: 482–483. [The Paleobiology Data Base, meant to take Sepkoski's approach into the 21st century.]

Sepkoski, J. J. 1981. A factor analytic description of the Phanerozoic marine fossil record. *Paleobiology* 7: 36–53.

Sepkoski, J.J. 1984. A kinetic model of Phanerozoic taxonomic diversity. Lll. Post-Paleozoic families and mass extinctions. *Paleobiology* 10: 246-267.

Sepkoski, J. J. 1993. Ten years in the library: new data confirm paleontological patterns. *Paleobiology* 19: 43–51.

Vermeij, G. J. 1987. *Evolution and Escalation*. Princeton: Princeton University Press. [The importance of predator/prey interactions in evolution. A fine set of essays, enormous list of references.]

Mass Extinctions

Brennecka, G. A. et al. 2011. Rapid expansion of oceanic anoxia immediately before the end-Permian mass extinction. *PNAS* 108: 17631–17634. Available at http://www.pnas.org/content/108/43/17631.full

Grasby, S. E. et al. 2011. Catastrophic dispersion of coal fly ash into oceans during the latest Permian extinction. *Nature Geoscience* 4: 104-107. Available at http://media.cigionline.org/geoeng/2011%20-%20Grasby%20-%20Catastrophic%20dispersion%20of%20coal%20fly%20ash%20into%20oceans%20during%20the%20latest%20Permian%20extinction.pdf

Knoll, A. H. et al. 2007. Paleophysiology and end-Permian mass extinction. *Earth and Planetary Science Letters* 256: 295-313. Available at http://pangea.stanford.edu/~jlpayne/Knoll%20et%20al%202007%20EPSL%20Permian%20Triassic%20paleophysiology.pdf

Kring, D. A. 2000. Impact events and their effect on the origin, evolution, and distribution of life. *GSA Today*: August 2000. Snappy summary of impacts and their consequences, with good reference list. Available at https://rock.geosociety.org/pubs/gsatoday/archive/sci0008.htm

McGhee, G. R. et al. 2004. Ecological ranking of Phanerozoic biodiversity crises: ecological and taxonomic severities are decoupled. *Palaeogeography, Palaeoclimatology, Palaeoecology* 211: 289–297. Available at http://earthsciences.ucr.edu/docs/McGhee_et_2004.pdf

McGhee, G. R. 2005. Modelling Late Devonian extinction hypotheses. Chapter 3 in *Developments in Palaeontology and Stratigraphy* 20, 37–50. Concentrates on impact scenarios.

Payne, J. L. and M. E. Clapham. 2012. End-Permian mass extinction in the oceans: an ancient analog for the Twenty-First Century? *Annual Review of Earth & Planetary Sciences* 40: 89–111.

Raup, D. M. and J. J. Sepkoski. 1982. Mass extinctions in the marine fossil record. *Science* 215: 1501–1503. Available at http://planet.botany.uwc.ac.za/nisl/Gwen's%20Files/Biodiversity/Chapters/Info%20to%20use/Chapter%206/RaupSepkoskiMass.pdf

Ruhl, M. et al. 2011. Atmospheric carbon injection linked to end-Triassic mass extinction. *Science* 333: 430-434. Available at http://973.geobiology.cn/photo/2011072478778489.pdf

Saunders, A. and M. Reichow 2009. The Siberian Traps and the end-Permian extinction: a critical review. *Chinese Science Bulletin* 54: 20–37. Available at http://www.le.ac.uk/gl/ads/SiberianTraps/PDF%20Files/The%20Siberian%20Traps%20and%20the%20End-Permian%20mass.pdf

Sengör, A. M. C. and S. Atayman 2009. The Permian extinction and the Tethys: an exercise in global geology. *Geological Society of America Special Paper* 448.

Shen, S-z. et al. 2011. Calibrating the end-Permian mass extinction. *Science* 334: 1367–1372.

Sobolev, S. V. et al. 2011. Linking mantle plumes, large igneous provinces and environmental catastrophes. *Nature* 477: 312–316.

Visscher, H. et al. 2011. Fungal virulence at the time of the end-Permian biosphere crisis? *Geology* 39: 883-886.

Question for Thought, Study, and Discussion

In the last paragraph of the chapter, I mention new biological inventions. All of them are fascinating. But here's the question. Lots of people worry about the possibility of intelligent life somewhere in the Universe. But intelligent life has to evolve, as it did here. Starting from a world inhabited only by the first cells, list (in order) some of the new biological inventions that were absolutely crucial along the way to intelligent life on Earth. For example, was vision one of them? Be prepared to defend your list.

SEVEN

The Early Vertebrates

In This Chapter

We now turn to following the history of vertebrates, partly because we are familiar with them, and because they include us. They evolved from soft-bodied invertebrates, which leads to some uncertainty about their earliest members. Even so, we can identify the earliest known vertebrate in early Cambrian rocks from China. The fossil record of early fish is reasonable, because these early fishes had strong bony plates covering their front ends. A variety of fish groups shows increasing swimming ability. All early fishes lacked jaws, but some time in the Silurian fishes evolved a jaw from parts of the gill system. The jawed fishes now dominate living fish faunas. Jaws allowed some fishes to become large predators, and the placoderms of the Devonian included some of the largest and most effective predators the seas have ever seen. Sharks are ancient fishes too, and their survival in today's seas reflects their array of splendid adaptations to marine carnivory. Some early fishes evolved to breathe air, and I discuss that at some length because it paved the way for fishes to leave the water.

Vertebrates

Vertebrates dominate land, water, and air today in ways of life that combine mobility and large size (more than a few grams). Only arthropods (insects on land and crustaceans in the sea) come close to competing for these ecological niches. As vertebrates ourselves, we have a particular interest in the evolutionary history of our own species and our

History of Life, Fifth Edition. Richard Cowen.
© 2013 by Richard Cowen. Published 2013 by Blackwell Publishing Ltd.

remote ancestors. It's hardly surprising that vertebrates should receive special treatment in this and almost every other book on the history of life.

It is easier for us to identify with vertebrates than with invertebrates. We can feel how ligaments, muscles, and bones work. We feed by using our jaws and teeth. We have sensory skin and good vision, and we sense vibrations in our ears. We walk, run, and swim. We have bodily sensations as we thermoregulate, and we understand by experience the bizarre system we have for getting oxygen and circulating it around the body. All vertebrates share some of these systems, and many vertebrates have them all. In contrast, most invertebrates have quite different body systems that are more difficult for us to identify with and to understand.

Our familiarity with vertebrate biology helps to make up for the rarity of vertebrate fossils. Vertebrates are rare even today in comparison with arthropods or molluscs, and vertebrate hard parts are held together only by skin, muscles, cartilage, and ligaments that rot easily after death. Even bones crumble and dissolve rather easily once they lose the organic matter that permeates them in life. Land vertebrates in particular live in a habitat that offers little chance of preservation. Bones are scattered and destroyed rather than buried and sheltered by sediment. Only the special interest in vertebrates shown by professional and amateur fossil collectors alike has compensated for the intrinsic poverty of vertebrate preservation. By now we have a very good idea of the major events in vertebrate history. (Explaining them is a different problem!)

Vertebrate Origins

Vertebrates evolved from invertebrates, which are simpler in structure. Vertebrates have a spine, a bony (or cartilaginous) column that contains a nerve canal and a notochord. The **notochord** is a specialized structure that looks like a stiff rod of dense tissue. It is a more fundamental character than the spine that surrounds it. It is a shared derived character that places vertebrates in the phylum Chordata, together with some soft-bodied creatures that have a notochord but do not have a head or a skeleton.

By using the stiffness of the notochord, a chordate without a spine can give its muscles a firm base to pull against, while retaining enough flexibility to allow a push against the water for efficient swimming. The notochord can store elastic energy that is released at the right moment to help swimming. I suspect that the evolution of the notochord, with this mechanism for energy storage and release, is the evolutionary novelty that promoted the success of soft-bodied chordates. It preceded by a long time the evolution of the skeleton of a typical vertebrate.

Urochordates and cephalochordates are two living groups of soft-bodied chordates that help to show us what a vertebrate ancestor might have looked like. Urochordates include **tunicates** (sea squirts), small boxlike creatures that live as adults in colonies fixed to the seafloor, and filtering

Figure 7.1 A colonial tunicate. This is *Botrylloides*, from the Mediterranean Sea. Image by Liza Gross, from her article (Gross 2007) in the open-access journal PLoS Biology. http://www.plosbiology.org/article/info:doi/10.1371/journal.pbio.0050098

food from the water (Fig. 7.1). But tunicate larvae swim actively, using the notochord and muscle fibers in a tail-like structure that is lost soon after they settle as adults. The tunicate *Ciona* has had its genome completely sequenced, and that genome looks much like that of vertebrates, but simpler.

Cephalochordates (Fig. 7.2) are marine creatures that filter small particles from seawater, which is also used for respiration. The notochord runs along the dorsal axis and is surrounded by packs of body muscle arranged in V-shaped chevrons. Alternate contractions of the muscle packs flex the body from side to side in a wave-like pattern that allows it to swim. Nerve tissue at the anterior end of the notochord marks the position of a primitive brain. In most of these characters, cephalochordates are much like fishes, even to the pattern of V-shaped muscles that is so obvious when one dissects a fish carefully in a laboratory or a restaurant.

Branchiostoma, the amphioxus (Fig. 7.2), is a typical cephalochordate. It lives and moves between sand grains and in open water, squirming and swimming in eel-like fashion with its muscle packs and notochord acting against one another.

It is difficult to sort out the relationships between tunicates, cephalochordates, and early fishes. Part of the problem is that they all probably evolved in the Ediacaran. In the astounding Cambrian Chengjiang fauna of China, slightly older than the Burgess Shale (Chapter 4), we find a cephalochordate, a tunicate, and *Haikouichthys* (Fig. 7.3): a creature with a definite head (which makes it a craniate, a basal fish). At the moment we have no clear way to decide

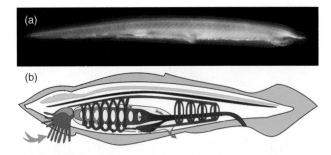

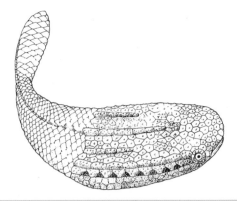

Figure 7.2 Photograph and diagram of a cephalo-chordate. a) *Branchiostoma* from the North Sea, about 2 cm long. Note the V-shaped muscle pattern. © Hans Hillewaart, used by permission. b) diagram of some body parts. Food and water are taken in (big blue arrow) through a mouth that bears tentacles (dark green). Water passes through a pharynx for respiration and exits at gill slits. Food is filtered from the water and passes through a gut to the anus. Blood (red) is circulated round vital organs. The body is stiffened by a notochord (brown), which lies next to a nerve chord with a bulge of nervous tissue at the front (yellow). Based on a diagram by PioM (Piotr Michal Jaworski), and placed into Wikimedia.

Figure 7.4 Reconstruction of an ostracoderm. *Astraspis*, from the Ordovician of Colorado, was about 13 cm (5 inches) long. Courtesy of David K. Elliott, Northern Arizona University.

Figure 7.3 Reconstruction of the earliest known fish, *Haikouichthys* from the Lower Cambrian Chengjiang Fauna of China. Art by Nobu Tamaru, and placed into Wikimedia.

which two of these three lineages are most closely related. All three possible combinations have been strongly argued.

We now have more than 500 specimens of *Haikouichthys*. The soft parts are difficult to interpret, but we can now say something about the first fishes from real fossils rather than from theoretical speculations. First, a lobe at the front has dark areas interpreted as eyes, and perhaps nasal sacs were present too. The mouth and gills lie immediately under this "head" region. Some structures are wrapped around the notochord, and these are probably cartilaginous vertebrae. If so, then *Haikouichthys* is a genuine vertebrate.

The notochord probably evolved as a structure that aided in swimming. But the physics of hydrodynamics dictates that swimming efficiency increases with body length.

As early chordates explored various ways of life, the more actively swimming species probably increased in body size. But there must come a body size for which efficient swimming requires more stiffness than a notochord can give, and some kind of cartilaginous or mineralized skeleton then becomes a cheap way of increasing efficiency. At the moment *Haikouichthys* is the first sign of this breakthrough in mechanical efficiency.

Ostracoderms

The earliest fishes with hard parts, from the Ordovician, did not have a bony internal skeleton. Instead, they evolved mineralized bony plates that covered some or all of their bodies, adding stiffness and giving rise to the term **ostracoderm** ("plated skin") for them. The plates of ostracoderms would have provided protection too, from possible predators and from abrasion by sand and rock surfaces. Solving the same problem in a different way, sharks today have an internal bony skeleton made of cartilage rather than bone, and they have a tough skin with strong fibers that stiffen the body considerably. Most other living fishes have a light scaly skin and an internal bony skeleton.

Bone is dense compared with cartilage, and the heavy plates of ostracoderms must have made ostracoderms relatively clumsy, with slow acceleration. They probably swam slowly along the sea floor, inside heavy bony boxes.

Astraspis from the Ordovician of Colorado is one of the best-preserved early fishes (Fig. 7.4). A headshield protected the anterior nerve center (which from this point can be called a brain), and also provided a stout nose cone for cutting through the water without flexing, and for probing into soft sediment. Behind the eyes were plates with multiple openings to allow water to flow out past the gills. The tail was short, stubby, symmetrical, and small, and these fishes probably swam well but not fast.

Figure 7.5 Many heterostracan fishes were adapted to scoop food from the sea floor, with the mouth on the underside of the head shield. This is *Anglaspis*. After Kiaer.

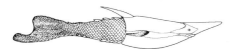

Figure 7.6 A reconstruction of *Pteraspis* shows the strong plated headshield and the flexible trunk and tail. I straightened out the nose, which is usually shown tilted slightly upwards.

The early fishes have left us a very skewed fossil record. We see mostly ostracoderms and their relatives, because their bony plates fossilize better. In fact, we see nothing of the small soft-bodied descendants of *Haikouichthys*. It is difficult to classify and construct a reliable cladogram for early fishes, so I will not provide one. Instead, I shall concentrate on the adaptations we can see evolving in early fishes to make them better swimmers and feeders in Paleozoic oceans.

Heterostracans

Heterostracans were the earliest abundant fishes (in the Silurian and Devonian). They had flattened headshields with eyes at the side, and they look well adapted for scooping food off the seafloor (Fig. 7.5). Some had plates around the mouth that could have been extended out into a shoveling scoop. The rigid head and the stiff, heavy-plated body imply that propulsion came mainly from the tail in a simple swimming style, with none of the control surfaces provided by the complex fins of modern fishes. Even so, the heterostracan way of life was successful, and their fossils are found all across the Northern Hemisphere.

Heterostracan fishes diverged quickly in Silurian times and their shapes evolved toward hydrodynamically more efficient shapes over time. **Pteraspids** (Fig. 7.6) were the most abundant heterostracans in Devonian times. They had beautifully streamlined armored headshields, with a sharp nose cone and a smooth curved shape that gave an upward motion to counteract the density of the armor. A spine projected backward over the lightly plated trunk, partly for protection and partly for hydrodynamic stability. Pteraspids had tails with the lower half longer than the upper; other things being equal, this too would have helped

(a)

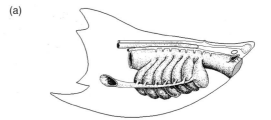

(b)

Figure 7.7 A diagram simplified from a reconstruction of amphiaspid gills by Larisa Novitskaya, showing their similarity to the exhaust system of an early supercharged racing car (this one is a 16-cylinder Bugatti!). Photograph by Gérard Delafond, and placed into Wikimedia.

the fish to counteract the weight of the headshield. The mouth lay under the head, and a ventral plate covered the gills. Water was taken in through the mouth, and the exit passages were neatly tucked toward the back of the headshield, much like the exhausts of a twin-jet fighter aircraft. In some forms the headshield was very flattened, for gliding through water as a delta-wing aircraft glides through air.

Another group of heterostracans, **amphiaspids**, are best known from Siberia. Larisa Novitskaya found specimens from which she could reconstruct the gills. Pteraspid and amphiaspid gill systems look like the exhaust systems of 1930s racing cars, sharing their design for efficient passage of fluids (Fig. 7.7).

But in all this successful evolution, heterostracans never evolved paired fins. Their swimming power came entirely from the trunk and tail, with perhaps a little help from the gill exhaust.

Osteostracans

Other things being equal, any swimming creature would benefit by evolving powerful swimming and better maneuverability. We have seen this already among the heterostra-

(a)

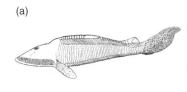

(b)

Figure 7.8 Osteostracan fishes. a) *Hemicyclaspis*, showing the sensory areas on each side of the headshield (after Stensio). b) *Cephalaspis*, a specimen that shows the body shape very well, with the two eyes pointing upward, and the dominance of the headshield relative to flexible trunk and tail. Photograph by Haplochromis, and placed into Wikimedia.

cans. An innovation came with the evolution of paired fins, leading to a Late Silurian radiation of new jawless fishes, the **osteostracans**. Osteostracans were like heterostracans in that they had a strongly plated headshield and a comparatively flexible body and tail that provided most of the propulsion.

The most important osteostracans, the **cephalaspids**, lived from Late Silurian to Late Devonian times. Their large solid headshields often had a large spine projecting forward and two spines extending backward at each corner (Fig. 7.8). Powerful paired fins were attached at the back corners of the headshield, just inside the protective spines. The body behind the headshield was laterally compressed, as in most living fishes, and small dorsal fins added stability. The cephalaspid tail was more versatile than the heterostracan tail, and it had some horizontal flaps that added new control surfaces.

Cephalaspids were bottom swimmers. The mouth was on the flat underside of the headshield. The eyes were small and close together on the top of the headshield (Fig. 7.8). In addition, cephalaspids had large sensory areas on each side of the headshield, covered with very small plates. These organs may have served as pressure sensors in murky water, though they may also have sensed electrical fields, as in living sharks.

Galeaspids

Galeaspids are a clade of small jawless fishes related to osteostracans. They were geographically confined to Eastern Asia, especially China and Vietnam. Galeaspids were successful, however, and range from Silurian to Devonian. They are unusual in having a central opening on the top side of the headshield (Fig. 7.9). The mouth and gill intakes are on the underside, as usual, and the eyes are small and set wide on the headshield, so the galeaspid opening is for something else.

A new galeaspid from the Silurian of China called *Shuyu* is so beautifully preserved that the internal structures of the head could be reconstructed from CAT scanning (Gai et al. 2011). The study showed that the skull opening funneled incoming water to a pair of structures that look very much like the paired nostrils of jawed fishes (Fig. 7.9).

(a)

(b)

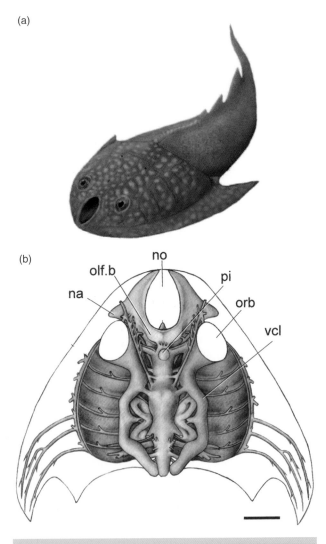

Figure 7.9 *Shuyu*, a galeaspid from the Silurian of China, about 3 cm (1 inch) long. It has a prominent opening in the center of the headshield. Art work by Brian Choo, used by permission. CT scans of the interior (Gai et al. 2011) showed a complex nasal structure inside the headshield. Courtesy Dr. Shikun Gai. The important labels are: no, nasal opening; ol.b, olfactory bulb, na, nasal sac, orb, eye. Scale bar, 2 mm.

As a jawed fish develops as an embryo, its nostril structures form first, before the jaw. Most likely, then, the jawless ancestors of jawed fishes evolved nasal structures before they evolved jaws. And if *Shuyu* is typical of galeaspids, then jawed fishes evolved from a galeaspid ancestor in the Silurian, most likely in China, perhaps even from *Shuyu* itself. This places galeaspids as closer to gnathostomes than osteostracans, reversing current thinking. It might also explain why a varied array of very early jawed fishes have been found at Chinese localities in the last ten years.

The Evolution of Jaws

By the Silurian, the jawless fishes were quite varied. Without jaws, they were confined to eating small particles, such as plankton from the surface, sediment on the seafloor, or soft, easily swallowed food such as worms or jellyfish. But somewhere among them were fishes in the process of a major breakthrough for the feeding ecology of fishes: **evolving jaws**.

Studies of anatomy and embryology suggest that the bones that form the vertebrate jaw evolved originally from the gill arches of jawless fishes. In living fishes, water is taken in at the mouth and passes backward past the gills, where oxygen and CO_2 are exchanged with the blood system (Fig. 7.10). Gills are soft, so they must be supported in the water current by thin strips of bone or cartilage called **gill arches** (Fig. 7.11). The more water passing the gills, the more oxygen can be absorbed and the higher the energy the fish can generate. Living fishes usually have pumps of some kind to increase and regulate the flow of water passing the gills. Most fishes use a pumping action in which they increase and decrease the volume of the mouth cavity by flexing the jaws. Tuna swim so fast that they create a ramjet action that forces water past the gills, just as the airscoops of some jet fighters funnel air into the turbines.

If jaws evolved from a gill arch, the evolution of the jaw was probably connected originally with respiration rather than feeding. Water flow over the gills of jawless fishes may have been impeded by their small mouths and by a slow flow of water past the gills, so their swimming performance may have been limited by oxygen shortage. Perhaps a joint evolved in the forward gill arch so that it flexed to open the mouth wider, pumping more water backward over the gills (Fig. 7.12), and transforming the gill arch into a true jaw.

Jawed fishes are **gnathostomes**, as opposed to the agnathans or jawless fishes. There is no general agreement on the details of their early evolution, mainly because they evolved and radiated very quickly in the late Silurian. I show a simplified cladogram of early-evolving gnathostome groups (Fig. 7.13).

The rapid radiation of gnathostomes into life in open waters did not drive agnathans to extinction, at least not quickly. Instead, during the late Silurian and the Devonian, gnathostomes added to fish diversity.

The evolution of jaws and a resulting extension of the potential food supply were keys to the tremendous evolu-

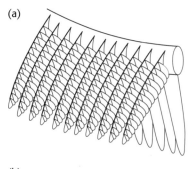

(a)

(b)

blood

water

water

Figure 7.10 How gills work in most living fishes. a) gills are arrays of thin platelike structures set in rows supported on a strong axis. b) oxygen-poor blood is pumped along a one-way system through each platelike structure. Water is pumped the other way, exchanging gases with the blood by shedding oxygen and taking up carbon dioxide.

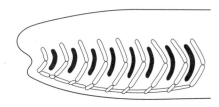

Figure 7.11 A jawless fish, showing its gill arches, and gill slits for outgoing water.

tionary success of jawed fishes. But teeth and jaws are only weapons: they must be applied to targets by a delivery system. The history of fishes since the Devonian has been largely one of increasing effectiveness in mounting and hinging the jaws, in the speed of strike, and in the hydrodynamics of propulsion and maneuverability. All these factors meant that the jawed fishes were able to extend their ecological range into the three-dimensional world of open water, as opposed to the largely bottom-feeding agnathans.

Acanthodians

The earliest jawed fishes are small Silurian forms called **acanthodians**. They are lightly built, not well preserved,

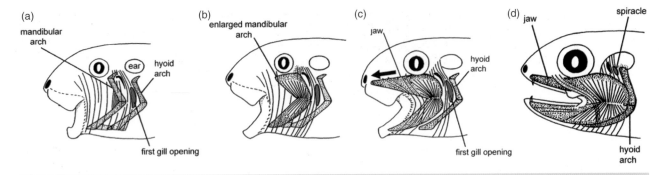

Figure 7.12   Evolving the fish jaw. a) an agnathan or jawless fish. The first gill arch, the mandibular arch, flexes a little to help water flow into the mouth and out through the first gill opening, b) the mandibular arch becomes larger and stronger as muscles flex it more strongly to improve the respiration of the fish. c) in an early gnathostome, the mandibular arch has evolved into a jaw, capable of biting down on a food item. d) the fully evolved jawed fish, with teeth forming along the structure of the jaws. This is a gradual transition, for good functional reasons, but it makes it relatively difficult to identify "the" first gnathostome fish. Simplified from the work of Mallatt (Mallatt 2008), by permission of Blackwell Publishing.

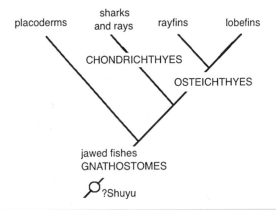

Figure 7.13   A simplified cladogram of early jawed fishes (gnathostomes). Evolving from something like *Shuyu*, the extinct placoderms are the basal group in this scheme. The extinct acanthodians are poorly known, and I have not shown them. They may be a "stem group" of various lineages clustered round the base of the gnathostome radiation. The cartilaginous fishes Chondrichthyes are a clade and so are the bony fishes, the Osteichthyes, which quickly diverged into rayfin fishes and lobefin fishes.

and not very well known. I mention them only because they are among the earliest known fishes with jaws, and one fortunate find may upset our whole understanding of the earliest jawed fishes and the structure of Figure 7.13.

Placoderms

Placoderms were abundant, worldwide fishes during Devonian times. Most placoderms had a well-developed head-

shield made of several plates, jointed to an armored girdle surrounding the front part of the trunk, making the fish very nose-heavy. The rest of the trunk was lightly scaled, and presumably the trunk and the long tail were flexible, for powerful swimming. There were several pairs of fins, indicating good control over movement. But the body was usually flattened to some extent, and the eyes were usually small and set on the upper side of the headshield.

Placoderms include a range of groups, though I shall only discuss the large powerful predators, the **arthrodires**, and the small box-like **antiarchs**. Arthrodires were powerful, streamlined fishes, but their great weight of armor and their generally flattened body shapes may have limited their swimming performance. Large pectoral fins aided stability and provided lift for the heavy armored head. The small eyes imply that they probably used other senses to a large extent, just as living sharks do. The jaws vary quite a lot, but some advanced placoderms had vicious sharp-edged tooth plates set into the jaw (Fig. 7.15); these were large carnivores up to 6 meters (20 feet) long. Others had large crushing tooth plates, perhaps for eating molluscs or arthropods. Ecologically, the rich late Devonian placoderm fauna included generalists and small precise pickers as well as the cutters and crushers. In fact, it is clear that the range of ecology in placoderms is similar in broad outline to that of today's bony fishes (Anderson 2008).

The arthrodires evolved a unique set of joints that operated the jaws, head shield, and trunk armor in a spectacular way. The head could be levered upward while the lower jaw dropped at the same time, quickly opening a wide gape that would have sucked prey toward the jaws just as they closed. Models show that the bite of a huge *Dunkleosteus* is one of the most powerful ever evolved (Anderson and Westneat 2007, 2009). Giant arthrodires include *Tityosteus*, the largest known Early Devonian fish, with a length of about 2.5 meters (8 feet). But *Tityosteus* is dwarfed by a Middle Devonian freshwater arthrodire *Heterosteus* and by the Late

Figure 7.14 The giant arthrodire *Dunkleosteus*, from the Late Devonian of North America, which was up to 6 m (20 feet) long. Art by Nobu Tamura, and placed into Wikimedia.

Figure 7.15 Reconstruction of the little antiarch *Bothriolepis*, from the Devonian of Canada. It is about 30 cm in length (about 1 foot). Image by Citron, and placed into Wikimedia.

Devonian *Dunkleosteus*, both of which grew to 6 meters (20 feet) long (Fig. 7.14).

Antiarchs are much more difficult to understand, but these "grotesque little animals" (as one famous paleontologist called them) were successful worldwide, mostly in freshwater environments. They were small, with headshields up to 50 cm (20 inches) long and a maximum known length just over a meter (3 feet). Their headshields were flattened against the bottom, with the eyes set close together high on the headshield. The mouth lay just under the snout. The body armor was long. Instead of pectoral fins, antiarchs had long, jointed appendages that look as if they were used for poling the fish along the bottom rather than swimming (Fig. 7.15). Antiarchs had small mouths and probably ate mud, filling an ecological role that had been taken by earlier jawless fishes. It's clear that they were slow, rather clumsy swimmers.

Placoderms had another unusual biological attribute: as far as we know, they all had live birth (Fig. 7.16), which also implies internal fertilization. Today's bony fish almost all spawn eggs and sperm into the water, but living sharks and rays have internal fertilization and live birth, and so does the coelacanth (see later in this chapter). The surprise discovery of embryos preserved inside Devonian placoderms (Long et al. 2009a, 2009b) was followed quickly by the realization that the sexes were dimorphic, with the males having pelvic fins modified into "claspers" that are important for mating (Ahlberg et al. 2009).

Placoderms became extinct at the end of the Devonian, and in view of their variety it is difficult to understand why. Possibly they were handicapped by their weight of armor, at least in comparison with the other fishes around them. One might compare them to the huge obsolete masses of steel and chrome that were the prestige American cars of the 1950s! The cartilaginous fishes and the bony fishes, each with their own adaptations, now dominate living fish faunas.

Figure 7.16 *Materpiscis*, a little placoderm from the Devonian of Australia, had live birth. Art by Alf Kuhlmann of Reel Pictures, image © Museum Victoria, used by permission.

Cartilaginous Fishes (Sharks and Rays)

Sharks and rays, and all their ancestors we have been able to identify, have cartilaginous skeletons rather than bone. This distinction dates back to the Early Devonian, when this group of fishes was just one of the many early successful lines that had recently evolved jaws.

The fossil record of sharks and rays is poor, because they rarely preserve well as fossils. They have cartilage rather than bone, and a tough skin rather than heavy scales. They do have formidable teeth, which are often well preserved as fossils, but teeth alone give only a vague idea about the entire fish. Occasionally a rare find of a body outline allows us to see that sharks have not changed a great deal in overall body shape during their evolution (Fig. 7.17).

Figure 7.17 An exceptional find allowed this reconstruction of the fossil shark *Akmonistion*, from the Carboniferous. Its shape is uncannily like that of many living sharks, except for the prominent dorsal structures that are associated in shark biology today with mating. Art work by Nobu Tamura, based on Coates and Sequeira (2001), and placed into Wikimedia.

Sharks have excellent vision and smell and an electrical sense, all of which combine to equip them well for hunting in all kinds of environments. They all have internal fertilization, and some have live birth. They are certainly not primitive. Sharks are simply a group of fishes that discovered a successful way of life several hundred million years ago.

Bony Fishes

The bony fishes, or Osteichthyes, evolved and radiated so fast in the late Silurian and early Devonian that we have very few fossils of the earliest forms. Almost as soon as we see them, they are divided into two major groups, the rayfins and the lobefins (Fig. 7.13). The critical fish faunas involved in the very earliest radiation were recently discovered in South China, which was an isolated mini-continent at the time. Fossils from this region will give us more information in the next few years.

We know already that air breathing was evolved by an early lineage of bony fishes. Some primitive rayfin fishes that survive today breathe air, with lungs. Living lobe-fins, lungfishes, do the same. But sharks, rays, and agnathan fishes today all breathe through their gills. The simplest evolutionary interpretation is that air breathing was evolved in early bony fishes, but has been lost in most rayfins, and in coelacanths.

Air Breathing

Why did early bony fishes evolve the ability to breathe air? The answer I used for many years involved low oxygen levels in water, but Colleen Farmer has suggested a better idea (Farmer 1997, 1999). I have used her work to write much of this section, though I have simplified it drastically.

Animal respiration has built-in universal features. Animals take in oxygen to burn their food in respiration, and they produce CO2 as a waste product. Carbon dioxide is toxic because it dissolves easily in water to form carbonic acid. Animals can tolerate only a small buildup of CO2 before passing it out of the system. (For example, it is high CO2 in our lungs that makes us want to breathe out, not shortage of oxygen.)

Gases are exchanged with the environment, whether it is water or air, as body fluids are passed very close to the body surface. For example, blood flows close under the lung surface in our own breathing. As long as the environment has higher O2 and lower CO2 than the body, diffusion acts to pass O2 in and CO2 out. The rate depends on several factors: the surface area and the thickness of tissue through which the gases must diffuse; the rate at which the external and internal fluids pass across the surface; and the concentrations of gases in the internal fluid and in the external medium. In normal fishes, CO2 and O2 diffuse in opposite directions across the gill surface (Fig. 7.10).

Respiration in Early Fishes

The earliest fishes probably had a respiration system like that of living cephalochordates (Fig. 7.2). Water is pumped into a basket-like structure, and food particles and oxygen are taken out of it. Oxygen is carried in a blood system pumped by a heart, and travels through the body tissues, delivering oxygen, until it reaches the heart again.

This system was inherited by later jawless fishes. They may have evolved sophisticated gills (Fig. 7.7), but their system had a basic flaw: blood arrived at the heart depleted in oxygen, because it had flowed all around the body first (Fig. 7.18). The more active the fish, the more likely it was to suffer heart failure! This is not an ideal piece of engineering, but you inherit what your ancestors give you.

Relatively slow-moving and rather small, jawless fishes flourished for millions of years with this system. (When you think about it, their success on Cambrian seafloors confirms that global oxygen levels were reasonably high, at least in shallow water.)

With the evolution of jaws and the radiation of jawed fishes, more lineages must have become more active foragers and predators, at larger body sizes. There must have been strong selective pressure to modify the ancient system to cope with the extra oxygen demands of a more active life.

Oxygen Intake

It is easier and cheaper to get oxygen from air rather than water. Water is hundreds of times denser and more viscous

(a)

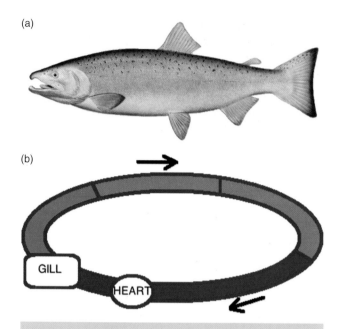

(b)

GILL

HEART

Figure 7.18 Respiration in a fish that uses gill breathing, such as an early fish, or the coho salmon shown here. A weak heart pumps blood to the gill, then oxygenated blood circulates round the body before it reaches the heart again. There is a considerable danger that the heart will become oxygen-deprived. Based on Farmer (1997).

Figure 7.19 Lack of oxygen in the waters of Greenwich Bay, Rhode Island, killed about a million fish in August 2003. Photograph by Tom Ardito, used by permission from the Insomniacs Research Group at Brown University. See http://www.geo.brown.edu/georesearch/insomniacs

than air, and even at best it contains less oxygen. Many gill-breathing animals have to pump external water across their gill surfaces at ten times the rate they pump their internal blood. Gills have to be designed to resist the leakage of dissolved body salts, and the tissues across which oxygen is exchanged cannot be as thin as they can in air, so gas exchange is rarely anywhere near 100% efficient.

Because oxygen diffuses 100,000 times more quickly in air than in water, oxygen-poor air is rare. But oxygen-poor *water* does occur quite often, especially in tropical regions, wherever warm freshwater or saltwater lakes, ponds, or lagoons are partly or completely isolated, especially in a hot season. Warm, rotting debris can quickly use up oxygen, especially if there is little or no natural water flow. Even if the effect is only seasonal, it may still be critical for fishes and other organisms living in the water. The water is stagnant, hot, and full of rotting debris, often teeming with bacteria that may also release toxic substances.

Even today, there are often natural **fish kills** in which massive mortality occurs among fishes (Fig. 7.19). Many fish kills are related to a lowering of oxygen in the environment: for example, in shallow pools, rivers, or lagoons that heat up too much. The immediate culprit is the environmental insult, of course, but the crisis is worsened because the fishes were using a gill system that could not handle the oxygen shortage.

Why would fishes swim into oxygen-poor water, where gill breathing is difficult? The food supply may be rich for

fishes that can tolerate it, and there are situations in which fishes might benefit from swimming into areas of warm, often oxygen-poor water near the surface.

Many carnivorous fishes today are bottom feeders, hunting for small prey that live on or in the surface of the sediment. In warm latitudes, the bottom waters are often much cooler than the surface waters, which are heated by the sun. Digestion can be very slow in cold-blooded animals, especially if they live in cold environments. That may be a critical factor holding back growth and development. In such cases, increasing the digestive rate by swimming into the warm surface water can produce faster growth, earlier maturity, and more successful reproduction.

But what happens to a fish that swims into surface water because it is warm, only to find that it is also oxygen-poor? Even if surface waters are generally low in oxygen, there is always a thin surface layer of water, about a millimeter thick, that gains oxygen from the air by diffusion. Many living fishes in tropical environments come to this surface layer to bathe their gills in the surface oxygen layer. They can breathe, but they have to solve other problems too. If they break the surface, their bodies extend out into the air, losing some buoyancy.

Some living fishes in this situation bite off bubbles of air and hold them in their mouths for positive buoyancy, to remain at the surface without active swimming. Some living species of gobies use this action to breathe. Once they have an air bubble, they can extract oxygen from it in the back of their mouths much more efficiently than at the gills. When the oxygen level in the mouth bubble falls, reducing its size and its buoyant effect, the fish must then get rid of the bubble and bite off another. Rhythmic air breathing might have evolved this way, as fishes get rid of CO2 from their mouth bubbles while they are still losing it at gill surfaces as well.

Oxygen intake in the mouth enriches the blood supply there, and a fish can store oxygen in an air bubble. An air bubble that takes up only 5% of body volume can increase oxygen storage by 10 times compared with a fish without a bubble. Therefore, bubble breathing doesn't mean that a fish is completely tied to the surface; it can make extended dives to the bottom. This is true today of all air-breathers with low metabolic rates, including crocodiles and turtles. Many water-dwelling insects use air bubbles too, and the wonderful diving-bell spider uses a silken dome to make a bubble-filled underwater home for itself (Seymour and Hetz 2011, Fig. 7.20).

In fishes that breathe air, freshly oxygenated blood flows directly to the heart (Fig. 7.21). Colleen Farmer suggests that this made an immense difference to early bony

fishes, enabling them to escape from the danger presented by an oxygen-starved heart (Fig. 7.18). Bony fishes then radiated dramatically as they became able to live more active lives in surface waters. Thus all rayfins (as far as we know) and all lobefins were capable of breathing air by Devonian time, because it made them more efficient fishes in the water.

Most living rayfins have now lost the ability to breathe air, and the structure that was once their lung has evolved into an enclosed gas-filled organ called the **swim bladder**, which helps them to maintain buoyancy in the water. This means that they reverted to the older system of Figure 7.18, which seems counter-productive. Yet it can only have happened for good functional reasons. Colleen Farmer suggests that most ray-fin lineages reverted to gill respiration (in Mesozoic times) after they became vulnerable to newly evolved aerial hunters at the surface: first pterosaurs, and then seabirds. This is speculative, of course: but air-breathers are vulnerable since they must come to the surface to breathe. Remember that 19th-century whalers relied on spotting whales "spouting" at the surface to locate and kill them. Submarines face the same problem: a submarine is most vulnerable at or on the surface.

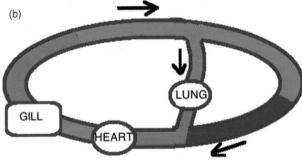

Figure 7.20 *Argyroneta aquatica* is a spider that lives in streams in Central Europe. It builds an underwater dome from silk, then fills it with air bubbles that it brings down from the surface. There it lives, breeds, and eats the insects it catches in the water and on the water surface. Photograph © Dr. Stefan Hetz of Humboldt University in Berlin, and used by permission.

Figure 7.21 Respiration in fishes that use air breathing as well as a gill. The living tarpon shown here has evolved to use its swim-bladder as a lung NOAA photograph. But in early bony fishes, some of the blood from the gill was circulated to the mouth, throat, or lung, where it was charged up with oxygen and delivered directly to the heart. This removed a major disadvantage of the earlier system shown in Figure 7.18. Diagram based on Farmer (1997).

Actinopterygians (Rayfin Fishes)

Actinopterygians, or rayfins, have very thin fins that are simply webs of skin supported by numerous thin, radiating bones (called rays). Typically, rayfins are lightly built fishes that swim fast or maneuver very well. They have dominated marine and freshwater environments of the world since the end of the Devonian. It is tempting to suggest that their evolutionary success largely reflects their mastery of swimming and feeding in open water.

In general, the evolution of the rayfin fishes resulted in a lightening of the bony skeleton and the scaly armor, both of which improved locomotion. Increasing sophistication and variation in the shape and arrangement of the paired fins led to patterns that were optimum for specialized sprinters, cruisers, or artful dodgers. In the most advanced rayfins, swimming has come to depend more and more on the tail fin rather than on body flexing, while the other fins are modified as steering devices and/or defensive spines. Even flying fishes had evolved by Triassic times.

The jaws and skull of rayfins were gradually modified for lightness and efficiency. In particular, intricate systems of levers and pulleys allow advanced fishes to strike at prey more effectively by extending the jaws forward as they close. The same system also allows more efficient ways of browsing, grazing, picking, grinding, and nibbling, all encouraging the evolution of the tremendous variety of living fishes.

Sarcopterygians (Lobefin Fishes)

Sarcopterygians, the lobefin fishes, are distinguished as a separate group because they evolved several pairs of fins that are stronger than any found in a rayfin fish. Other differences in scale and skull structure confirm the separate evolution of these groups. They separated as early as the Silurian: the recently discovered fish *Guiyu* (Fig. 7.22) from the Silurian of southern China is the earliest known lobefin (Zhu et al. 2009). The major lobefin groups diverged during the Devonian in shape, structure, and ecology, into three major clades: coelacanths, lungfishes, and a group named tetrapodiforms which includes the ancestors of all tetrapods, including us (Fig. 7.23).

The central part (the lobe) of a lobe fin is sturdy and contains a series of strong bones, while the edges have radiating rays as in ray fins (Fig. 7.24). A lobe fin must beat more slowly than a ray fin of the same area because there is more mass to accelerate and decelerate, but the resultant stroke is more powerful. Furthermore, and fundamental to later vertebrate history, a lobe fin that imparts a powerful stroke to the water has to have some kind of support at its base, just as an oar has to be stabilized in a rowlock. Therefore, lobefin fishes have internal systems of bones and muscles that help to tie one dorsal and two ventral pairs of lobe fins to the rest of the skeleton (Fig. 7.24). These ventral linkages evolved to become the pectoral and pelvic girdles

Figure 7.22 *Guiyu*, from the Silurian of Southern China, is the best-known early osteichthyan (bony fish), but it is also clearly a sarcopterygian (lobefin). That means that the rayfin fishes had already diverged from the lobefins, and we can hope that good specimens of their ancestors will be found soon. Art work by Nobu Tamura, based on Zhu et al. 2009, and placed into Wikimedia.

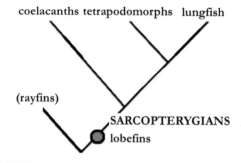

Figure 7.23 A cladogram of early lobefin fishes (sarcopterygians). They radiated in the Devonian from a Silurian ancestor like *Guiyu*. The Devonian forms were all powerful swimmers, unlike all their living descendants.

of land vertebrates, but of course that was not why they evolved: they evolved originally to allow early lobefin fishes to swim more effectively. All Devonian lobefins seem to have been effective swimmers and predators.

Lungfishes and coelacanths still survive, but only as rare and unusual fishes. Two species of coelacanth survive as small populations in South African and Indonesian waters, and one species of lungfish lives in each of the three southern continents: Australia, South America, and Africa. All these living lobefins have such unusual biology and ecology that they must be interpreted with caution. They are much evolved from their Devonian ancestors in structure and in habits, so they may not be very good guides to the biology of those ancestors.

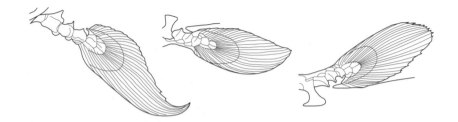

Figure 7.24 From left to right: the anterior ventral, the posterior ventral, and the posterior dorsal lobe fins of the living coelacanth *Latimeria*. The anterior (or pectoral) fin is significantly larger than the other two, but the posterior dorsal fin is just as large as its ventral counterpart and has just as strong an internal bony skeleton. (After Millot and Anthony.)

Figure 7.25 Drawing of *Latimeria*, the last surviving coelacanth. Coelacanths grow to about 5 feet long (close to 2 meters), and may live to be 200 years old. Originally drawn by Robbie Cada for Fishbase (www.fishbase.org), and placed by him into the public domain.

Figure 7.26 *Dipterus*, a Devonian lungfish from Scotland. Even in this crushed specimen, the powerful lobe-fins, like those of the living coelacanth (Figure 7.24) contrast with those of a rayfin. Photograph by Haplochromis, and placed into Wikimedia.

Coelacanths

A living coelacanth, *Latimeria* (Fig. 7.25), was unexpectedly discovered in 1938. Coelacanths had been known as fossils for decades, but it was thought that they had died out after the Cretaceous. We know now that they are very rare, very long-lived, and probably endangered. Living coelacanths are lazy swimmers, do not have lungs, and do not breathe air. The females bear live young, as many as 26 at a time, which develop internally from very yolky eggs.

Lungfishes

Living lungfishes are medium-sized, long-bodied fishes found in seasonal freshwater lakes and rivers in tropical areas. They seem best designed for rather slow swimming. Living lungfishes can breathe air, allowing them to survive periods of drought or low oxygen in seasonal lakes and rivers in tropical climates. Lungfishes probably survive today because they can tolerate environments that would kill most other fishes. The African lungfish can even tolerate a dry season in which its river dries up. It digs a burrow, seals itself inside, and estivates (turns its body metabolism to a very low level) until the rainy season sends water down the river and into the burrow, reviving it.

Figure 7.27 *Protopterus*, a living lungfish from Africa. Its pathetically weak fins can beat synchronously, especially the pelvic fins, to allow it to push effectively against the substrate. From an old engraving. For details and for movies, see King et al. 2011.

Lungfishes have evolved considerably to their present anatomy, biology, and ecology. The first lungfishes were marine fishes, and look as if they were much more active swimmers than their living descendants (Fig. 7.26, 7.27). Living lungfishes are descended from a clade of Devonian ancestors that evolved the ability to live in fresh water, where they evolved changes in teeth and jaws that mark a shift in feeding from other fishes to molluscs and crustaceans. (Living forms have flattened teeth shaped like plates, for crushing their prey.) But lungfishes had evolved burrowing for dealing with drought by Permian times, because many specimens have been found fossilized in their burrows!

Tetrapodomorpha

Tetrapodomorphs are the sister group of lungfishes (Fig. 7.23). The old name for them, rhipidistians, does not meet modern standards of cladistic precision. They include the ancestors of the land-going vertebrates, and are discussed in more detail in Chapter 8.

While placoderms were the dominant fishes of the Devonian, at least in size, the lobefins were most successful in shallow waters around coasts and in inland waters, but were hardly dominant. After the Devonian the rayfin fishes came to be the most successful group, with their combination of lightness and maneuverability, while lobefins were gradually confined to unusual habits and habitats. Perhaps in the process of being squeezed, ecologically speaking, a lineage of Late Devonian lobefins evolved adaptations that allowed them to expand in an unexpected direction—toward life in air.

Further Reading

Ahlberg, P. E. 2009. Birth of the jawed vertebrates. *Nature* 457: 1094–1095. Available at http:/staging.instructables.com/files/orig/FFU/DED8/FRMKL3CW/FFUDED8FRMKL3CW.pdf

Ahlberg, P. et al. 2009. Pelvic claspers confirm chondrichthyan-like internal fertilization in arthrodires. *Nature* 460: 888–889.

Anderson, P. S. L. and M. W. Westneat. 2007. Feeding mechanics and bite force modelling of the skull of *Dunkleosteus terrelli*, an ancient apex predator. *Biology Letters* 3: 76–79. Available at http://171.66.127.192/content/3/1/77.full

Anderson, P. S. L. and M. W. Westneat. 2009. A biomechanical model of feeding kinematics for *Dunkleosteus terrelli* (Arthrodira, Placodermi). *Paleobiology* 35: 251–269. Available at http://biosync.fieldmuseum.org/tdikow/tdikow/sites/default/files/DunkPaleoBio.pdf

Anderson, P. S. L. 2009. Biomechanics, functional patterns, and disparity in Late Devonian arthrodires. *Paleobiology* 35: 321–342. Available at http://eis.bris.ac.uk/~glpsla/page6/assets/Anderson2009.pdf

Anderson, P. S. L. et al. 2011. Initial radiation of jaws demonstrated stability despite faunal and environmental change. *Nature* 476, 206–209. [Devonian disparity among gnathostomes.]

Coates, M. I. and S. E. K. Sequeira. 2001. A new stenacanthian chondrichthyan from the Lower Carboniferous of Bearsden, Scotland. *Journal of Vertebrate Paleontology* 21: 438–459. [Akmonistion.] Available at http://pondside.uchicago.edu/oba/faculty/coates/CoatesSequeira2001Akmonist.pdf

Farmer, C. 1997. Did lungs and the intracardiac shunt evolve to oxygenate the heart? *Paleobiology* 23: 358–372. Available at http://biologylabs.utah.edu/farmer/publications%20pdf/1997%20Paleobiology23.pdf

Farmer, C. G. 1999. Evolution of the vertebrate cardiopulmonary system. *Annual Reviews of Physiology* 61: 573–592. Available at http://www.biologia.ufrj.br/labs/labpoly/Farmer1999.pdf

Gai, Z. et al. 2011. Fossil jawless fish from China foreshadows early jawed vertebrate anatomy. *Nature* 476: 324–327. [The galeaspid *Shuyu*.] Available at http://www.biologie.uzh.ch/index/nature10276.pdf

Janvier, P. 1996. *Early Vertebrates*. Oxford: Clarendon Press.

King, H. M. et al. 2011. Behavioral evidence for the evolution of walking and bounding before terrestriality in sarcopterygian fishes. *PNAS* 108: 21146–21151. Available at http://www.ncbi.nlm.nih.gov/pmc/articles/PMC3248479/

Long, J. A. et al. 2008. Live birth in the Devonian period. *Nature* 453: 650–652.

Long, J. A. et al. 2009. Devonian arthrodire embryos and the origin of internal fertilization in vertebrates. *Nature* 457: 1124–1127. Available at http://pro.unibz.it/staff2/fzavatti/corso/img/nature07732.pdf

Mallatt, J. 2008. The origin of the vertebrate jaw: neoclassical ideas versus newer, development-based ideas. *Zoological Science* 25: 990–998.

Nikaido, M. et al. 2011. Genetically distinct coelacanth population off the northern Tanzanian coast. *PNAS* 108: 18009–18013. Available at http://www.pnas.org/content/108/44/18009.full.pdf+html

Shu, D-G. et al. 2003. Head and backbone of the Early Cambrian vertebrate Haikouichthys. *Nature* 421: 526–529. Available at http://www.bio.pku.edu.cn/userfiles/File/lifm//shudegan/Nature2003-Haikouichthys.pdf

Zhu, M. et al. 2009. The oldest articulated osteichthyan reveals mosaic gnathostome characters. *Nature* 458: 469–474, and comment by M. I. Coates, pp. 413–414. [*Guiyu*.]

Questions for Thought, Study, and Discussion

Summarize Colleen Farmer's idea that bony fishes would operate more efficiently if they were air-breathers. (You might need to draw diagrams to persuade yourself!) It seems that many Devonian fishes breathed air. But now most of them don't. What happened to their lungs, and why did they evolve to become what seems to be less efficient?

EIGHT
Leaving the Water

In This Chapter

Leaving the water is difficult for both plants and animals, for a variety of reasons associated with food or nutrients, exposure to extremes of heat and cold, respiration, and reproduction. I treat plants first because the earliest land plants were probably Ordovician in age. By the end of the Devonian the land plants had evolved into the world's first forests, probably providing habitat for the first land animals.

These were mostly arthropods at first, but air-breathing fishes had evolved into tetrapod-like animals by the end of the Devonian. We have many of the morphological steps in this process preserved in fossils, but we do not have a full understanding of the ecology and biology of the earliest tetrapods.

Problems of Life in Air

Plants, invertebrates, and finally vertebrates evolved to live on land in the Middle Paleozoic. There were major problems in doing so, related not so much to the land surface as to exposure to air. Many marine animals and plants spend their lives crawling on the seafloor, burrowing in it, or attached to it. As a physical substrate, the land surface is not very different. But land organisms are no longer bathed in water. There are predictable consequences for the evolutionary transitions involved, many of them based on the laws of physics and chemistry.

Organisms weigh more in air without the buoyant effect of water, so support is more of a problem. Air may be very humid, but it is never continuously saturated, so organisms living in air must find a way to resist desiccation. Tiny organisms are particularly sensitive to drying out in air, because they have relatively large surface areas but cannot

History of Life, Fifth Edition. Richard Cowen.
© 2013 by Richard Cowen. Published 2013 by Blackwell Publishing Ltd.

hold large reserves of fluid. Therefore reproductive stages and young stages of plants and animals are very sensitive to drying. Temperature extremes are much greater in air than they are in water, exposing plants and animals to heat and cold. Oxygen and carbon dioxide behave differently as gases than they do when dissolved in water, so respiration and gas exchange systems must change in air. The refractive index of light is lower in air than in water, and sound transmission differs too, so vision and hearing must be modified in land animals.

There are also ecological consequences. Seawater carries dissolved nutrients, but air does not, so some organisms, especially small animals and plants, have a food supply problem in air. It's unlikely that the same food sources would be available to an animal that crossed such an important ecological barrier, so invasion of the land would often be associated with a change in feeding style.

All the major adaptations for life in air had to be evolved first in the water, as adaptations for life in water. Only then would it have been possible for organisms to emerge into air for long periods. We must reconstruct a reasonable sequence of events during the transition, then test our ideas against evidence from fossil and living organisms.

The Origin of Land Plants

We have no idea when plants first colonized land surfaces. Plants must have emerged gradually into air and onto land from water, and the first "land" plants must have been largely aquatic, living in swamps or marshes.

Almost all the major characters of land plants are solutions to the problems associated with life in air. Land plants grow against gravity, so they have evolved structural or hydrostatic pressure supports (**hard cuticles** or **wood**) to help them stay upright. They cannot afford evaporation from moist surfaces, so they have evolved some kind of **waterproofing**. **Roots** gather water and nutrients from soil and act as props and anchors. **Internal transport systems** distribute water, nutrients, and the products of photosynthesis around the plant. Even so, all these adaptations for adult plants are useless unless the **reproductive cycle** is also adapted to air. Cross fertilization and dispersal require special adaptations in air. All these adaptations must have evolved in a rational and gradual sequence. But because the first stages would have been soft-bodied water plants, the fossil record of the transition is difficult to find.

A scenario for the evolution of land plants was presented by John Raven (Raven 1984). Water-dwelling plants, probably green algae, were already multicellular. Green algae grow rapidly in shallow water, bathed in light and nutrients. One might think that cells in a large alga are comparatively independent of one another. In the water, each cell has access to light, water, nutrients, and a sink for waste products. But the fastest growing points of algal fronds need more energy than the photosynthesis of the cells there can supply, so some green algae have evolved a transport system between adjoining cells to move food quickly around the plant. They presumably do this because they can then grow more rapidly.

Raven's scenario begins with green algae living in habitats that were subject to temporary drying. The algae might already have evolved to disperse spores more effectively by releasing them into wind instead of water. Spores, even in algae, are reasonably watertight and could easily have been adapted for release into air from special spore containers (**sporangia**) growing high enough to extend out of the water on the uppermost tips of otherwise aquatic plants. As plant tissue extended into air, photosynthesis increased because light levels in air are higher than they are in water, especially at each end of the day, and are free from interference by muddy water. Furthermore, CO_2 is more easily extracted from air than it is from water.

As plants grew out into air, some tissues were no longer bathed in the water that had provided nutrients and a sink for waste products. Internal fluid transport systems between cells became specialized and extended. Photosynthesis was concentrated in the upper part of the plant that was exposed to more light. Photosynthesis fixes CO_2, so there had to be continual intake of CO_2 from the air. However, plant cells are saturated with water, but air is usually not, so the same surfaces that take in CO_2 automatically lose water. Sunlight heats the plant, encouraging evaporation. The water loss had to be made up by transporting water up the stem to the photosynthesizing cells.

Water is transported much more effectively as liquid than as vapor. Early land plants evolved a simple piping system called a **conducting strand** of cells to carry water upward. The conducting strand, found in living mosses, is powerful enough to prevent water loss in small, low plants, if soil water is abundant. But mosses quickly dry up if soil water is in short supply.

Early land plants began to evolve a **cuticle** (a waxy layer) over much of their exposed upper surfaces. The cuticle helps the plant through alternating wet and dry conditions. In wet times it acts as a waterproof coating. It prevents a film of water from standing on the plant that could cut off CO_2 intake. In dry times it seals the plant surface from losing water by evaporation. A cuticle may also have added a little strength to the stem of early plants, and its wax probably helped to protect the plant from UV radiation and from chewing arthropods.

But the cuticle also cut down and then eliminated, from the top of the plant downward, the ability to absorb waterborne nutrients over the general plant surface. Nutrients were taken up more and more at lower levels of the plant, eventually taking place on specialized absorbing surfaces at the base (**roots**) which probably evolved from the runners that these plants often used to reproduce asexually. As roots grew larger and stronger, they helped to anchor and then to support the plant. Roots also extended deep into the soil to extract water, and, often with the help of fungi and bacteria, to extract nutrients.

As cuticle evolved, it sealed off CO_2 uptake over the general plant surface, so plants evolved special pores called **stomata** where CO_2 uptake could be concentrated. If it is too hot or too dry, stomata can be closed off by **guard cells** to control water loss. As CO2 uptake was localized, plants evolved an **intercellular gas transport system** that led from the stomata into the spaces between cells, improving CO2 flow to the photosynthesizing cells. The same system was also used to solve an increasingly important problem. As roots enlarged, more and more plant tissue was growing in dark areas where photosynthesis was impossible; yet those tissues needed food and oxygen. Soils are low or lacking in O2, especially when they are waterlogged. The intercellular gas transport system feeds O2 from the air down through the plant to the roots, sometimes through impressively large hollow spaces.

Later plants evolved **xylem**, an improved piping system for better upward flow of water from the roots. Xylem is made of very long dead cells arranged end to end to form long pipes up and down the stem. Even a narrow xylem can transport water much faster than can normal plant tissue, and, once begun, the evolution of xylem was probably a rapid process that immediately gave plants greater tolerance of dry air. Plants that have xylem are called **vascular plants** (Fig. 8.1).

Xylem cells are dead, so xylem transport is passive, driven entirely by the suction—or negative pressure—of evaporation from the upper part of the plant, and it takes place at no cost to the plant. The forces generated can be very large, so the long narrow walls of xylem cells may tend to collapse inward. Xylem cell walls came to be strengthened by a structural molecule, **lignin**. Once lignin had evolved, it was used later to strengthen the roots and stem

as plants grew taller and heavier. Later still it also became a deterrent to animals trying to pierce and chew on plant tissues.

As plants became increasingly polarized, with nutrient and water being taken up at the roots and photosynthesis taking place in the upper parts, the xylem and gas transport systems improved, but neither of them could transport liquid downward. This problem was solved by the evolution of another transport system called **phloem** that works like the cell-to-cell transport system of green algae. Phloem cells carry photosynthate from photosynthesizing cells to growing points such as reproductive organs and shoots, and to tissues such as roots that cannot make their own food.

Throughout the process, the advantage that encouraged plants to extend into air in spite of the difficulties involved was the tremendous increase in available light. Marine plants are restricted to the narrow zone along the shore where light has to penetrate sediment-laden, wave-churned water. Growth above water increases light availability. Furthermore, competition for available light tended to encourage even more growth of plant tissues above the water surface, and more effective adaptations to life in air (Fig. 8.1). Once plants could grow above the layer of still air near the water surface, spores could be released into breezes. Greater plant height and the evolution of sporangia on the tips of branches were both adaptations for effective dispersal.

The Earliest Land Plants

The earliest spores that belonged to land plants come from Middle Ordovician rocks (Fig. 8.2 and Wellman et al. 2003). They look very much like the spores of living liverworts, and there are fragments of their parent plants preserved with them. All evidence from living plants suggests that the simplest ones fall into a natural group called **bryophytes** (liverworts, mosses, and hornworts), and that liverworts (Fig. 8.3) are the most basic of the three (Fig. 8.4). Fortunately the fossil record is perfectly consistent with this (at the moment).

Late Silurian and Early Devonian Plants

Well-preserved land plants are not found in rocks older than Late Silurian. Though it is Devonian in age, *Aglaophyton* (Fig. 8.5a) probably has a grade of structure that evolved in Silurian times. It grew to a height of less than 15 cm (about 6 inches). It had most of the adaptations needed in land plants (cuticle, stomata, and intercellular gas spaces), but it did not have xylem, only a simple conducting strand, so it was not a fully evolved vascular plant.

The Late Silurian fossils almost certainly include vascular plants. *Cooksonia* (Fig. 8.5b) was only a few centimeters high and had a simple structure of thin, evenly branching stems with sporangia at the tips, and no leaves. But it also

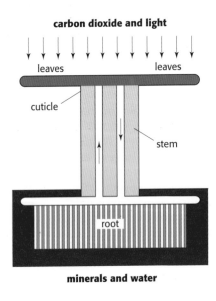

carbon dioxide and light

leaves   leaves

cuticle

stem

root

minerals and water

↑ Xylem transports water and minerals upward from the roots to the leaves.

↓ Phloem transports food around the plant from the leaves where it is made.

Figure 8.1   The basic land plant.

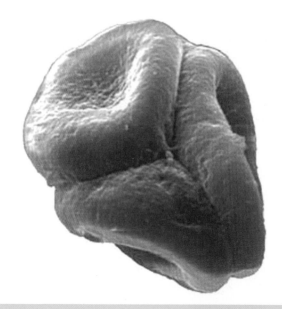

Figure 8.2 One of the earliest known spores from a land plant, from the Ordovician of Libya. Courtesy Charles Wellman of Sheffield University, UK.

Figure 8.3 Living liverworts. They prefer damp shady places, probably much like their Ordovician ancestors. This is *Conocephalum* from Scotland. Photograph by Lairich Rig, and placed into Wikimedia.

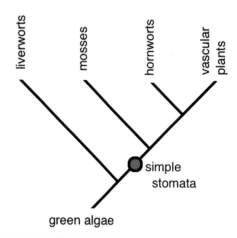

Figure 8.4 There is general agreement on the evolution of early land plant groups. Evolving from green algae, the liverworts, mosses, and hornworts are successive clades at the base of land plants, before the evolution of the vascular transport system. Simple stomata, however, evolved as early as the mosses.

either in standing water or in dense clusters, aided by the fact that they reproduced largely by budding systems of rhizoids for asexual, clonal reproduction, as strawberries do today (Fig. 8.5). This style of reproduction not only gave mutual support to individual stems, but, by "turfing in," a cloned mass of plants could help to eliminate competitive species. Plants like this could have grown and reproduced very quickly, a way of escaping the consequences of relatively poor adaptations for living in air.

Rhynia (Fig. 8.5c), like *Cooksonia*, is a genuine vascular plant. It is named for a famous plant fossil bed, the Rhynie Chert of northern Scotland. Plants were occasionally flooded by silica-rich water from hot springs, and preserved perfectly with all their tissues (Fig. 8.6).

At some point in the Early Devonian, a lineage of small plants evolved thick-walled cells called **tracheids** within their xylem. The thickened cell walls allowed the secondary xylem to transport liquid under greater pressure, increasing its efficiency, and made the tracheids relatively rigid. In some living plants the tracheid system is very large and quite rigid, and we know it now as wood. Late Devonian plants were able to grow to greater heights, competing with one another for light, and therefore for living space.

Later Devonian Plants

Structural advances are seen in later Devonian and Early Carboniferous plants. Successive floras all lived in lowland floodplains, which have a good fossil record. It looks as if we are seeing waves of ecological and evolutionary replacement on all levels, from individual plants to world floras, as structural innovations allowed each plant group to outcompete its predecessor.

had central structures that were probably xylem rather than simple conducting strands. Later species of *Cooksonia* from the earliest Devonian have definite strands of xylem preserved, and cuticles with stomata, so they probably had intercellular gas spaces and were better adapted for life in air.

Early Devonian land plants were dramatically more diverse. They grew up to a meter high, although they were slender (1 cm diameter). For support they must have grown

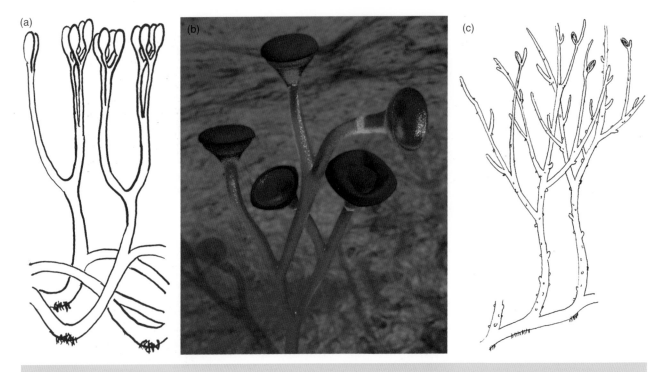

Figure 8.5   Some Silurian and Devonian plants. a) *Aglaophyton*, from the Devonian, has not evolved a vascular system. Reconstruction by Grienstiedl, based on Edwards (1986), and placed into Wikimedia. b) a new reconstruction of *Cooksonia*, the earliest vascular plant, to show its funnel-shaped sporangia in various stages of development. Image made by Smith609 and modified by Peter Coxhead: placed into Wikimedia. c) reconstruction of *Rhynia*, a vascular plant from the Devonian of Scotland, by Grienstiedl, and placed into Wikimedia.

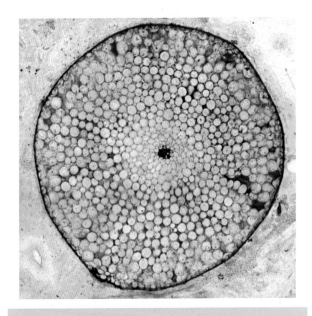

Figure 8.6   A microscopic examination of the stem of *Rhynia* shows its cell structure preserved beautifully. The xylem is the small dark area in the center of the stem, and the phloem is the large zone of circular tubes surrounding it. Image by Plantsurfer, and placed into Wikimedia.

For example, *Rhynia* had only 1% of its stem cross-section made of xylem (Fig. 8.6). Other Devonian plants had 10%, and the whole stem was more strongly built. Plants could grow taller (up to 2 meters high) and compete for light more efficiently than *Rhynia*. Other improvements in reproduction and light gathering, through the evolution of leaves rather like those of living ferns and through more complex branching, also aided plant efficiency.

Vascular plants diverged into two great groups in the Devonian, the **lycophytes** and all the others. Lycophytes today are very small plants, but in the Devonian and Carboniferous, some of them grew to be very large trees, which I shall discuss later. The other group (all other vascular plants) needs a name: it is euphyllophytes ("plants with nice leaves") (Fig. 8.7), because their leaves have interesting vein structures, while lycophyte leaves have very simple veins. Oddly, true leaves, roots and wood evolved independently in each group, perhaps as many as several times. The fossil record is not good enough to see the details. By Middle Devonian times there were many fernlike plants with well developed leaves.

Trees are woody and large, so they stand out in the fossil record: sometimes literally, because they may be preserved still upright in life position. Fossil tree trunks from the Middle Devonian of New York suggest plants over 10 meters (30 feet) high, with woody tissue covered by bark. Once plants reached these heights, shading of one species

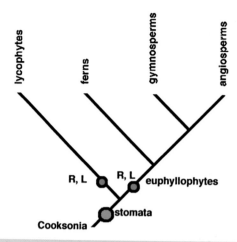

Figure 8.7 Cladogram of early vascular plants. Stomata evolved after Cooksonia, but true roots and leaves (R, L) evolved separately in lycophytes and euphyllophytes. Gymnosperms include conifers, gingkos, and cycads.

Figure 8.8 This image of me taking a photo of a late Devonian *Archaeopteris* will be taken next year from my time machine. I will be checking the 1962 reconstruction of Charles Beck which connected the fossil foliage of *Archaeopteris* (Figure 8.9) with the fossilized wood of its tree trunk.

by another would have led to fairly complex plant communities. We see the first large forests by the late Devonian. *Archaeopteris* grew to be 30 meters high (100 feet), and was the dominant tree in forests that were globally widespread (Fig. 8.8, Fig. 8.9).

Plants evolved **seeds** (rather than spores) in the Late Devonian, too. A Late Devonian seedlike structure called *Archaeosperma* looks as if it belongs to a tree very much like *Archaeopteris*. This was a great advance: all previous plants had needed a film of water in which sperm could swim to fertilize the ovum, but seed plants can reproduce away from water.

Thus by the end of the Devonian, all the major innovations of land plants except flowers and fruit had evolved. Forests of seed-bearing trees and lycopods had appeared, with understories of ferns and smaller plants.

The increasing success of land plants, especially their growth to the size of trees, must have produced ever-larger amounts of rotting plant material in swamps, rivers, and lakes, leading to very low O_2 levels in any slow-moving tropical water (O_2 is used up in decay processes). At the same time, the increasing photosynthesis by land plants drew down atmospheric CO_2 and increased atmospheric oxygen.

All this probably helped to encourage air breathing among contemporary freshwater arthropods and fishes, and it led to better preservation of any fossil material deposited in anoxic swamp water. Some coals are known from Devonian rocks, but truly massive coal beds formed for the first time in Earth history in the Carboniferous Period, which was named for them.

The dominant process in Devonian plant evolution seems to have been selection based on simple efficiency—in size and stability, photosynthesis, internal transport, and

reproductive systems. Plant groups replaced one another as innovations appeared. Perhaps the most interesting part of this story of early plants is the rate at which innovations appeared. There is no obvious reason why the process should not have gone faster, or slower. The innovations we have discussed should have given immediate success whenever they appeared. But it took the length of the Devonian (about 50 m.y.) for land plants to evolve to seedplants. Even "obvious" innovations may take time to evolve and accumulate.

The evolution of seeds seems to have been the foundation for success in the Early Carboniferous. Seed plants invaded drier habitats, and seed dispersal by wind (rather than water) became important; we have winged seeds from Late Devonian rocks. Seed dispersal allowed some plants to specialize as invaders into new areas, avoiding the increasingly dense and competitive habitats in wetlands and along rivers.

Comparing Plant and Animal Evolution

Whether one counts spores or plant macrofossils, there is a striking increase in land plant diversity from Silurian to Middle Devonian time, when a diversity plateau was reached that extended into the Carboniferous. A second

Figure 8.9 A frond from the top of an *Archaeopteris* tree. Image by RC.

For example, CO2 uptake must be accompanied by the loss of water vapor, since the plant is open to gas exchange. Many plant adaptations are responses to the problem of water conservation. Because it is so basic a part of their biology, an innovation here could provide a new plant group with an overwhelming advantage. Other plant systems such as light-gathering are equally likely to be improved by innovation.

Plant distributions are sensitive to climate. If climate changes, plants must adapt, migrate, or become extinct. In extreme circumstances, there may be no available refuges. Thus, the tree species of Northwest Europe were trapped early in the ice ages between the advancing Scandinavian glaciers to the north and the Alpine glaciers to the south, and were wiped out. In contrast, similar species in North America were able to move their range south along the Appalachians, then north again as the ice retreated.

On the other hand, plants are well adapted to deal with temporary stress, even if it is catastrophic to animals. Plants readily shed unwanted organs such as leaves or even branches in order to survive storms and extreme weather. Many weeds die, to overwinter as seeds or bulbs. Even when plants are removed, by fire or drought, the soil is always rich with seeds, so that mass mortality of full-grown plants does not mean the end of the population. Plants are the dominant biomass in communities recovering after volcanic eruptions, tropical storms, or other catastrophes have devastated an area. In many land communities, the removal of dominant trees by storm, fire, or human agency is followed by the rapid growth of species that are very good at colonizing disturbed areas.

The First Land Animals

As plants extended their habitats into swamps and on to riverbanks and floodplains, they would have provided a food base for animal life evolving from life in water to life in air. The marine animals best preadapted to life on land were arthropods. They already had an almost waterproof cover and were very strong for their size, moving on sturdy walking legs. The incentive to move out into air might have been the availability of organic debris washed ashore on beaches, or perhaps the debris left on land by the first land plants. (Foraging crabs are obvious members of many beach communities.) Plant debris, whether it's on a beach or on a forest floor, tends to be damp; it provides protection from solar radiation and is comparatively nutritious. Thus it is not surprising that the earliest land animals were arthropods that ate organic debris, and other arthropods that ate them.

Different arthropods probably moved into air by different routes. The easiest transition would seem to have been by way of estuaries, deltas, and mudflats, where food is abundant and salinity gradients are gentle.

The earliest land arthropods are known from Late Silurian trace fossils of their footprints, rather than the animals themselves. Very small arthropods have been found at

increase in Carboniferous land plant diversity was followed by a long period of stability. A third, Late Mesozoic expansion in land plants and animals raised diversity to current levels.

The pattern looks rather like the pattern of Sepkoski's three major faunas in the oceans. But the radiations among land plants and marine animals did not occur at the same time, so they were not directly linked. Extinctions among plants are different from those among animals, which suggests that plants and animals may respond to quite different extinction agents. Andrew Knoll suggested three major factors (Knoll 1984):

- Plants are more vulnerable to extinction by competition.
- Plants are more vulnerable to climatic change.
- Plants are less vulnerable to mass mortality events.

These differences reflect basic plant biology. All plants do much the same thing. They are all at the same primary trophic level, so they cannot partition up niches as easily as animals can. A new arrival in a flora may be competitively much more dangerous than a new arrival in a fauna.

several Early Devonian localities. Most of them (mites and springtails) were eating living or dead plant material, and in turn were probably eaten by larger carnivorous arthropods such as early spiders. Larger arthropods are usually found in tiny fragments, but it is clear that some were large by any standard. We have pieces of a scorpion that was probably about 9 cm (over 3 inches) long, and a very large millipede-like creature, *Eoarthropleura*, which probably lived in plant litter and ate it. At 15–20 cm (6–8 inches) long, this was the largest terrestrial animal of the Early Devonian. But it is preserved in fragments, which does not make it photogenic . . .

This early terrestrial ecosystem did not include any vertebrates as permanent residents, but no doubt the entire food chain, including fishes in the rivers, lakes, and lagoons, benefited from the increased energy flow provided by plants and their photosynthesis.

Tetrapodomorphs

The invasion of the air by plants, and by invertebrates that exploited them for food and shelter, led to a large increase in organic nutrients in and around shorelines. In the Devonian we see the first signs that fishes were beginning to exploit the newly enriched habitats near the shore and near the surface. But we must not imagine that vertebrates adapted quickly to life in air, or that they readily left the water.

We saw in Chapter 7 that lobefin fishes evolved into different ways of life by the end of the Devonian. Lungfishes came to specialize in crushing their prey, small clams and crustaceans. If we can judge by the last surviving species of coelacanth, this group came to hunt in the water by stealth, followed by a quick dash. The tetrapodomorphs seem to have been the Late Devonian lobefins best adapted to hunting fishes in shallow waters along sea coasts and into brackish shoreline lagoons and freshwater lakes and rivers. They look more active than coelacanths and were probably fast-sprinting ambush predators.

Tetrapodomorphs had long, powerful, streamlined bodies with strong lobe fins and tail (Fig. 8.10), adapted for strong swimming. They had long snouts, especially the larger ones. Perhaps as a result, they evolved a skull joint that allowed them to raise the upper jaw as well as, or instead of, lowering the lower jaw as they gaped to take

prey. This could have had two important effects related to life in shallow water. First, the snout movement would have changed mouth volume, perhaps allowing extra water to be pumped over the gills without moving the lower jaw. Second, tetrapodomorphs could have caught prey in shallow water by raising the snout without dropping the lower jaw. Crocodiles do exactly the same thing as they take prey in shallow water. Some tetrapodomorphs may have been able to chase prey right up to or even beyond the water's edge. Their powerful ventral lobe fins, set low on the body, may have allowed them to drive after prey on, over, or through shallow mud banks, thus making rapid trips over surfaces that could be called "land."

The main sprinting propulsion in tetrapodomorphs came from the tail. The lobe fins were set on the dorsal side of the body as well as ventrally (Fig. 8.10, Fig. 8.11). In deeper water, tetrapodomorphs could attack prey from any angle. But in shallow water the ventral fins took on additional importance. The pectoral ventral fins could be used against the bottom as supports, strengthening the posture of the anterior trunk and acting as props in chasing; the pelvic ventral fins acted to grip and push on the substrate so that maximum effort could be expended against it, adding to the thrust. Lobe fins evolved toward limbs, not as an adaptation for walking, but to become a more efficient fish.

Tetrapodomorphs evolved an adaptation for air breathing that you and I still have. If you breathe through your nose, air reaches your lungs through a passage called a **choana** that runs from your nostrils, through your sinuses, and through the back of your throat. That same passage evolved in the earliest and most basal of the tetrapodomorphs, *Kenichthys* from the Middle Devonian of China. This is clear evidence that all tetrapodomorphs are closer to tetrapod ancestry than any other lobefins.

Well-known Devonian fish called *Osteolepis* and *Eusthenopteron* (Fig. 8.10, Fig. 8.11, Fig. 8.12, Fig. 8.13) are further along that same lineage, and have long been recognized as the most likely tetrapod ancestors. The skull bones, the pattern of bones in the lobe fins, and the general size, shape, and geographic distribution of these fishes are close to

Figure 8.11 *Eusthenopteron foordi*, from the Devonian of Canada. Photograph by Ghedoghedo, and placed into Wikimedia.

Figure 8.10 An early tetrapodomorph, *Osteolepis* from the Devonian of Scotland. Art work by Nobu Tamura, and placed into Wikimedia.

Figure 8.12 Reconstruction of *Eusthenopteron*, from the Devonian of Canada. Strong ventral fins could have been used to help swimming in very shallow water. Art by Nobu Tamura, and posted on Wikimedia.

Figure 8.13 In *Eusthenopteron*, the dorsal lobe fins were attached to the spine just as firmly as the ventral fins were, and presumably played just as important a role in swimming. In those later tetrapodomorphs that evolved toward life in shallow water and excursions out into air, the ventral fins evolved to become limbs because they happened to be placed where they could push on the substrate. (After Jarvik.)

Figure 8.14 A series of cartoon fishapods, from Middle Devonian to early Carboniferous, arranged along a cladogram to show the evolutionary transition from tetrapodomorph (first three) to tetrapod (Last three). From basal to derived, they are *Eusthenopteron*, *Panderichthys*, *Tiktaalik*, *Acanthostega*, *Ichthyostega*, and *Pederpes*. Diagram by Maija Karala, and placed into Wikimedia.

those of the earliest tetrapods. The lobe fins have bony elements corresponding to a 1–2–several–many pattern. Our limbs do the same: our arms have humerus; radius + ulna; wrist bones (carpals); hand bones; and our legs have femur; tibia + fibula; ankle bones (tarsals); foot bones. We and all other tetrapods share the same pattern, inherited from tetrapodomorphs.

We have a picture, then, of varied Devonian tetrapodomorphs, all hunters and most adapted to shallow-water habitats. None was adapted to be active out of water for any length of time. These creatures are informally called **fishapods** (Fig. 8.14) because increasingly their strong ventral fins begin to look like feet.

From Tetrapodomorph to Tetrapod

Tetrapodomorph locomotion in shallow water and on shallow mudbanks would have been improved by stronger fins, especially stronger fin edges. Land locomotion consisted at first of the same undulatory twisting that salamanders still have, with the fins acting simply as passive pivots (Fig. 8.15). The fins gradually exerted stronger traction on the substrate, which may have encouraged the multiple rays

in the fins to become fewer and stronger until toed feet evolved. In the process, the pectoral fins came to support the thorax, while the pelvic fins came to be better suited to push the body forward. The pelvic fin evolved a hinge joint at a "knee" and a rotational joint at an "ankle," a pattern that persisted into tetrapods. This difference was inherited by all later vertebrates: elbows flex backward, knees flex forward. As the pectoral and pelvic girdles evolved better linkage with the fins, the fins evolved gradually to become clearly defined limbs.

Other changes also took place as tetrapodomorphs evolved into tetrapods. A leathery skin evolved to resist water loss, and senses improved for an air medium. Ecologically, tetrapods and tetrapodomorphs divided up the habitat as they diverged. Derived tetrapodomorphs (evolving tetrapods) spent more and more time at and near the water's edge, sunning and basking, while basal tetrapodomorphs remained creatures of open water.

I use the word tetrapod to describe an animal that has feet rather than fins (note that this is a *stem* definition of a tetrapod). It is clear that there was a gradual transition

(a)

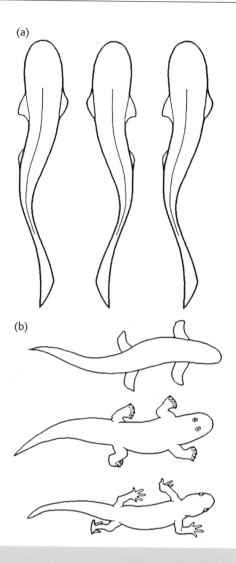

(b)

Figure 8.15 Locomotion from fish to tetrapod. a) fishes swim by undulating the body significantly, while the head swings less. b) the same basic body movements are used by a tetrapodomorph fish swimming or squirming over a mud bank, by an early tetrapod crawling, and by a salamander walking on dry land. No sudden or large shifts in locomotory mechanism were required for the transition, even though the fish has fins and the salamander has feet.

Figure 8.16 *Tiktaalik*, a tetrapodomorph from the Late Devonian of Arctic Canada. The skull is about 20 cm (8 inches) long. Photograph by Ghedoghedo, and placed into Wikimedia.

Figure 8.17 *Tiktaalik*, reconstruction by Zina Deretsky, presumably in conjunction with the scientific team that discovered *Tiktaalik*. Courtesy National Science Foundation.

between the two structures. The bones of tetrapod limbs, all the way to the toes, are coded by the same sets of genes that once coded for the bones in the fins of their gnathostomes. Subtle changes in those genes, and subtle changes in the bones they code for, shape the evolutionary sequences that lead to goldfish fins, tetrapod legs, bird wings, bat wings, and dolphin flippers. Obviously, if we had a lot of fossils we could see those changes happen in detail among Devonian fishapods. But we already have the basic sequence (Fig. 8.14).

At the moment, the tetrapodomorph that looks closest to tetrapods is *Tiktaalik*, from the Late Devonian of Arctic

Canada (Fig. 8.16, Fig. 8.17) (Daeschler et al. 2006). It had a flatter and broader skull than its predecessors, and it was connected to the spine in such a way that the head was mobile. *Tiktaalik* has a neck, in other words, allowing it to seize prey with a slashing sideways grab. It has pectoral fins that could be positioned for swimming/pushing in locomotion, or could act as support props for the heavy skull and thorax (Shubin et al. 2006), presumably to make air breathing easier while lying at the water's edge. The ribs were broad and overlapping, as part of the same way of life. *Tiktaalik* was found in rocks that were laid down in broad coastal river systems.

Limbs and Feet: Why Become Tetrapod?

Why would tetrapodomorphs, as fast-swimming predators in the water, have evolved lobe fins that increasingly looked and operated like the tetrapod limb? How would a tetrapodomorph have benefited from an ability to push on a resistant substrate, rather than using a swimming stroke in water? Why take excursions out into air, rather than simply breathe air at the water surface? In other words, why would a tetrapodomorph become a tetrapod? Evolutionarily, it can have resulted only from a chain of events that produced an improved tetrapodomorph.

The old story about this transition was that an ability to withstand air exposure helped a tetrapodomorph find another pool of water if the one it lived in dried up. This idea is probably wrong. Animals in the Florida Everglades around drying waterholes stay with the little supply there is, rather than striking off into parched country in the hope of finding more: it's simply a better bet for survival.

Basking?

The evolution of strong, low-slung lobe fins on tetrapodomorphs probably helped them to hunt small prey in shallow water by poling their bodies through and over mudbanks. The fins became powerful enough to support the weight of the fish, at least briefly, while it gasped and thrashed its way along. The brief exposures to air would not have been long enough to pose much danger of drying out, but they would have pre-adapted tetrapodomorphs for longer periods of exposure.

If some tetrapodomorphs evolved the habit of sunning themselves on mudbanks to warm up their bodies, their digestion would have been faster than in the water. Other things being equal, they would have grown faster, matured earlier, and reproduced more successfully than their competitors did. Basking behavior would have been effective even if the fish exposed only its back at first, supported mainly by its own buoyancy. But such effectiveness would have encouraged longer and more complete exposure. Some fishes, and many living amphibians and reptiles (including alligators and crocodiles), bask while they digest (Fig. 8.18).

As a basking, air-breathing tetrapodomorph became more exposed, more of its weight would have rested on the ground, threatening to suffocate it by preventing the thorax from moving in respiration. The pectoral fins in particular would therefore have become stronger, to take more and more of the body weight during basking (Fig. 8.17). Part of the shoulder girdle originally evolved to brace the gill region, and part to link with the pectoral fins. So the pectoral fins of tetrapodomorphs were still strongly linked with the skull and backbone.

Basking behavior may have made a more competitive fish, but we would still have recognized it as a tetrapodomorph. What other factors might have encouraged its evolution into a completely new kind of creature in a completely new environment?

Reproduction?

The most vulnerable parts of the life cycle of a fish are its early days as an egg and hatchling. If some tetrapodo-

Figure 8.18   Basking crocodiles. These are Indian marsh or "mugger" crocodiles. (They open their mouths because they have no sweat glands in their leathery skin.) Photograph by Kmanoj, and placed into Wikimedia.

morphs could make very short journeys—even a meter or so to begin with—over land, or over very shallow water, they would have been able to find small, warm pools, lagoons, ponds, and sheltered backwaters nearby to spawn in. There would have been fewer predators in these side pools than in open water, and eggs and young would have survived better there. In much the same way and for the same reasons, salmon struggle to swim far upstream to spawn, and many freshwater fishes swim into seasonally flooded areas to breed.

Isolated warm ponds would also have been ideal breeding grounds for small invertebrates such as crustaceans and insects, which would have formed a rich food supply for the young tetrapodomorphs. Then, reaching a size at which they could handle larger prey and that would give them some protection against being eaten themselves, the young tetrapodomorphs could make their way back to the main stream and take on their adult way of life as predators on fishes. Among young crocodiles today, the greatest cause of death (apart from human hunting) is being eaten by an adult crocodile. Crocodiles provide intensive parental care while their young are small. Iguanas tend to separate juvenile and adult habitats. Tetrapodomorphs perhaps solved the same problem by arranging for their young to spend time away from other adults.

The First Tetrapods

The best-known and securely identified early tetrapods are *Ichthyostega* and *Acanthostega* from almost complete skel-

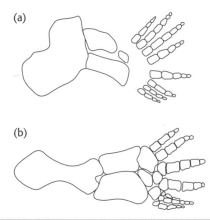

Figure 8.20 The massive skull of *Acanthostega*. Photograph by Ghedoghedo, and placed into Wikimedia.

Figure 8.19 Early feet and toes, from two tetrapods from the Late Devonian of Greenland. A), the front limb of *Acanthostega*. The limb clearly has (eight) toes, but it looks more like a functional flipper rather than a walking foot. Below, the hind limb of *Ichthyostega*. There are seven toes (well, six and a half), and it is more like a foot than a flipper. (After Clack and Coates.)

Figure 8.21 A reconstruction of *Acanthostega*, showing it as a tetrapod that was not as well adapted to going on land as its contemporary *Ichthyostega*. Art by Nobu Tamura, and placed into Wikimedia.

etons from the Late Devonian of Greenland, and *Tulerpeton*, which is a pile of bones from Russia that includes bones from more than one animal. Other less well preserved "stem tetrapods" have been found within a 2-m.y. time period close to the Devonian/Carboniferous boundary (363 Ma). Small arthropods and plants were not a suitable food supply for these early tetrapods, which were all large (more than a meter long). These animals ate fishes in the water. After that, there are interesting variations.

The stem tetrapods had many digits. The number varied, but it was not five: *Acanthostega* had eight toes and *Ichthyostega* had seven (Fig. 8.19); *Tulerpeton* had six. The number of examples is limited, but the loss of toes seems to be linked to the relative use of the foot in pushing on the bottom. *Tulerpeton* could have walked quite well on land, *Acanthostega* was much more adapted to life in water, and *Ichthyostega* was somewhere in between.

Acanthostega (Fig. 8.20, 8.21) seems to have had the most fish-like biology of the first tetrapods. It still had functional gills, for example. Its forelimbs were rather weak, its ribs did not curve round to support its weight well, and its eight-toed lower limbs were still somewhat flipper-like. It may have been best adapted to eating fish in weed-choked shallows, and it may not have been able to support its weight for long (or at all!) out of water.

Ichthyostega had a massive skeleton (Fig. 8.22) but was otherwise very much like Late Devonian tetrapodomorphs in spine, limb, tooth, jaw, palate, and skull structure, and probably in diet and locomotion. Like them, it had a tail fin, but unlike them it had a strong rib cage and limbs and feet rather than lobe fins. *Ichthyostega* solved the problem of supporting the chest for breathing on shore by having a massive set of ribs attached to the backbone.

Figure 8.22 Reconstruction of the skull of *Ichthyostega*, from the Late Devonian of Greenland: low and massive, with many long fish-eating teeth. Photograph by FunkMonk, and placed into Wikimedia.

Figure 8.23 Elephant seals fighting on the California coast. I do not think for a moment that *Ichthyostega* looked or fought like this, but I do want to make the point that air-breathing aquatic animals can and do move to land sometimes for display, fighting, mating, and breeding. Photograph by Dawn Endico, and placed into Wikimedia.

These ribs (I suggest) were adaptations for excursions into air.

As an adult, *Ichthyostega* was probably much like a living crocodile in ecology. New specimens show that its hind feet are not well designed for walking, and so far its front feet are unknown (though they were placed on very strong bones). Jenny Clack compares its potential for land movement with that of the living elephant seal (though at smaller size). Elephant seals basically haul themselves around on the beach, using the strong front feet for propulsion and the hind feet as props and skids. Nevertheless, unwary tourists have been damaged by the sudden and rapid charge of a big male elephant seal driven to rage by twittering intruders into its "territory" (Fig. 8.23).

The aquatic hunting of *Ichthyostega* was aided by a unique ear structure. A large air-filled pocket in the skull probably amplified any underwater sound reaching it, then transmitted the signals through a long thin stapes bone to the inner ear. No other tetrapod has anything quite like it. However, all the stem tetrapods could have come out into air and on to land for reasons other than food, such as the digestive and reproductive advantages I have mentioned, and the new information about *Ichthyostega* is certainly compatible with that. After all, elephant seals leave the water for display, fighting for dominance, mating, and breeding.

Further Reading

Plants

Beerling, D. 2007. *The Emerald Planet: How Plants Changed Earth's History*. Oxford University Press.

Cleal, C. J. and B. A. Thomas. 1999. *Plant Fossils: The History of Land Vegetation*. Woodbridge, England: Boydell Press. Very nicely written, with beautiful photographs. Covers Paleozoic plants very well.

Heckman, D. S., et al. 2001. Molecular evidence for the early colonization of land by fungi and plants. *Science* 293: 1129–1133.

Knoll, A. H. 1984. Patterns of extinction in the fossil record of vascular plants. In M. H. Nitecki (ed.). *Extinctions*, pp. 21–68. Chicago: University of Chicago Press.

Raven, J. A. 1984. Physiological correlates of the morphology of early vascular plants. *Botanical Journal of the Linnean Society* 88: 105–126.

Wellman, C. et al. 2003. Fragments of the earliest land plants. *Nature* 425: 282–285, and comment, pp. 248–249. Available at http://eprints.whiterose.ac.uk/106/1/wellmanch1.pdf

Animals

Ahlberg, P. et al. 2008. *Ventastega curonica* and the origin of tetrapod morphology. *Nature* 453: 1199–1204. Available at http://www.diva-portal.org/smash/record.jsf?searchId=2&pid=diva2:221254

Boisvert, C. A. 2005. The pelvic fin and girdle of *Panderichthys* and the origin of tetrapod locomotion. *Nature* 438: 1145–1147.

Boisvert, C. A. et al. 2008. The pectoral fin of *Panderichthys* and the origin of digits. *Nature* 456: 636–638. Available at http://www.diva-portal.org/smash/record.jsf?searchId=1&pid=diva2:221262

Clack, J. A. 2002. *Gaining Ground: The Origin and Evolution of Tetrapods*. Bloomington: University of Indiana Press. This is not a light-weight book in size or level, but it is very well written, and gives a complete picture of research up to 2002.

Clack, J. A. et al. 2003. A uniquely specialized ear in a very early tetrapod. *Nature* 425: 65–69. The wonderful inner ear of *Ichthyostega*.

Daeschler, E. B. et al. 2006. A Devonian tetrapod-like fish and the evolution of the tetrapod body plan. *Nature* 440: 757–763. [*Tiktaalik*.] Available at https://www.com.univ-mrs.fr/~boudouresque/Publications_DOM_2006_2007/Daeschler_et_al_2006_Nature.pdf

Davis, M. C. et al. 2007. An autopodial-like pattern of Hox expression in the fins of a basal actinopterygian fish. *Nature* 447: 473–476. Available at http://science.kennesaw.edu/~mdavi144/Publications_files/Davis2007b.pdf. See blog by Carl Zimmer, http://scienceblogs.com/loom/2007/05/23/old_hands_and_new_fins_1.php

Downs, J. P. et al. 2008. The cranial endoskeleton of *Tiktaalik roseae*. *Nature* 455: 925–929.

Heatwole, H. and R. L. Carroll. (eds.) 2000. *Amphibian Biology* Volume 4 *Palaeontology*. Chipping Norton, N.S.W., Australia: Surrey Beatty & Sons. [The first seven chapters are especially relevant to early tetrapods.]

Janvier, P. 1996. *Early Vertebrates*. Oxford: Clarendon Press.

Janis, C. M. and C. Farmer. 1999. Proposed habitats of early tetrapods: gills, kidneys, and the water–land transition. *Zoological Journal of the Linnean Society* 126: 117–126.

Shear, W. A. 1991. The early development of terrestrial ecosystems. *Nature* 351: 283–289.

Shear, W. A. et al. 1996. Fossils of large terrestrial arthropods from the Lower Devonian of Canada. *Nature* 384: 555–557.

Questions for Thought, Study, and Discussion

Fishes must have evolved to live in air (that is, on land) for good reasons. Summarize those reasons. Remember that evolution cannot make huge jumps: major changes have to occur in small steps. Now, if living in air was such a good idea, why was it only one lineage of fishes that successfully managed that new way of life. After all, the Devonian was the "Age of Fishes". There were many other candidates, you would think.

NINE
Tetrapods and Amniotes

In This Chapter

The early tetrapods diversified in the early Carboniferous, though their fossil record is patchy and confined to rather special habitats. The maximum size increased and the skeletons became stronger, though all these creatures would have been clumsy on land. The ecological diversity is surprising, however, with some early tetrapods evolving to look like little water snakes, or crocodile-like predators, or small and large lizards. Somewhere in this diverse array, some early tetrapods evolved a shelled egg with specialized compartments to foster the growth of an advanced hatchling.

This is the amniotic egg, the basis for the dramatic radiation into truly land-going vertebrates (reptiles, birds, and mammals) that followed. An amniotic egg must be laid in air, and it hatches into air, so the final link with a water existence is broken. This evolutionary innovation took place in giant swampy forests that covered much of the tropical Earth, and formed many of the coalfields we are mining today. The forest ecosystem was rich in diversity, with plants, insects, reptiles, and amphibians all thriving.

Early Tetrapods

Once the first tetrapods evolved, they radiated quickly into a great variety of sizes, shapes, and ways of life. Although this is a poorly known part of the terrestrial vertebrate record, it is also an exciting area in which research is constantly turning up new fossils. In this chapter I will give a progress report on the story as we see it now, and some of the problems that these early land animals faced and solved.

The earliest tetrapods, *Acanthostega* and *Ichthyostega* (Chapter 8) spent their adult lives in water, with (in my scenario) only occasional journeys into air for basking and spawning. One could describe this way of life as amphibious, but that does not make the animals amphibian in formal terms. The early tetrapods have left living descendants that are divided sharply into amphibians (Amphibia) and amniotes (Amniota: reptiles, birds and mammals). Many groups of early tetrapods became extinct, and are

History of Life, Fifth Edition. Richard Cowen.
© 2013 by Richard Cowen. Published 2013 by Blackwell Publishing Ltd.

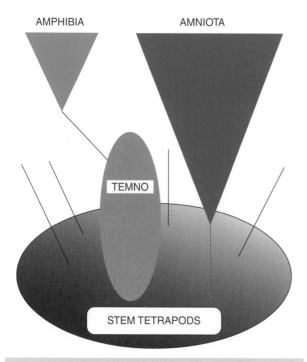

Figure 9.2 *Pederpes*, an early tetrapod from the Early Carboniferous of Scotland. About 1 meter (3 feet) long. Image by Dmitry Bogdanov, and placed into Wikimedia.

Figure 9.1 Diagram of the evolution of early tetrapods. Many lineages of "stem tetrapods" flourished in the Carboniferous, and many eventually died out. Living amphibians evolved in the Permian or Triassic from temnospondyl ancestors, and the lineage that led to living amniotes (reptiles, birds, and mammals) probably originated from a stem tetrapod early in the Carboniferous. The details of these lineages are still not worked out.

difficult to classify even when we have good skeletons preserved. Amphibia and Amniota are defined in a crown-group manner, so we are looking backward for the earliest ancestor of each group. That leaves a mass of early tetrapods that are neither Amphibia or Amniota: these are stem tetrapods (Fig. 9.1).

Ecologically, early tetrapods were the first large animals to exploit the environment in and around the water's edge. Their variety reflects different adaptations to different habitats and different ways of life. Some were dominantly terrestrial, some aquatic, and some genuinely amphibian. Naturally, there were variations even within each group. I shall describe the various groups, then try to set them into their ecological roles in the late Paleozoic world.

Living amphibians are all small-bodied and soft-skinned, and in these respects are quite unlike early tetrapods. They are newts and salamanders, frogs and toads, and caecilians, which are burrowing legless amphibians. Living amphibians are usually classed together as "smooth amphibians" or Lissamphibia, though in a crown-group definition this is the same as saying Amphibia. This clade is very much derived and probably did not evolve until Late Permian or even Triassic times. The biology of living amphibians is

fascinating, but may not be any guide to the origin, the paleobiology, or the classification of the early tetrapods of the Late Devonian and Carboniferous. In exactly the same way, living amniotes have been evolving for 300 m.y. or so, and are equally poor indicators of the ways of life of their distant ancestors. Altogether, evolutionary patterns among early tetrapods are difficult to work out.

Several tetrapod groups were evolving in parallel in the Late Devonian, some like *Acanthostega* toward a more aquatic life than *Ichthyostega*, some like *Tulerpeton* toward a more terrestrial life (Chapter 8). The early tetrapods radiated quickly into many lineages. Two of them were the ancestors of Amphibia and Amniota, so they still survive, and the others are closely or distantly related to these two clades. The fossil record is biased, because it favors large animals over small, and it favors preservation in water rather than on land. We simply do the best we can with it.

Pederpes is a meter-long tetrapod (Fig. 9.2), from early Carboniferous rocks of Scotland (age around 350 Ma). Its feet have only five toes, unlike earlier tetrapods. It still looked and probably behaved like a small crocodile, spending most of its time in the water. It is one of only a handful of tetrapods from Early Carboniferous rocks, and is the first tetrapod with feet that are genuinely adapted for walking on land.

A really good collection has come from rocks dating to about 335 Ma at East Kirkton in Scotland, where there was a complex tropical delta environment at the time, including shallow pools fed by hot springs. No fishes were found in the same levels as the tetrapods, possibly because the pools were too hot for them to live in. The tetrapods are probably members of an early community of animals that lived in the rivers and swamps near the pools, walked by them and sometimes fell into them or were washed into them. These animals included scorpions and millipedes, the earliest known harvestman, and, of course, several tetrapod groups. These included two groups of larger tetrapods, **temnospondyls** (Fig. 9.3) and **anthracosaurs**, but other tetrapods were smaller in body size and varied in ecology. Temnospondyls led (eventually) to living amphibians. The anthracosaurs led (eventually) to living amniotes (Fig. 9.1).

To show how difficult the East Kirkton fossils can be, consider the fossil *Eucritta melanolimnetes*, "the creature from the black lagoon," which has a skull rather like an

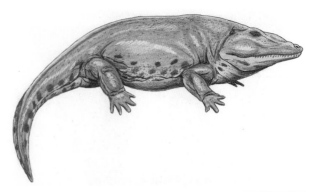

Figure 9.3 *Eryops* from the Early Permian of Texas is a poster child for large powerful temnospondyls. It is about 2 meters (6 feet) long, and massively built. Photograph © Joshua Sherurcij, and placed into Wikimedia.

Figure 9.5 *Eryops*, a Permian temnospondyl about 2 m (6 feet) long. This one looks contented and well fed! See Figure 9.3 for the reality of its skeleton. Image by Dmitry Bogdanov, and placed into Wikimedia.

Figure 9.4 *Eucritta*, a tetrapod about 25 cm (10 inches) long, from the Early Carboniferous of Scotland. It has a puzzling combination of characters. Image by Dmitry Bogdanov, and placed into Wikimedia.

fully preserved feet look strong and well adapted for walking. Temnospondyls probably had a biology much like that of crocodiles, and they were fish-eaters as adults (see *Eryops*, Fig. 9.3, Fig. 9.5). The jaw was designed to slam shut on prey, and the skull was therefore strongly built.

More terrestrially adapted temnospondyls had very massive, strong skeletons capable of supporting them on land, even when they were rather small (Fig. 9.3). Some temnospondyls even became more terrestrial during their lifetimes. For example, young *Trematops* had a jaw designed for eating small, soft food items, but adults had a carnivorous jaw and a lightly built skeleton capable of rapid movement. As adults they were probably land-going predators.

As part of their adaptation to life in air, temnospondyls had an ear structure that could transmit airborne sound. The stapes bone is strong and seems to have conducted sound to an amplifying membrane that sat in a special notch in the skull. Other early tetrapods, including early amniotes, did not evolve such an advanced system; but it is rather like the system that living frogs and toads have.

Some Triassic temnospondyls were giant marine animals like *Mastodonsaurus* (Fig. 9.6). This group ranged globally, from Russia to Antarctica. Temnospondyls survived into the Cretaceous in parts of Gondwana. The last big temnospondyl *Koolasuchus*, up to 5 m (16 feet) long, survived in rivers in Australia, no doubt hunting like a crocodile. (Real crocodiles could not have tolerated the cool temperatures of Cretaceous Australia, but living amphibians can live in cold water as long as they spawn in spring or summer.)

At some point in the Permian, a lineage of temnospondyls evolved into the ancestors of living amphibians. While most temnospondyls were large, the amphibian ancestors were small-bodied, and probably specialized on small-bodied prey, especially insects. The closely related little Permian genera *Gerobatrachus* and *Doleserpeton* have each been proposed as the ancestor of frogs and salamanders. *Gerobatrachus* has the advantage of the irresistible nickname "frogamander" bestowed on it by its describer Jason

anthracosaur and a body rather like a temnospondyl, with a "reptile-like" palate. It looks rather like a salamander, but of course it is not one (Fig. 9.4). *Eucritta* adds complexity to the puzzle, rather than simplifying it.

Ancestors of Living Amphibians: Temnospondyls

Temnospondyls are the largest and most diverse group of Carboniferous tetrapods: 40 families and 160 genera have been described altogether. There are over 30 skeletons of temnospondyls in the East Kirkton collections.

Temnospondyls were large, with teeth like those of osteolepiforms and *Ichthyostega*. The most common temnospondyl at East Kirkton, *Balanerpeton*, is about 50 cm (20 inches) long. It has heavy bones for its size, and its beauti-

Figure 9.6 The skull of *Mastodonsaurus*. A huge temnospondyl. Image by Ghedoghedo, and placed into Wikimedia.

Figure 9.8 *Oestocephalus* is a little aistopod, about 15 cm long (6 inches), from the Carboniferous of North America. Aistopods typically have over 100 vertebrae. Image by Nobu Tamura, and placed into Wikimedia.

Figure 9.7 *Gerobatrachus*, the so-called frogamander, from the Permian of Texas. It is a possible temnospondyl ancestor of living amphibians. Image by Nobu Tamura, and placed into Wikimedia.

Figure 9.9 *Microbrachis* is a microsaur, about 15 cm (about 6 inches) long, from the Late Carboniferous of the Czech Republic. It still had gills as an adult. Image by Nobu Tamura, and placed into Wikimedia.

Anderson, and an artist's rendition that makes it look cute (Fig. 9.7). But it is only one specimen, and a juvenile one at that. There are many well-preserved adult specimens of *Doleserpeton*, and they have been well studied, so the weight of evidence is with *Doleserpeton* at the moment (Sigurdsen and Green 2011).

Small But Interesting Groups of Early Tetrapods

We do not yet know how to classify some of the small, mainly aquatic early tetrapods. **Aistopods** like *Oestocepha-* *lus* (Fig. 9.8) were small, slim tetrapods that had lost almost all trace of their limbs. They probably lived rather like little snakes, perhaps hunting insects and worms among leaf litter. They do not preserve well and their fossils are quite rare. **Microsaurs** were small, with weakly calcified skeletons (Fig. 9.9). Their remains are usually fragmentary, and they include many juvenile forms.

Figure 9.10 *Diplocaulus* is a horned nectridean, about a meter long (3 feet) from the Permian of Texas. It is difficult to imagine any other function for the horns than hydrodynamic control during swimming. Image by Nobu Tamura, and placed into Wikimedia.

Figure 9.11 *Seymouria* was one of the few anthracosaurs well adapted for terrestrial life. It may be close to the origin of amniotes. Photograph by Ryan Somma, and placed into Wikimedia.

Nectrideans are better preserved and understood. They had a short body and a long, laterally flattened tail that made up two-thirds of their total length and was probably used for swimming. The vertebrae were linked in a way that allowed extremely flexible bending. Nectrideans probably swam like salamanders.

Horned nectrideans are fascinating. They had flat, short-snouted skulls with the upper back corners extended backward on each side. In early forms the extensions were quite small, but later they evolved to look like the swept-back wings of a jet fighter, as in *Diplocaulus* (Fig. 9.10).

Anthracosaurs

Anthracosaurs are the other large and diverse group of early tetrapods. *Tulerpeton* may be the earliest one (Chapter 8), though we need more complete specimens to confirm this. There are certainly two anthracosaurs in the East Kirkton fauna. Most anthracosaurs were adapted for life primarily in water, as long-snouted and long-bodied predators, presumably crocodile-like fish-eaters, with jaws designed for slamming shut on prey. Their limbs were not very sturdy, but they may have been very good at squirming among dense vegetation in and around shallow waters. It's unlikely that they had the speed and power to compete with tetrapodomorphs in open water, even though some were quite large, up to 4 meters (13 feet) long.

A few anthracosaurs were smaller, slender animals, adapted to terrestrial life. These tetrapods seem to be the closest relatives of early reptiles (Fig. 9.1), even though their ears remained adapted for low-frequency water-borne sound. *Seymouria* (Fig. 9.11) and *Diadectes* are well-known members of this large clade.

The consensus is that amniotes evolved somewhere near or in anthracosaurs, but no one hypothesis is strong. The problem is that amniotes began small, whereas most anthracosaurs were large. Since a major shift in habitat and ecology was probably involved here, convincing evidence is going to be difficult to find.

Amniotes and Amniota

The word **amniote** means a tetrapod that forms eggs inside a membrane. The word **Amniota** is a formal name for the clade of tetrapods that includes all living amniotes (mammals, reptiles, and birds) and their common ancestor. This is an uneasy combination of terms. The common ancestor of living amniotes may or may not have been the first tetrapod to lay an amniotic egg (how would we know?). But if we use a crown-based method of classification, we have to accept its awkward aspects along with its power.

The Amniotic Egg

Living amphibians differ from living amniotes in several characters of the skeleton that can be recognized in fossils, and in other characters that affect the soft parts and cannot be recognized in fossils. The major soft-part character of living amniotes is that they have eggs surrounded by a membrane, rather than the little jelly-covered eggs of fishes and amphibians. This fundamental difference in biology needs special attention because it was so important in the evolution of tetrapods into entirely terrestrial habitats.

How did the amniotic egg evolve, and who evolved it? (Perhaps the earliest amniotes laid amphibian-style eggs, and the amniotic egg evolved later in the lineage. Perhaps some early tetrapods laid amniotic eggs before the common ancestor of living amniotes appeared. This is not an easy argument to grasp at first, but only someone uneasy about the process of evolution would have any logical problem with it.)

Amphibians have successfully solved most of the problems associated with exposure to air. But their reproductive system was and is linked to water, and it remains very fish-like. Almost all amphibians spawn in water and lay a great number of small eggs that hatch quickly into swimming larvae. The eggs do not need any complex protection against drying, because if the environment dries, the larvae are doomed as well as the eggs. Thus, selection has acted to encourage the efficient choice of suitable sites for laying eggs, rather than devices to protect eggs. Both fishes and amphibians may migrate long distances for spawning, and favored sites are often disputed vigorously.

Living reptiles have a different system. Their juveniles hatch into air as competent terrestrial animals, often miniature adults. Yet the stages of embryological development are strikingly similar to those of amphibians. The difference is that reptiles develop for a longer time inside the egg, which in turn means that the egg must be larger and must provide more food and other life-support systems. Reptiles typically lay far fewer eggs than amphibians of comparable body size, so they have evolved more complex adaptations to ensure greater chances of survival for each individual egg.

A reptile (**amniotic**) egg is enclosed in a tough membrane covered by an outer shell made of leathery or calcareous material. The membrane and shell layers allow gas exchange with the environment (water vapor, CO_2, and oxygen) for the metabolism of the growing embryo, but they also resist water loss. Reptiles lay eggs on land, so eggs are not supported against gravity by water. Instead, the shell gives the egg strength, protects it, holds it in a shape that will allow the embryo room to grow freely, and buffers it against temperature change and desiccation. Birds' eggs are much like reptile eggs, usually with harder shells, and in most mammals the whole egg (without a shell) is nurtured internally so that the embryo emerges from the amnion at the time it emerges from the mother ("live birth").

Inside the amniotic egg, the embryo is nourished by a large **yolk**, and special internal sacs act as gas-exchange and waste-disposal modules. The most fundamental innovation, however, is the evolution of another internal fluid-filled sac, the **amnion**, in which the embryo floats. The amniotic egg acts toward the embryo like a spacecraft nurturing an astronaut in an alien environment: it has food storage, fuel supply, gas exchangers, and sanitary disposal systems (Fig. 9.12).

Because the embryo inside an amniotic egg is encased in membranes, and often inside a shell, the female's eggs must be fertilized before they are finally packaged. Internal fertilization must have evolved along with the amniotic egg.

The evolution of the amniotic egg broke the final reproductive link with water and allowed tetrapods to take up truly terrestrial ways of life. Its evolution demanded changes in behavior patterns and in soft-part anatomy and physiology. The transitional forms either evolved into or were outcompeted by more advanced animals, so they are now extinct and unavailable for direct study.

The amniotic egg was probably evolved by an early tetrapod that looked like a little reptile. I present here a reasonable scenario for the evolution of amniotes. It may be supported or contradicted by future evidence, or replaced by a simpler or more elegant story.

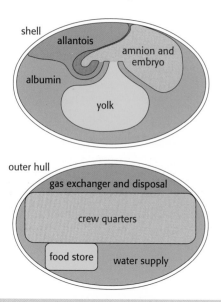

Figure 9.12 The amniotic egg is in many ways analogous to a spacecraft.

With the increasing ability of tetrapods of all kinds to make forays onto land, their breeding grounds became much less secure. Sites that had once been safe refuges for young animals gradually became more susceptible to raiders. The same evolutionary pressures that I suggested in Chapter 8 for the origin of tetrapods from fishes now drove some tetrapods to seek still safer refuges for breeding and the development of their young. In so doing they evolved into the first amniotes, and were preserved in environments quite different from those of their ancestors.

Forays farther from water became more practicable where flourishing plant life provided a myriad of damp hiding places, in and around Carboniferous swamps. Small tetrapods could have found small, sheltered, hidden places that were damp enough to foster egg development but not obvious enough to attract predators. They may have had behavior patterns like those of many living amphibians, particularly tree frogs and tree-dwelling salamanders.

The major problem in laying unshelled eggs away from larger water bodies is drying, at spawning time and during development. A crude sort of egg membrane would have been a partial solution to the developmental problem. Further refinement of the system is then fairly easy to imagine. Internal fertilization was probably a preliminary solution to the problem of desiccation while spawning. (Some living amphibians have independently evolved a crude kind of internal fertilization.)

Most frogs and toads undergo a complex development after hatching. A drastic metamorphosis from tadpole to adult involves not only a major anatomical reorganization but a major change in life style. The problems associated with this kind of amphibian reproduction can be solved, sometimes in spectacular ways. Tree frog eggs often hatch into tadpoles in places where there is little water. Some

Figure 9.14 Life reconstruction of the early amniote *Hylonomus* as an analog of a living lizard. Its fossils were found inside upright fossil tree trunks in Late Carboniferous rocks of Nova Scotia, Canada. Art by Nobu Tamura, and placed into Wikimedia.

Figure 9.13 The little golden dart-poison frog *Colostethus* lives in bromeliad plants in the South American cloud forest. It carries its tadpoles to the water pools formed by bromeliad flower cups, where they develop in a miniature pond. Photograph by Godfrey R. Bourne. Courtesy National Science Foundation.

frogs carry their tadpoles one by one to little pools in bromeliad plants (Fig. 9.13); some carry them in pouches on their bodies, where the young develop into miniature adults; and in some Australian frogs the females swallow their eggs after they are fertilized and hatch and develop the young internally in the digestive tract (the females don't feed while they incubate!).

However, early tetrapods may have had a much more direct development, hatching as miniature adults. A few frogs today lay large eggs, 10 mm across, which hatch into miniature adults. These eggs show no sign of evolving toward amniotic eggs, but they show that some living amphibians can lay large eggs that then develop without a complex metamorphosis. Presumably, as the amniotic egg evolved, the reproductive problems faced today by living frogs were avoided by simply allowing the embryo to develop longer and longer inside the egg. Longer development could, of course, have evolved gradually along with increased size and complexity of the egg.

As long as an egg does not dry out, it may have a better oxygen supply in damp air than in water. An egg laid in water, especially shallow warm water, may be exposed to lethal anoxic conditions.

This story provides a unifying theme that links the evolution from fish to amniote through the early tetrapods. Throughout, evolutionary change is linked with successful reproduction. As a by-product, successful animals are encouraged to enter new habitats. As they do so, they evolve ways of exploiting those habitats, and new ways of life become not only possible but encouraged. Simple themes that explain many facts are always satisfying: but they are only stories until they are tested against evidence.

Why Were the First Amniotes Small?

The large early tetrapods had a long, heavy skull, with a jaw designed for slamming shut on larger prey, which was then swallowed whole or in large pieces. There was little chewing, and the jaw muscles generated little pressure along the tooth line.

However, some Carboniferous tetrapods, and all early amniotes, were small. All early amniotes were about the same size as living lizards, and much like them in body proportions, posture, and jaw mechanics (Fig. 9.14). They were probably like them in ecology too. They had a notably small skull with a short jaw well suited to hold, chew, and crush small, wriggling prey, and to shift the grip for repeated bites. The small head was set on a neck joint that allowed very swift three-dimensional motion. Why small size?

Animals that spent longer time on land did not do so simply because they were seeking places to breed, but because there were potential food supplies there. Carboniferous forests were rich in worms, insects, and grubs. Young or small animals would have been best suited for foraging after this kind of food. Worms, insects, and grubs are small, though highly nutritious; they are easy to seize, process, and swallow; and they can be found among cracks and crevices in a maze of plant growth in a complex three-dimensional forest. Most of these potential prey items are slow-moving, and a successful predator need not have been quick and agile at first. But small and light-bodied animals could have quickly evolved greater agility as their repertoire of prey extended to the expanding number of large insects in Carboniferous forests.

Small body size may also have been favored by a thermoregulatory effect. Animals encounter greater temperature extremes in air than they do in water, and small animals can shelter more easily from chilling or overheating among vegetation, in cracks and crevices, or in hollow tree trunks, than can animals the size of *Ichthyostega*. Small bodies are also quicker and easier to heat by basking in the sun. Again, this suggests that terrestrial and/or arboreal excursions would have most benefited juvenile or small animals.

A scenario of amniote evolution, mostly due to Robert Carroll, is that they evolved on a forest floor covered with rotting material, leaf litter, fallen branches, and tree stumps, ideal places for prey to hide and amniotes to search (Fig. 9.15).

The scenario is reasonable enough to people used to temperate forests. But tropical forest floors are clean. The shade of the forest canopy is so thick and continuous that no vegetation grows at ground level, except along river banks where water barriers break the continuity of the canopy, or where storms (or people) have carved an open track of fallen trees. Fallen branches, leaves, bodies, and other pieces of organic debris are broken down and recycled so quickly on the ground by fungi and insects that vertebrates find it hard to make a living there. In contrast, the canopy and the river banks teem with small vertebrates: reptiles, amphibians, mammals, and birds in the canopy, and fishes in the water.

The best candidates for first amniotes are found in the Late Carboniferous. Many tree stumps and tree trunks have been fossilized upright in life position in rocks associated with coal forests (Fig. 9.16 and Fig. 9.17), and amniotes have been found in Nova Scotia, Canada, preserved inside some of the hollow stumps. *Hylonomus* (Fig. 9.14) is one example. This may not be a freak of preservation. The amniotes may have lived inside hollow tree trunks, as little

Figure 9.15 Reconstruction of a Carboniferous swamp forest. Art by Mary Parrish, under the direction of Tom Phillips and William DiMichele, and used by permission. See http://www.mnh.si.edu/ETE/ETE _Research_Reconstructions_Carb_step1.html for the steps in reconstructing the environment. Trees were tall but often had shallow roots and weak structure, so they were frequently felled by storm and flood. A rich fauna of insects, spiders, and other arthropods lived in this ecosystem, and I think it is likely that small early reptiles lived as much in the tree tops as under them. This scene has much growth on and near the forest floor because it is near natural open spaces around water bodies.

Figure 9.16 a) *Sigillaria*, a fossil tree found still standing upright in rocks to the north of the town of Stanhope, County Durham, England. It was removed and re-mounted in 1862 in Stanhope, next to the Church of St. Thomas. Photograph © Andrew Curtis, and placed into Wikimedia. b) a famous engraving from J. W. Dawson's 1868 account of the geology of Nova Scotia, Canada, the first scientific account of tree fossilized upright. (Coal miners had been finding them for centuries.)

Figure 9.17 Another upright fossil tree at Joggins, Nova Scotia. Geological hammer for scale. Photograph by M. C. Rygel, and placed into Wikimedia.

insectivorous mammals do today in tropical rain forests, or perhaps they sheltered in the hollow stumps during storms or were washed into them in floods.

Whichever suggestion one prefers, I would argue that amniotes were feeding in the canopy forest in the Late Carboniferous. Vertical climbing is easy with a small body size, so small Carboniferous vertebrates could have been tree dwellers, as many salamanders are today. Trees offer damp places in which to lay eggs, and rich insect life high in the canopy forest provides abundant food. Even today, salamanders (and spiders) are the top carnivores in parts of the Central and South American canopy forest. The rich fossil record of Late Carboniferous insects, scorpions, spiders, and amniotes may reflect the ecosystem of the canopy rather than the forest floor.

Carboniferous Land Ecology

Little is known yet about the land ecology of Early Carboniferous times; the East Kirkton fossils are the best-known tetrapod fauna from this time, and they are preserved in an unusual setting. All the Devonian and Carboniferous tetrapods so far discovered lived close to the equator.

The evidence is much better when we turn to the Late Carboniferous. Late Carboniferous coalfields have been intensely studied for economic reasons, yielding a lot of information that gives us a good picture of the flora and global paleoecology of the time. Swamp forests in tropical lowlands were dominated by lycopods, and the vegetation

and organic debris that were deposited in oxygen-poor water formed thick accumulations of peat, now compressed and preserved as giant coal beds stretching from the American Midwest to the Black Sea.

By the time all this carbon was buried, there may have been high levels of oxygen in the seas and atmosphere of the Carboniferous. The evolution of flight in insects, a very fuel-intensive activity, may have been made possible by a richly oxygenated atmosphere, but at the moment both the data and the inference are very speculative.

There were no herbivores among early land arthropods, possibly because of lignin (Chapter 8). This universal substance in vascular plants is formed through biochemical pathways that include toxic substances which are often stored in cell walls and dead plant tissue. From the Silurian to the Late Carboniferous, lignin and its associated biochemistry probably made vascular plants invulnerable to potential herbivores.

But eventually, of course, both invertebrates and vertebrates made the breakthroughs that allowed direct herbivory. Bacteria and fungi can break down the toxins in dead plants, and it's possible that symbiosis (Chapter 3) with one or both allowed some animals to eat living plant material for internal enzyme-assisted digestion. Also, early land plants evolved larger sporangia and seeds (Chapter 8) that were very nutritious and low in toxins, and therefore more liable to attack by arthropods. Insects quickly evolved the anatomy to feed on the reproductive tissues of plants. Seeds may have evolved not only for better waterproofing of the embryo but also to deter insect predation.

The first insect is Devonian, but the dominant fact of early insect evolution is the explosive radiation of winged insects in the Late Carboniferous, about 325 Ma. Some had mouthparts for tearing open primitive cones, and their guts were sometimes fossilized with masses of spores inside. Others had piercing and sucking mouthparts for obtaining plant juices. Overall, it seems that leaf eating was rare among early insects; instead, they ate plant reproductive parts, sucked plant juices, or ate other insects. Gigantic dragonflies were flying predators on smaller arthropods; Late Paleozoic dragonflies were the largest flying insects ever to evolve, with wingspans up to 60 cm (Fig. 9.18).

Explosive evolution had occurred among land-going invertebrates by the Late Carboniferous, much of it linked with the evolution of herbivory among insects: 137 genera of terrestrial arthropods are recorded from the Mazon Creek beds of Illinois, including 99 insects and 21 spiders, with millipedes present also. Most of the living groups of spiders had evolved by the Late Carboniferous, with only the sophisticated orb-web spiders missing. Centipedes were important predators.

Millipedes are important forest recyclers today, feeding on decaying plant material. They include flattened forms that squirm into cracks in dead wood and literally split their way in, reaching new food and making space for shelter and brood chambers at the same time. Carboniferous millipedes reached half a meter in length, and a giant relative, *Arthropleura*, reached 2.3 m (7 feet) long, and

WAYNE MARDER
COLLEGE
LEARNING RESOURCES

Figure 9.18 Reconstruction of the giant Carboniferous dragonfly *Meganeura*, which had a wingspan up to 60 cm (2 feet). Diagram by Dodoni, and placed into Wikimedia.

50 cm (18 inches) across. The gut contents of *Arthropleura* suggest that it ate the woody central portion of tree ferns.

Most early vertebrates, however, were carnivorous. Fishes, small tetrapods, and giant dragonflies all ate insects, and in turn were eaten by larger carnivores. For land vertebrates, then, Carboniferous swamp plants provided shelter and cover, but not food: herbivory by vertebrates evolved late, as we shall see in the next chapter.

Further Reading

Anderson, J. S. et al. 2008. A stem batrachian from the Early Permian of Texas, and the origin of frogs and salamanders. *Nature* 453: 515–518. [*Gerobatrachus*.] Available at http://www.naherpetology.org/pdf_files/988.pdf

Bickford, D. 2002. Male parenting of New Guinea froglets. *Nature* 418: 601–602.

Carroll, R. L. 2001. The origin and early radiation of terrestrial vertebrates. *Journal of Paleontology* 75: 1202–1213. Nice review. Available at http://www.usfca.edu/fac_staff/dever/tetropodevol.pdf

Clack, J. A. 2002. *Gaining Ground: The Origin and Evolution of Tetrapods*. Bloomington: University of Indiana Press. This is not a light-weight book in size or level, but it is very well written, and gives a complete picture of research up to 2002.

Heatwole, H. and R. L. Carroll. (eds.) 2000. *Amphibian Biology. Volume 4. Palaeontology*. Chipping Norton, Australia: Surrey Beatty and Sons.

Labandeira, C. C. 1998. Early history of arthropod and vascular plant associations. *Annual Reviews of Earth & Planetary Sciences* 26: 329–377. Concentrates most on Late Paleozoic associations, but reviews the whole story lightly. Available at http://si-pddr.si.edu/dspace/bitstream/10088/5965/1/AR_Earth_Planet_1998.pdf

Perry, D. 1986. *Life Above the Jungle Floor*. New York: Simon and Schuster. An account of the fauna inside hollow tree trunks reads like a visit to the Pennsylvanian, except for the bats. The section on coal forests is good, the sections on dinosaurs are not so sound.

Robinson, J. et al. 2005. The braincase and middle ear region of *Dendrerpeton acadianum* (Tetrapoda: Temnospondyli). *Zoological Journal of the Linnean Society* 143: 577–597. A CT scan shows that the stapes bone is very much like that of living amphibians, implying that temnospondyls were their ultimate ancestors.

Question for Thought, Study, and Discussion

Summarize the biology of the early tetrapods on land, including the need for them to lay their eggs in water. How much of the land surface of the world could they survive in? Then think about the early amniotes, who laid their eggs on land. Clearly they could succeed in different environments. But could they expand over a lot of other land environments? Summarize the problems in geographic and climatic expansion of the early amniotes.

ASKHAM BRYAN
COLLEGE
LEARNING RESOURCES

TEN
Early Amniotes and Thermoregulation

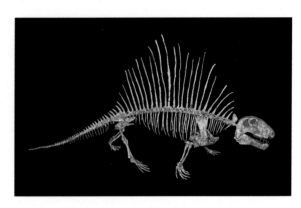

In This Chapter

The radiation of amniotes continued into the Permian, and by the end of the Permian land vertebrates had reached all the continents, and from pole to pole. The dominant amniotes were the synapsids (mammal ancestors). I discuss the pelycosaurs at length: an early group of synapsids that reached large body sizes as predators and as browsing herbivores. Any vertebrates that are fully adjusted to life on land mist be adapted toward changing temperatures, and there has been a lot of discussion about thermoregulation in pelycosaurs. In particular, pelycosaurs like Dimetodon probably used a great bone-supported sail on their backs to help regulate their body temperature. Finally, I discuss Triassic synapsids, which were more varied than Permian ones.

The Amniote Radiation

The radiation of amniotes was probably encouraged by ecological opportunities away from water bodies. But away from water, microenvironments have lower humidity, more exposure to solar radiation and to colder nights, less vegetation and shelter, and greater temperature fluctuations. Some degree of temperature control or thermoregulation is needed to live in such habitats, and the varied responses of reptiles to environmental and physiological challenges are major themes in their evolutionary history.

Tetrapods emerged onto land and the first amniotes evolved in warm, humid, tropical regions along the southern shores of the great northern continent Euramerica. Life away from such swamps and forests demands adaptations for dealing with seasons, where temperature, rainfall, and food supply vary much more and are less predictable than in the tropics. In many ways, such challenges to early land

History of Life, Fifth Edition. Richard Cowen.
© 2013 by Richard Cowen. Published 2013 by Blackwell Publishing Ltd.

vertebrates were simply extensions of the problems involved in leaving the water. In this chapter we shall follow the early history of amniotes and discuss the adaptations that allowed them their great terrestrial success.

Amniotes came to be dominant large animals in all terrestrial environments in Permian times. The radiation probably began in Euramerica, because hardly any land vertebrates are known from Siberia, from East Asia, or from the whole of Gondwana before Middle Permian times. Seas and mountain ranges may have blocked land migrations; or problems of thermoregulation may have confined land vertebrates to the tropics of Euramerica until the Middle Permian. The invasion of other continents and/or climates was accompanied by a spectacular evolution of varied body types. Since amniotes rather than amphibians radiated so successfully, perhaps it was their solution to thermoregulatory problems that allowed them to invade regions in higher latitudes.

Three major groups of amniotes had diverged by the Late Carboniferous and Early Permian. The earliest amniotes had **anapsid** skulls (they had no openings behind the eye) (Fig. 10.1a). This character was inherited from fishes and early tetrapods. The two major amniote clades have derived or advanced skull types, in which there are one or two large openings behind the eye socket. **Synapsids** (with one skull opening behind the eye socket) (Fig. 10.1b), diverged first, from an anapsid ancestor that we have not yet identified. Synapsids dominated Late Paleozoic land faunas. They include the Late Paleozoic pelycosaurs and their descendants, the therapsids and mammals. Synapsids never evolved the water-saving capacity to excrete uric acid rather than urea, a character that all other surviving amniote groups share.

Diapsids are amniotes with two skull openings behind the eye socket (Fig. 10.1c). They include the dominant land-going groups of the Mesozoic (including dinosaurs and pterosaurs) and all living amniotes except mammals (that is, reptiles and birds). Turtles have no skull openings, so are technically anapsid. But their ancestors were most likely diapsids that lost the two skull openings. Turtles evolved in Late Triassic times from diapsid ancestors we have not yet identified.

The earliest well-known diapsid is *Petrolacosaurus* (Fig. 10.1c), which looked like and probably lived like a lizard (Fig. 10.2) (but then the earliest amniotes did too: see Figure 9.14). Compared with later diapsids, *Petrolacosaurus* had a heavy ear bone, the stapes, that could not conduct airborne sound. As in most early tetrapods, the massive stapes probably transmitted ground vibrations through the limb bones to the skull.

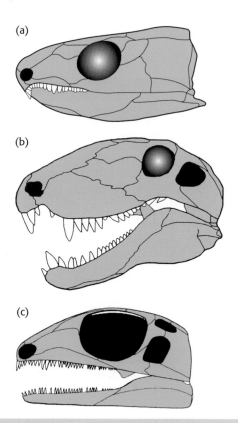

Figure 10.1 Three different skull types among amniotes are defined by the number of holes in the skull behind the eye socket. a) anapsid (no holes), represented by *Captorhinus*. b) synapsid (one hole), represented by *Dimetrodon*. c) diapsid (two holes), represented by *Petrolacosaurus*. Images a and b are © Dalton Harvey and used by permission.

Pelycosaurs

The first diapsid evolved in the Late Carboniferous, but the major radiation of diapsids took place much later, in the Triassic (Chapter 11). The dominant Late Carboniferous and Permian reptiles were synapsids, including five of the six other amniotes found with *Petrolacosaurus*.

Early synapsids are classed together as **pelycosaurs**, the most famous of which are the sail-backed Permian forms such as *Dimetrodon* (Fig. 10.1b). They were already the most important group of fully terrestrial tetrapods. Over 50% of Late Carboniferous amniotes were pelycosaurs, and

Figure 10.2 Reconstruction of the earliest well-known diapsid, *Petrolacosaurus* from the Late Carboniferous of Kansas. It was about 40 cm long (16 inches), and looked externally like a modern lizard. Image by Nobu Tamura, and placed into Wikimedia.

over 70% of Early Permian amniotes. After that they appear to decline, but only because one clade of them had evolved into the dominant therapsids of the Late Permian.

Despite their variety, early pelycosaurs are rare. The earliest one is *Archaeothyris*, found in the fossil tree trunks of Nova Scotia. It was small and lizardlike, but it had the characteristic synapsid skull. All early reptiles, in summary, were small insect-eaters. The pelycosaurs were the first to evolve to larger size, and perhaps because of that evolved into groups that were more abundant as fossils, varied in diet, and more widespread geographically than the other reptiles: for a while, at least.

Pelycosaur Biology and Ecology

Locomotion

Pelycosaurs are well enough known that we can reconstruct how they walked. The massive front part of the body was supported by a heavy, sprawling fore limb. The lighter hind limb had a greater range of movement, although it was also a sprawling limb. There was no well-defined ankle joint, and the toes were long and splayed out sideways as the animal walked. Thus, the feet provided no forward thrust but simply supported the limbs on the ground. The fore limbs were entirely passive supports that prevented the animal from falling on its face, while the hind limbs provided all the forward thrust in walking with powerful muscles that rotated the femur in the hip joint.

Think of two children (or adults) playing "wheelbarrow" (Fig. 10.3). The propulsion and steering are both from the rear, and the wheelbarrow is stable only as long as the leader stays stiff. Spinal flexibility is important to many swimming animals, particularly those that actively pursue fishes. But in pelycosaurs, a strong stiff backbone prevented the body from collapsing in the middle under its own weight, and allowed thrust from the hind limb to be converted directly into forward motion. Therefore, most pelycosaurs were predominantly terrestrial animals. If they swam, they were slow swimmers that hunted by stealth rather than speed. Only one pelycosaur (*Varanosaurus*) had a really flexible spine, and it may have been almost entirely aquatic.

Carnivorous Pelycosaurs

Early pelycosaurs were all carnivorous: they all have the pointed teeth and long jaws of predators. Two groups remained completely predatory. **Ophiacodonts** became quite large. *Ophiacodon* itself was 3 meters (10 feet) long and probably weighed over 200 kg (450 pounds). Many ophiacodonts have long-snouted jaws with many teeth set in a narrow skull. The hind limbs tended to be longer than the fore limbs.

Ophiacodonts may have hunted fishes in streams and lakes of the swamps and deltas of the Late Carboniferous and Permian (Fig. 10.4), although they were perfectly capable of walking on drier, higher ground, and like crocodiles, their prey may well have included terrestrial animals

Figure 10.3 Wheelbarrow race: photograph by John Trainor and placed into Wikimedia.

Figure 10.4 *Ophiacodon*, 2 meters (7 feet) or more in length, from the Early Permian of Texas. Reconstruction as a fish-eating pelycosaur by Dmitry Bogdanov, and placed into Wikimedia.

coming down to the water to drink. Their general lack of spinal flexibility (except in *Varanosaurus*) may suggest that they were slow swimmers, possibly eating more tetrapods than fishes.

Sphenacodonts were specialized carnivores on land. Many of their skull features betray the presence of very strong jaw muscles, and the teeth were very powerful. They were unlike typical early amniote teeth in that they varied in shape and size and included long stabbing teeth that look like the canines of mammals. The sphenacodont body was narrow but deep, and the legs were comparatively long. Both of these characters suggest that sphenacodonts were reasonably mobile on land.

The earliest sphenacodont was *Haptodus* (Fig. 10.5), a little less than a meter long and fairly lightly built. Similar forms existed throughout the Permian, but later sphenacodonts were much larger. The group is best known from spectacular fossils of *Dimetrodon* (Fig. 10.1c). *Dimetrodon* had vertebrae extended into spines projecting far above the backbone (Fig. 10.6). (I'll discuss these structures later.)

Evolution within carnivorous pelycosaurs reflected their prey capture. The jaws slammed shut around the hinge, with no sideways or front-to-back motion for chewing. With this structure, a long jaw made it easier to take hold of prey, but the force exerted far from the hinge was not very great. Small prey could perhaps be killed outright by slamming the jaw on them.

In ophiacodonts, which may have hunted in water for fish, the difficult part of feeding would have been seizing the prey; their teeth were subequal in length in a long, narrow jaw. Most fish-eaters swallow their prey whole.

In sphenacodonts, which were terrestrial carnivores, the head was bigger and stronger. Long, stabbing teeth were set in the front of the jaw (Fig. 10.1b). Struggling prey could be held between the tongue and some strong teeth set into the palate, and could be subdued by powerful crushing bites from the teeth at the back of the jaw. Robert Carroll suggested that the success of pelycosaurs in the Carboniferous and Permian, compared with diapsids, was due to their massive jaw muscles, which were strong enough to hold the jaws steady against the struggles of large prey. Carnivorous pelycosaurs thus could become large predators, not simply small insectivores.

Vegetarian Pelycosaurs

Carroll's suggestion cannot be the whole story, because there were also vegetarian pelycosaurs. Caseids and edaphosaurs were the first abundant large terrestrial animals, and were among the first terrestrial herbivores. They had similar body styles, presumably because they were similar ecologically. They had about the same range of body size as the carnivorous sphenacodonts, but they had smaller, shorter heads that gave more crushing pressure at the teeth. There were no long canines, and the teeth were short, blunt, and heavy. In addition, smoothing of the bones at the jaw joint allowed the lower jaw to move backwards and forwards slightly, grinding the food between upper and lower teeth. Caseids ground their food between tongue and palatal teeth, while edaphosaurs had additional tooth plates in their lower jaw that they used to grind food against palatal teeth. Vegetation is low-calorie food compared with meat, so herbivores need a large gut to contain a lot of food (Fig. 10.7). As one would expect, the bodies of all these vegetarian pelycosaurs were wide to accommodate a large gut. The limb bones were short but heavy.

Caseids were more numerous than edaphosaurs. They included *Cotylorhynchus*, which was over 3 meters long (10 feet), and weighed over 300 kg (650 pounds) (Fig. 10.7). Caseids had small heads for their size, which perhaps implies that they did not chew very much, and perhaps had powerful digestive enzymes or gut bacteria to help break down plant cellulose.

The earliest edaphosaur *Ianthasaurus* was not a vegetarian but a small insect eater (Fig. 10.8). It had a small sail on its back and sharp pointed predatory teeth. Vegetarian edaphosaurs evolved in the Early Permian. They are best known from *Edaphosaurus* itself (Fig. 10.9), which carried a very large sail made of vertebrae extended into spines, rather like *Dimetrodon*.

Figure 10.5 *Haptodus*, the earliest sphenacodont, from the Late Carboniferous of Kansas. Reconstruction by Dmitry Bogdanov, and placed into Wikimedia.

Figure 10.6 *Dimetrodon*, the best-known sphenacodont, from the Permian of Texas. Photograph by H. Zell, and placed into Wikimedia.

How Does Herbivory Evolve in Tetrapods?

Most plant material is difficult to digest. Vertebrates can break down cellulose only if they chew it well and have some way of enlisting fermenting bacteria as symbionts

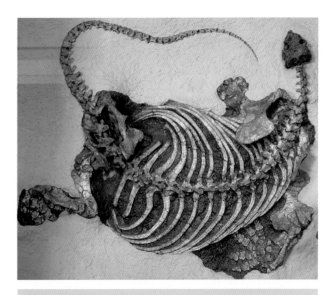

Figure 10.7 *Cotylorhynchus*, a big caseid from the Permian of Texas. The small head and capacious gut mark it as a vegetarian, even without looking at the teeth. Photograph by Ryan Somma, and placed into Wikimedia.

Figure 10.9 *Edaphosaurus* from the Permian of Texas was a big vegetarian pelycosaur, about 3 meters (10 feet) long, and weighing 300 kilograms (over 600 pounds). Photograph © Ken Angielczyk/Field Museum, used by permission.

Figure 10.8 *Ianthasaurus* is the earliest edaphosaur, from the Late Carboniferous of Kansas. It was a small insect eater not a vegetarian, about 15 cm long (6 inches). Image by Dmitry Bogdanov, and placed into Wikimedia. (He forgot to draw the rock that *Ianthasaurus* was sunning itself on!)

protein or sugar, especially the reproductive parts, but only a small animal can selectively feed on plant parts.

In other words, there are only two possible evolutionary pathways toward herbivory. One of them begins with animals that are small, active, and selective in their food gathering, eating high-calorie foods such as juices, nectar, pollen, fruits, or seeds from plants. Examples today are small mammals, hummingbirds, and insects. If an animal then enlists gut bacteria as symbionts, however, the diet can contain more and more cellulose, and a larger vegetarian can evolve, as in many mammal groups, including leaf-eating monkeys and gorillas. Large birds can also be herbivores: the extinct moas of New Zealand are good examples. *Ianthasaurus* suggests that edaphosaurs evolved herbivory this way.

The other pathway begins at rather large body size with rapid and rather indiscriminate feeding, possibly omnivory, so that a large volume of low-calorie food can be processed. Bearlike mammals are examples of a group in which some members have evolved away from a carnivorous way of life toward omnivory and then to a completely vegetarian diet, as in pandas.

Because vegetarianism depends so much on body size, diets must change with growth. Most living reptiles and amphibians change their diet as they grow. Food requirements and opportunities change as they reach greater size and can catch a different set of prey. Among living reptiles, small and young iguanas are carnivorous or omnivorous, while large iguanas are largely vegetarian but take meat occasionally. Living amphibians today are almost all small and carnivorous.

The giant Carboniferous coalfields (Chapter 9) contain rock sequences in which many beds consist almost entirely of carbon formed from plant debris such as leaves, trunks,

(Chapter 3) to aid digestion. Living vegetarians do this: for example, cattle and many other grazers have bacteria in a stomach compartment called the **rumen** (so they are called ruminants). Horses and rabbits have gut bacteria lower in the digestive tract. Any vertebrate that begins to eat comparatively low-protein plant material must process large volumes of it, and so must have a rather large food intake at a rather large body size. Some plant material is high in roots, spores, and pollen, plus half-rotted and unrecognizable fragments. Carboniferous coalfields have been studied so intensively that we can reconstruct their plant communities very well; we can tell, for example, that some plants had spread away from the rivers and lakes into so-called uplands—probably not very high above sea level but with distinctly drier air and soil than the lowland swamps.

The rich floras first provided a food base for insects, but large terrestrial herbivores appeared in the Late Carboniferous of Euramerica. The anthracosaur *Diadectes*, for example, ranged up to as much as 4 meters (13 feet) long as an adult, and synapsids of this size were also common. Large herbivores appeared at the same time as a major change in plants, when upland plants replaced the coal swamp forests.

Why were tetrapods relatively slow to evolve herbivory? First, because the wet tropical forest in which they evolved is a poor habitat for ground dwelling herbivores. As in today's tropical forests, most leaves were in the canopy, and leaf litter was broken down quickly by fungi and arthropods. The first vertebrate herbivores could not have found much green material on the floor of the coal forests (Fig. 9.15). Vertebrates could not have evolved herbivory until they could survive well on the forest margin, away from the watery habitats most likely to be preserved.

Second, any large-bodied vegetarian eats large volumes of low-calorie plant material and needs gut bacteria to help digest the cellulose. Gut bacteria work well only in a fairly narrow range of temperature, so an additional requirement for the first successful large-bodied vegetarians was some kind of thermoregulation.

Thermoregulation in Living Reptiles

Body functions are run by enzymes, which are sensitive to temperature. Other things being equal, enzymes work best at some optimum temperature; any other body temperature implies a loss of efficiency—in digestion, in locomotion, in reaction time, and so on. Birds and mammals have a sharp peak of efficiency that drops off radically with a small rise or fall in body temperature. Reptiles are called cold-blooded, but in fact they take on the temperature of their surroundings, so can be warm or cold. Their bodies can function over quite a range of internal temperature, but they also have an optimal temperature, and reptiles try to control it at that level by **behavioral thermoregulation**.

Generally, reptiles try to maintain their body temperatures at the highest level that is consistent with safety and cost. Although it takes energy to stay warm, the higher activity levels that are possible at higher temperatures give greater hunting or foraging efficiency, greater food intake, faster digestion, and faster growth: remember the section on basking in Chapter 8. As long as the climate is warm and food supply is abundant enough to fuel a reptile, thermoregulation that produces or maintains warm body temperature gives a net gain in reproductive rate and so is selectively advantageous. The same principles should apply to all cold-blooded vertebrates.

Body size is a vital factor in thermoregulation. Small bodies have a low mass with a relatively large surface area. Small reptiles bask in the sun, sit in the shade, hide in burrows or in leaf litter, or exercise violently (often with push-ups) to change their body temperatures. Their small mass allows them to respond quickly to temperature changes by behavioral means, giving them sensitive control over their body processes. Large reptiles have a natural resistance to temperature change because of their mass: it takes a long time to heat them up or cool them down (just as it takes a long time to boil a full kettle of water).

Behavioral thermoregulation is more energy-consuming and much less responsive for larger reptiles than for smaller ones. So large reptiles today live in naturally mild tropical climates with even temperatures day and night and season to season (like the large monitor lizards of Indonesia, Australia, and Africa), or they live near or in water, which buffers any changes in air temperature (like crocodiles and alligators, which even so are never found far outside the tropics). There are no large lizards at high latitudes.

Thermoregulation in Pelycosaurs

The spectacular pelycosaur *Dimetrodon* was large, over 3 meters long. It evolved very long spines on some of its vertebrae, forming a row of long vertical spines along the backs of these creatures (Fig. 10.6). In life, the bones were covered with tissue to form a huge vertical sail (Fig. 10.10). Most people think that the sail was used for thermoregulation.

Here is the simplest version of the story. *Dimetrodon* was too large to hide from temperature fluctuations (in a crevice or tree-stump or burrow, for example). It probably used its sail to bask in the early morning and the late afternoon, turning its body so that the sail intercepted the sunshine. By pumping blood through the sail, it could collect

Figure 10.10 Reconstruction of *Dimetrodon* to show the spines covered with tissue to form a "sail". Art by Dmitry Bogdanov, and placed into Wikimedia.

solar heat and transfer it quickly and efficiently to the central body mass (solar panels work this way to heat water). Once warm and active, *Dimetrodon* would face no further problem unless it overheated. It could shed heat from the sail by the reverse process, turning the sail end-on to the sun. At night, heat would be conserved inside the body by shutting off the blood supply to the sail.

The sail, as an add-on piece of solar equipment, allowed rapid and sensitive control over body temperature. Enzyme systems could have been fine-tuned to work at high bio-chemical efficiency within narrow temperature limits, and the animal could have foraged even in environments where air temperatures fluctuated widely. The activity levels, loco-motion, and digestive systems of *Dimetrodon* were all improved. Smaller reptiles that lived alongside them would have been able to heat up quickly in the morning, simply because they were smaller, and it would have been impor-tant for *Dimetrodon* to be equally active at that time, for effective hunting.

Some pelycosaurs did not have a sail at all, and the small pelycosaur *Ianthasaurus* had only a small sail (Fig. 10.8). Young *Dimetrodon* had a small sail too. The area of the sail was related to body size, which makes sense if it was used for behavioral thermoregulation like living lizards. A large body warms and cools slowly. Birds and mammals burn large amounts of food in a built-in, high, internal metabo-lism that allows them to be continuously "warm-blooded". The sail, as add-on solar technology, may have made *Dimetrodon* into a super-pelycosaur, but it didn't make it a mammal.

But *Edaphosaurus* has a sail too (Fig. 10.9), which was also thought to be a thermoregulatory device. Christopher Bennett (1996) pointed out that *Edaphosaurus* had knobs on the bony spines on its sail. Testing a model in a wind tunnel, Bennett found that the knobs would have generated eddies in breezes blowing past the sail. This would have no effect on solar collection by the sail, which is a radiation effect, or on its cooling by radiation, but it would make it a better cooling device by increasing convection over the skin. (Moving air cools bodies better than it heats them, as we all know from personal experience in breezes and winds.)

More recent work on the bone structure of *Edaphosaurus* spines suggests that there was no system of canals through the bone for the blood vessels that would be needed for efficient thermoregulation, either dominantly heating, or dominantly cooling (Huttenlocker et al. 2011). Most likely, then, *Edaphosaurus* had a sail for some other reason. The most likely alternative is for display. To make the story even more complicated, Tomkins et al. (2010) used a theoretical model to argue that none of the pelycosaurs had a sail that was the best design for thermoregulation. Instead, they argued, all sails were for display.

This is an uncomfortable situation. The simple resolu-tion, I suspect, is that the sails were used for both functions. The sails were distinctive enough to signal the species they belonged to, perhaps the gender and/or age, but there was also enough thermoregulation going on to add to the evo-

Figure 10.11 A male fiddler crab displaying his impressive right claw, in the sunshine on a Louisiana beach. Image by Junglecat and placed into Wikimedia.

lutionary advantage of the sails, and to help to defray the cost of building them.

There is a perfectly good living analog, too. Fiddler crabs are famous because the male grows one of his front claws to enormous size. He waves it to attract females and to intimidate male rivals, though if there is a fight between two males, the crab with the larger claw always wins. Recent research has found that the enormous surface area of the claw also gives male crabs enough thermoregulatory ability that they can stay longer displaying outside their burrows, even in bright sunshine on the beach (Fig. 10.11). Here too, simple measurements of the claws, like simple measure-ments of pelycosaur sails, do not fit the specifications of a perfect thermoregulatory structure. And in both cases, the function is display, but with important thermoregulatory benefits.

If pelycosaurs with sails thermoregulated, then other pelycosaurs (*Cotylorhynchus*, for example, Figure 10.7) probably thermoregulated too, in behavioral ways that left no traces on the skeleton. After the Permian, we see little sign of thermoregulatory devices as advanced pelycosaurs evolved into therapsids. There is indirect evidence, however, that therapsids had limited thermoregulation; but that evi-dence is presented in another chapter.

Permian Changes

Shifting continental geography resulted in major biogeo-graphic changes in the Permian (Chapter 6). The large southern continent Gondwana moved north to collide with Euramerica, and by the Middle Permian these blocks formed a continuous land mass. A little later, Asia crashed into Euramerica from the east, buckling up the Ural Mountains to complete the assembly of the continents into the global supercontinent Pangea.

These tectonic events put an end to the wet climates that had fostered the system of large lakes, swampy deltas, and shorelines along the south coast of Euramerica, where the Carboniferous coal forests had flourished. Permian floras, in contrast, were dominated by gymnosperms, mostly gingkos, conifers, and cycads. Compared with other Late Paleozoic plants, conifers were better adapted for drought resistance, and they probably evolved in much drier uplands, because they are rare on lowland floodplains. Tree-sized lycopods disappeared from coal swamps in the Late Carboniferous as climates became drier, and Permian conifers extended into lowlands and replaced them.

The Invasion of Gondwana

Geological evidence from Gondwana shows that a huge ice sheet was centered on the South Pole (Fig. 10.12) in Late Carboniferous and Early Permian times. Ice sheets moving northward scoured rock surfaces and deposited stretches of glacial debris on a continental scale.

The continental collisions that formed Pangea allowed land animals to walk into Gondwana. But pelycosaurs, which had always had a narrow tropical distribution, remained in the tropical areas, much reduced in diversity; Late Permian pelycosaurs are found only in North America and Russia. Instead, Gondwana was invaded by their descendants, the synapsid reptiles called **therapsids**.

Thermoregulation in Therapsids

Therapsids lived mainly at middle or high latitudes rather than in tropical regions: almost all therapsid clades evolved in Gondwana and spread outward from there. The restriction, or adaptation, of therapsids to drier and more seasonal habitats may have encouraged their success in southern Gondwana, away from the tropics and toward higher latitudes. Thousands of specimens of therapsids have been collected from Late Permian rocks in South Africa, for example, and someone with time on his hands estimated that these beds contain about 800 billion fossil therapsids altogether! There are literally dozens of species, and we have a good deal of evidence about their environment. The glaciations were over and vegetation was abundant, with mosses, tree ferns, horsetails, true ferns, conifers, and a famous leaf fossil, *Glossopteris*. The climate may well have been mild considering that South Africa was at 60° S latitude (Fig. 10.12). But it must have been seasonal, so the supply of plant food would have been seasonal too.

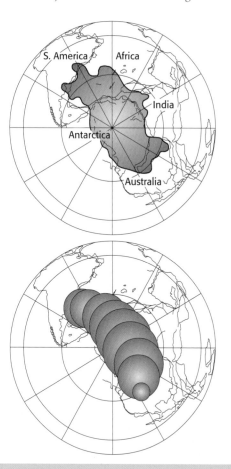

Figure 10.12 Traces of an ice age are widespread over the continents that formed Gondwana in the Late Paleozoic. The edges of the ice sheets can be marked with confidence, but not the precise time at which the ice reached those edges. The simplest explanation is that an ice cap formed over the south pole of the time. Over millions of years, Gondwana slowly drifted over the pole, and the edges of the migrating ice cap left an irregular trace. The traces ended as the last little Permian ice cap melted.

When we find large extinct synapsids at such high latitudes, we can be reasonably sure that they were unlike living reptiles in their metabolism. Their thermoregulation may have been more sophisticated than simple behavioral reactions. We know that mammals evolved from late therapsids in the Late Triassic. Did Permian therapsids already have a mammalian style of thermoregulation, with automatic internal control, a furry skin, and a high metabolic rate? We have too little evidence to say, but the scrappy evidence available suggests that the answer is no.

Therapsids had sprawling fore limbs and did not move very efficiently compared with later reptiles and mammals. Unlike other reptiles, many therapsids had short, compact, stocky bodies, with short tails: good adaptations for conserving body heat. They may also have had hair or thick hides for conserving heat, but there is no way of detecting that from their fossils. All this suggests that therapsids did not have a large energy budget. They may have had some moderate form of internal temperature control, but nowhere near as good as that in living mammals.

Therapsid Evolution

The evolutionary history of therapsids has not yet been properly worked out, and the classification is still changing rapidly. Part of the problem is that the therapsids radiated very quickly into several groups, and by Late Permian time they had spread globally all over Pangea. Therapsids as a group had larger skull openings than pelycosaurs did, indicating that they had more powerful jaws. The whole skull was strengthened and thickened. Therapsids also had much better locomotion than pelycosaurs. There is little doubt that therapsids evolved from a lineage of sphenacodont pelycosaurs, as relatively small- to medium-sized carnivores. Figure 10.13 shows one possible scheme of therapsid evolution.

Dinocephalians

Dinocephalians were the first abundant therapsids. They moved much better than pelycosaurs. Their spine was quite stiff, and limb length and stride length were longer than in pelycosaurs. The forelimbs were still sprawling, but the hind limbs were set somewhat closer to the vertical, accentuating the wheelbarrow mode of walking we have already described for pelycosaurs.

Dinocephalians became very large. They ranged from Russia to South Africa. They had large skulls and, like all therapsids, they had strong canines. They also had well-developed incisors that seem to have been both efficient and important in feeding. Dinocephalians had unusual front teeth: their upper and lower incisors, and sometimes the canines too, interlocked along a line when the mouth closed, forming a formidable zigzagged array that would bite off pieces of food (animal or plant) as well as piercing and tearing (Fig. 10.14).

The earliest dinocephalians, the anteosaurs, were carnivores with skulls up to a meter long. Like sphenacodonts, they killed prey mainly by slamming the long, sharp front teeth into them, then tearing and piercing. Apparently the back teeth were not used very much; they were fewer and smaller than in sphenacodonts.

Most other dinocephalians, the tapinocephalians, look carnivorous at first sight because they have large canines and incisors in the front of the jaw (Fig. 10.14). But they had a broad, hippo-like muzzle, a large array of flattened back teeth, and massive bodies with a barrel-like rib cage that must have contained a capacious gut (Fig. 10.15).

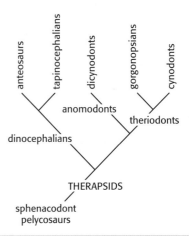

Figure 10.13 One possible hypothesis of the evolution of the major groups of therapsids. The only surviving clade of therapsids is the cynodonts, a group of advanced therapsids that evolved toward mammals in the Late Triassic.

Figure 10.14 *Estemmenosuchus* from the Late Permian of Russia was a vegetarian tapinocephalian member of the dinocephalians, in spite of the formidable appearance of its teeth. The incisors were spaced and were likely used to seize and tear off leaves. The canines and the great bony cheek plates would have been used in display and/or fighting. (After Chudinov.)

Figure 10.15 Therapsids had compact stocky bodies and short tails. *Keratocephalus* is a tapinocephalian from the Middle Permian of South Africa. It was about 3 meters (10 feet) long, and weighed perhaps 500 kg (1000 pounds).

Figure 10.16 Looking inside the mouth of a hippo. Photograph by Aqwis and placed into Wikimedia.

These animals may have been omnivorous, but more likely the incisors were cropping, cutting teeth used on vegetation, and the canines were fighting tusks, not carnivorous weapons. (Look inside the mouth of a hippo sometime! (Fig. 10.16).) The jaw exerted most pressure when closed, for efficient chewing rather than slamming.

Some tapinocephalians were particularly bizarre, with horns; some of them, probably males, had great bony flanges on the cheeks (Fig. 10.14). All tapinocephalians had thick skull bones, sometimes up to 11 mm (half an inch) thick. Herbert Barghusen suggested that individuals butted heads, presumably to establish dominance within a group. Large vegetarians today tend to fight by head butting or pushing, while carnivores today are quick and agile and tend to use claws and teeth as they fight. Early therapsids, even the carnivores, were heavy and clumsy; they had sprawling limbs that were so committed to supporting the body that they could not have used claws as weapons.

Advanced Therapsids

The rapid evolution of therapsids brought a new wave of advanced forms across the world in the Late Permian. These are the theriodonts and anomodonts. Theriodonts were all carnivores, with low flat snouts and very effective jaws. **Gorgonopsians** are theriodonts named for their ferocious appearance. They were the dominant large carnivores of the Late Permian, and look as if they specialized on large prey. Their sabertoothed killing action (Fig. 10.17) clearly involved a very wide gape of the jaw and a slamming action that drove the huge canines deep into the prey. The incisors were strong, but the back teeth were small and must have been practically useless. The snout was rather short, but deep in order to hold the roots of the canines. The limbs

were long and fairly slender, and gorgonopsians may have been comparatively agile. The skull is only 50 cm (20 inches) long in the largest known gorgonopsian, so limb joints could therefore be more lightly built, and the whole locomotion was improved. The hind limb could be swung into an erect position, stride length was greater, and the foot was lighter, altogether indicating greater speed. They didn't go in for much chewing, but simply tore large chunks from prey that was too large to eat in one bite (sharks, crocodiles, and Komodo dragons do that too). Their front teeth had serrations on them to slice through muscle and tendon.

Anomodonts

Anomodonts evolved in the Late Permian, and very quickly radiated to become the most important herbivores of the Late Permian. They were the first truly abundant worldwide herbivores, with a great variety of sizes and specializations. They make up 90% of therapsid specimens and much of the therapsid diversity preserved in Late Permian rocks.

The earliest dicynodonts were already so specialized as herbivores at their first appearance that they show no close resemblance to other therapsid groups and are difficult to classify. Early dicynodonts already had a secondary palate, separating the mouth from the nostrils so that they could breathe and chew at the same time.

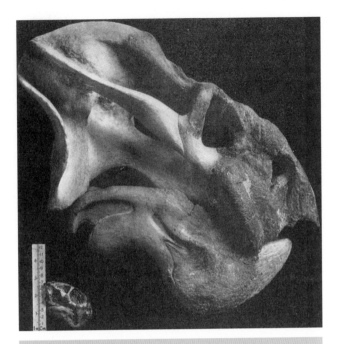

Figure 10.18   The box-like skull of two dicynodonts from the Permian of South Africa. The large skull is *Odontocyclops* and the small one is *Diictodon*. From the American Museum of Natural History digital library, in the public domain.

Figure 10.17   *Ivantosaurus* is an early theriodont from Russia with saberteeth about 10 cm (4 inches) long. The character continued into the gorgonopsians. (After Chudinov.)

At their peak in the Late Permian, dozens of species of dicynodonts were living in Gondwana, and they survived long into the Triassic. They differed from other therapsids in having very short snouts, and they had lost practically all their teeth except for the tusklike upper canines, which were probably used for display and fighting rather than for eating. Because there were no chewing teeth, the jaws must have had some sort of horny beak (like that of a turtle) for shearing off pieces of vegetation at the front and grinding them on a horny secondary palate while the mouth was closed. The jaw joint was weak, and moved forward and back in a shearing action instead of sideways or up and down. As part of this system, the jaw musculature was unusual, set far forward on the jaw, and took up a good deal of space on the top and back of the skull. These unusual jaw characters had their effect on the whole shape of the skull, which was short but high and broad, almost boxlike. The extensive muscle attachments resulted in

the eyes being set relatively far forward on a short face (Fig. 10.18). Dicynodonts look as if they cropped relatively tough vegetation with their beaks, and then ground it up in a rolling motion in the mouth. As in other herbivores, the body was usually bulky, with short, strong limbs.

The success of dicynodonts is astonishing. Most dicynodonts were rather small, though they ranged from rat-sized to cow-sized. Presumably the fact that the horny feeding structures of dicynodonts were replaced continuously throughout life had a great deal to do with their success. Reptiles with teeth replace them throughout life, but intermittently, so it is difficult for them to achieve continuously effective tooth rows. Other therapsids evolved effective cutting and grinding teeth, but teeth do wear out with severe and prolonged use.

Dicynodont jaws varied a lot, presumably because of their diet. There were dicynodonts with cropping jaws and with crushing jaws (perhaps for large seeds), and many browsers and grazers. Some dicynodonts were specialized for grubbing up roots, and some for digging holes, although they remained vegetarian. In a spectacular discovery in South Africa, skeletons of a little dicynodont were found at the bottom of sophisticated spiral burrows (Fig. 10.19).

The extent of specialization among dicynodonts suggests that the climate was reasonably mild and food supply reasonably reliable at the time, in spite of the high latitude and inevitable seasonal changes. Most Permian dicyno-

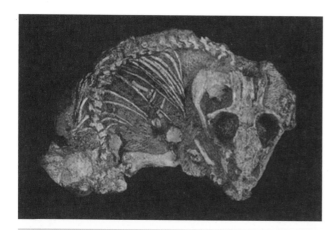

Figure 10.19 One of several specimens of the little Permian dicynodont *Diictodon* that have been found fossilized inside their burrows. Courtesy Dr. R. M. H. Smith of the South African Museum.

Figure 10.20 The dicynodont *Lystrosaurus*, which survived the Pemian-Triassic extinction and flourished in cool climates in Gondwana. Photograph by Rama and placed into Wikimedia.

donts were small, with skulls about 20 cm (8 inches) long. Possibly many of them were small so that they could burrow to avoid seasonal changes in temperature and food supply.

Dicynodonts declined abruptly at the end of the Permian, but a few lineages persisted, often in great numbers. The best-known dicynodont of all is a very specialized Early Triassic form, *Lystrosaurus* (Fig. 10.20). It has been found in India, Antarctica, South Africa, and South China; its strange distribution helped early structural geologists to identify and then piece together the scattered fragments of Gondwana.

Synapsids and Their Diapsid Replacements

With their generally large bodies, their radiation into herbivores and carnivores of varying sizes, and their experimentation with horns, fangs, and fighting, a Late Permian therapsid community viewed from a long distance would not seem totally strange to a modern ecologist, especially one familiar with the large mammals of the African savanna. However, the comparison would not stand close examination. All therapsids moved in a slow clumsy fashion, especially the larger ones. Slow motion is fine as a way of life if it applies to prey and predators too, otherwise both would eventually die out (think about it).

At the end of the Permian and into the Early Triassic, synapsids were the dominant land vertebrates. But beginning in the earliest Triassic, they gradually lost that dominance to diapsid reptiles. Paradoxically, a Late Permian community would not have looked as foreign as a Triassic one. Triassic land ecosystems did not evolve to look more mammal-like. Instead, therapsids were replaced by archosaurian diapsid reptiles, which had evolved from quite different Permian ancestors. We look at "the Triassic Turnover" in Chapter 11.

Further Reading

Bennett, S. C. 1996. Aerodynamics and the thermoregulatory function of the dorsal sail of *Edaphosaurus. Paleobiology* 22: 496–506.

Huttenlocker, A. K. et al. 2011. Comparative osteohistology of hyperelongate neural spines in the Edaphosauridae (Amniota: Synapsida). *Palaeontology* 54: 573–590.

Sues, H.-D. and R. R. Reisz. 1998. Origins and early evolution of herbivory in tetrapods. *Trends in Ecology & Evolution* 13: 141–145. [Review.] Available at http://www.erin.utoronto.ca/~w3reisz/pdf/SuesReisz1998.pdf

Question for Thought, Study, and Discussion

Explain carefully the history of ideas about the sail on the back of Dimetrodon. Now think about baby Dimetrodons. They were small, and they did not have a sail. Did they have any thermoregulation? And if not, why not? And then think about all the other pelycosaurs in the Permian world, and ask the same question.

ELEVEN
The Triassic Takeover

In This Chapter

As we saw in Chapter 6, there was a huge extinction at the end of the Permian. In the following Triassic period, synapsid amniotes were slowly replaced by an impressive array of diapsid reptiles, the ancestors of lizards and snakes, dinosaurs and birds, and crocodiles. Other diapsids returned to living in freshwater or in the ocean, and a few gliding diapsids are known. The reason for the diapsid take-over is not clearly established, but some of it is related to improvements in locomotion, thermoregulation, and respiration in various diapsids, that are not matched by the synapsids. I discuss research on living diapsids that allows us to interpret the biology of their distant ancestors. I conclude by outlining the history of diapsid groups in the Triassic, with the dinosaurs evolving just before the end of the period.

Diapsids

Chapter 10 may have given the impression that the only significant evolution among Permian and Triassic amniotes took place among synapsids. This, of course, is not true. Permian amniote faunas were dominated ecologically and numerically by synapsids, first by pelycosaurs and then by therapsids. But a good deal of evolution was going on among diapsid reptiles, and in the Triassic they came to replace synapsids as the dominant land vertebrates. The replacement was so dramatic that it has come to be a debating ground for the general question of replacement of one vertebrate group by another. The variety of Triassic diapsids leads to a mass of unfamiliar names, and I have tried to keep the list as simple as I can. Questions and some possible answers relating to the diapsid takeover form the major themes of this chapter.

Basal Diapsids

Diapsids are probably descended from *Petrolacosaurus* (Fig. 10.2), and most basal diapsids were basically lizard like in size, structure, and behavior. But basal diapsids (Fig. 11.1) also evolved into some interesting ways of life as early as the late Permian, especially in Gondwana, and without competing with the synapsids for large-bodied ways of life. For example, *Coelurosauravus* (Chapter 13) was a glider in

History of Life, Fifth Edition. Richard Cowen.
© 2013 by Richard Cowen. Published 2013 by Blackwell Publishing Ltd.

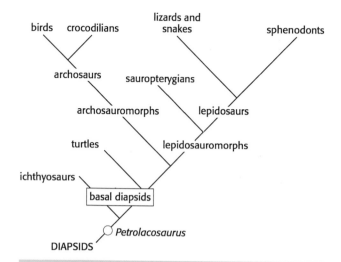

Figure 11.1 Cladogram of major groups of diapsids. The placement of the marine reptiles is unclear, though they probably arose early in diapsid evolution, perhaps, as shown here, from "basal" diapsids. Turtles are particularly difficult. Genetically, they are very close to archosaurs, but the placement of early diapsids cannot be done genetically, and has to be based on morphology.

Figure 11.2 *Hovasaurus*, a swimming diapsid from the Permian of Madagascar. Art by Nobu Tamura, and placed into Wikimedia.

the forest and *Hovasaurus* was aquatic. Both are from the Permian of Madagascar, which was part of Gondwana at the time.

Three hundred specimens of *Hovasaurus* make it one of the best-known Permian diapsids. Overall, it was lizard like, perhaps only 30 cm (1 foot) long from snout to vent. But the tail was exceptionally long, strong, and deep (Fig. 11.2), so the whole animal was close to a meter (3 feet) in length. The tail had at least 70 vertebrae and certainly looks like a swimming appendage. Inside the fossils, the abdominal cavity consistently contains a mass of small quartz pebbles. The pebbles often have a characteristic shape, tapering at both ends. Presumably they were swallowed by the animal during life. They are too small to be food-grinding pebbles and too far back in the abdomen to have occupied the stomach in life. Probably they were contained in a specially adapted abdominal sac. *Hovasaurus* almost certainly swallowed the stones as ballast for diving. Living Nile crocodiles do the same thing, and the extinct plesiosaurs may have done so too. This means that a very early diapsid had evolved a relatively sophisticated adaptation to an aquatic way of life.

The basal diapsids probably include the ancestors of two major groups of marine reptiles, the ichthyosaurs and turtles that I will discuss in Chapter 14 (Fig. 11.1).

The radiation into major diapsid clades on land began in the latest Permian but was truly spectacular in the Triassic. The diapsid takeover from synapsids during that time was an astonishing series of events.

There are two major surviving clades of diapsids: the **lepidosaurs** (lizards, snakes, and the tuatara of New Zealand) and the **archosaurs** (crocodiles and birds) (Fig. 11.1). We use a crown-group definition of the clades Lepidosauria and Archosauria: all those diapsids that are more closely related to the living survivors than to anything else. As usual, there are extinct clades that branched off below the base of these crown groups, and these are placed in larger clades called Lepidosauromorpha and Archosauromorpha. The scheme is logical though the names are clumsy.

Lepidosauromorphs

On land, the crown-group lepidosaurs have been the dominant group of small-bodied reptiles since the Mesozoic. They consist of three major clades (Fig. 11.1). **Squamata** are the numerous and diverse smaller living reptiles, including lizards and snakes. **Sphenodontia** include only one living form, the tuatara, *Sphenodon* (Fig. 11.3, Fig. 11.4), an outwardly lizard like animal that survives today only on a few islands off the coast of New Zealand (Chapters 17 and 21). Its skull characters show that it is not a true lizard. Sphenodonts are known as far back as the Triassic. And third, the extinct reptiles belonging to the **Sauropterygia** belong here too, branching off before the crown lepidosaurs, in the Triassic. These marine reptiles include placodonts and plesiosaurs.

Archosauromorphs

Archosauromorphs include the largest aerial and terrestrial animals that have ever lived, and they rose to dominate land ecosystems by the Late Triassic. They include a number of

Figure 11.3 A beautiful Victorian engraving of the tuatara, *Sphenodon*, from New Zealand. From Lydekker.

Figure 11.4 Henry is a teaching tuatara in Invercargill, New Zealand. Photograph by KeresH, and placed into Wikimedia.

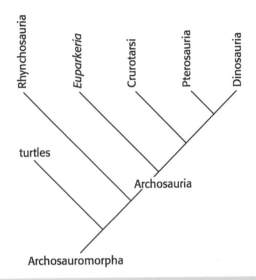

Figure 11.5 Cladogram of archosauromorphs. Turtles may be basal archosauromorphs rather than basal diapsids. Crurotarsi is the clade that includes living crocodilians and extinct forms that are closely related to them.

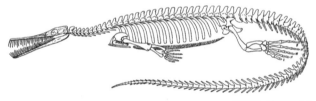

Figure 11.6 *Mesosaurus*, a little fish-eating reptile from the Permian of Gondwana. About 1 meter (3 feet) long. After McGregor. The discovery of *Mesosaurus* in fresh-water sediments in Brazil and South Africa helped to confirm that South America and Africa had once been joined in the supercontinent Gondwana.

groups that will receive significant discussion (Fig. 11.5). **Turtles** are certainly diapsids and may be basal archosauromorphs (the point is still not settled). **Rhynchosauria** are basal archosauromorphs that were the dominant large herbivores for a brief period during the Triassic.

The more we look at early archosauromorphs, the more we realize how fast and how early they radiated. Archosauria themselves evolved in the Early Triassic, so the other varied archosauromorph groups must already have split off. It is tempting to guess that the great radiation started immediately after the great Permo-Triassic extinction (and may have in some way been caused by it), but we do not yet have the fossils to test that idea.

Archosauria includes two great clades. One is the crocodile-like Crurotarsi (crocodiles, alligators, and related extinct forms), and the other is the bird-like Ornithodira (pterosaurs, plus dinosaurs and their sub-group the birds). The pterosaurs evolved true flapping flight much earlier than birds did, and they dominated the skies throughout the Mesozoic (Chapter 13). The Dinosauria are discussed in Chapter 12. Birds (Chapter 13) are derived diapsid, archosauromorph, archosaurian, dinosaurs.

The Triassic Diapsid Takeover: The Pattern

Large pelycosaurs dominated the tropical regions of Euramerica in the Early Permian. The only amniote outside this area was aquatic, the little fish-eating *Mesosaurus* (Fig. 11.6), which lived in and around the African and Brazilian parts of Gondwana.

By the Late Permian, land animals could walk into Gondwana. Therapsids had replaced pelycosaurs as the dominant land reptiles. New advanced therapsids dominated the Late Permian, particularly dicynodonts.

Gondwana had rich Triassic faunas, and land animals were free to disperse throughout Pangea. Therapsid diversity dropped sharply in the Permo-Triassic extinction, although the species that did survive were widespread and numerous. Dicynodonts were extraordinarily abundant at larger sizes, and cynodonts were medium-sized herbivores. There were few therapsid predators: most of them were small- and medium-sized cynodonts such as *Cynognathus* (Chapter 15). Some of the early archosauromorphs were small and carnivorous, although therapsids outnumbered them 65 to 1 at first. But by the end of the Early Triassic, some archosauromorphs were 5 meters (16 feet) long, with massive skulls a meter long. In South Africa, *Euparkeria* was a fast, lightly built carnivore that in retrospect is very close to the ancestry of the Archosauria (Fig. 11.7).

Therapsids, especially dicynodonts, were the dominant herbivores well into the Late Triassic. But Middle Triassic diapsids showed marked improvements in running ability over earlier forms, and by the end of the Middle Triassic, rhynchosaurs became abundant vegetarians alongside the dicynodonts. There were even greater changes among the carnivores. Diapsid carnivores of various sizes became abundant from the start of the Triassic. Among therapsids, cynodont carnivores were at most medium-sized but were still abundant and diverse.

Therapsids and rhynchosaurs declined distinctly in the Late Triassic, though they were still important ecologically. By the latest Triassic most of the therapsids had disappeared, along with rhynchosaurs and many other archosauromorphs. The vegetarians of the latest Triassic were almost all prosauropod dinosaurs; diapsid carnivores were larger, more diverse, and more mobile than before, and they were joined by the first theropod and ornithischian dinosaurs. The first mammals were few and small.

Finally, at the end of the Triassic, dinosaurs quickly overwhelmed terrestrial ecosystems throughout the world, replacing thecodonts and therapsids alike in every medium- and large-bodied way of life, to form a land fauna dominated by dinosaurs that lasted throughout the Jurassic and Cretaceous Periods.

The replacement of therapsids by archosaurs was worldwide. It was rapid but not sudden (the Triassic is about 50 m.y. long). Clearly, the replacement does not look as if the diapsids were dramatically superior to the therapsids: it took too long. But we can show clearly that archosaurs were superior to the synapsids and other Triassic animals in respiration and locomotion. This difference provides all the reason we need if we wish to explain the Triassic decline of the therapsids by diapsid competition—except, of course, for the shifty, vicious, nocturnal therapsid survivors, the mammals.

Respiration, Metabolism, and Locomotion

David Carrier put together some simple but powerful ideas about the links between respiration, locomotion, and physiology (Carrier 1987, Carrier and Farmer 2000).

Fishes have limitations maintaining high levels of exercise, as we saw in Chapter 7, because of the basic features of the body structure of a fish. Even so, many sharks swim all their lives without rest. Gill respiration gives reasonable oxygen exchange in normal aquatic circumstances.

The evolution of lungs helped to improve fish performance in low-oxygen situations (Chapter 7), but even so, early tetrapods moving about in air (on land) had potential difficulties. In early tetrapods, the shoulder girdle and the fore limbs in particular, powered in part by the muscles of the trunk, are largely used for supporting and moving the body over the ground. Amphibians and living reptiles still look awkward as they move. In walking and running, the trunk is twisted first to one side and then the other. As the animal steps forward with its left front foot, the right side of the chest and the lung inside it are compressed while the left side expands (Fig. 11.8). Then the cycle reverses with the next step. This distortion of the chest interferes with and essentially prevents normal breathing, in which the chest cavity and both lungs expand uniformly and then contract. If the animal is walking, it may be able to breathe between steps, but sprawling vertebrates cannot run and breathe at the same time. In the first edition I called this problem **Carrier's Constraint**, and the name seems to have stuck.

Animals can run for a while without breathing: for example, Olympic sprinters usually don't breathe during a 100-meter race. Animals can generate temporary energy by *anaerobic glycolysis*, breaking down food molecules in the blood supply without using oxygen. But this process soon builds up an oxygen debt and a dangerously high level of lactic acid in the blood. Mammalian runners (cheetahs and humans, for example) often use anaerobic glycolysis even though they can breathe while they run; it's a useful but essentially short-term emergency boost, like an afterburner in a jet fighter.

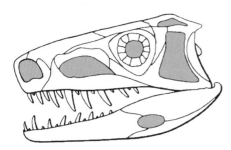

Figure 11.7 *Euparkeria* from the Late Permian of South Africa seems to be close to the ancestor of all later archosaurs. After Ewer.

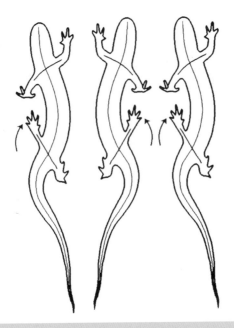

Figure 11.8 Lizard locomotion. David Carrier pointed out that the sprawling locomotion of a lizard or salamander forces it to compress each lung alternately as it moves (see text).

Figure 11.10 Toads can walk slowly and breathe at the same time, but they do not run. Instead, they hop to avoid Carrier's Constraint. Photograph by Iric, and placed into Wikimedia.

Figure 11.9 This lizard has stopped to take a breath. Photograph in Zion National Park by Thomas Schoch, and placed into Wikimedia.

Living amphibians and reptiles, then, can hop or run fast for a short time, first using up the oxygen stored in their lungs and blood, then switching to anaerobic glycolysis. They cannot sprint for long, however. If lizards want to breathe, they have to stand still (Fig. 11.9). Lizards run in short rushes, with frequent stops. By attaching recorders to the body, Carrier showed that the stops are for breathing, and that lizards don't breathe as they run. Therefore, all living amphibian and reptilian carnivores use ambush tactics to capture agile prey: chameleons and toads (Fig. 11.10) flip their tongues at passing insects, for example.

The giant varanid lizard, the ora or Komodo dragon, which eats deer, pigs, and tourists (most notably, the Swiss Baron Rudolf von Reding on 18 July 1974), goes a little way toward solving Carrier's Constraint by pumping air into its lungs from a throat pouch; but that only gives it a small improvement in performance. The Komodo dragon has a short sprinting range, but it prefers to ambush prey from 1 meter away.

Amphibians and most living reptiles have a three-chambered heart, which has usually been regarded as inferior to the four-chambered heart of living mammals and birds. But the three-chambered heart is useful to a lizard. Lizards run to catch food or to get away from danger, so they must use their resources most efficiently at this time. In a run, it is useless and perhaps dangerous for the lizard to waste energy pumping blood to lungs that cannot work. The lizard thus uses all the heart and blood capacity it has to circulate its store of oxygen around the whole body. The price it pays is a longer recharging time when it has to resupply oxygen to the blood, but it is usually able to do this at a less critical moment.

Early tetrapods all had sprawling gaits and faced a great problem. Their respiration and locomotion used much the same sets of muscles, and both systems could not operate at the same time. Imagine the laborious journey of *Ichthy-*

ostega from the water to its breeding pools, with a few steps and a few gasps repeated for the whole journey. One can understand why so many early tetrapods remained adapted to life spent largely in water, and why many early amniotes often looked amphibious. *Eryops*, for example, swam with its tail (Fig. 9.5), so would have had no major difficulty in devoting its rib-cage muscles to taking deep breaths at the surface.

When we see land animals such as pelycosaurs, with stiffened backbones and teeth designed for carnivorous and vegetarian diets rather than fish eating, we have to conclude that Carrier's Constraint had at least partially been solved. It's no good, for example, to raise metabolic rate by solar thermoregulation if there is no reliable oxygen supply to the tissues.

I suggest that the secret of the pelycosaurs was the stiffening of the backbone. They simply did not twist the body much as they moved. They had long bodies and relatively short limbs compared with lizards, and in any case a short step would not have rotated the trunk very much or distorted the lungs. The stiffening of the body also meant that most of the fore limb rotation was taken up at the shoulder joint, rather than being transmitted to the trunk. Furthermore, pelycosaurs had wheelbarrow locomotion, and the front limbs were mainly reactive support props, so the muscles operating them did not exert forces on the chest wall except to support the shoulder joint. On the other hand, the driving muscles of the pelvic girdle attached far from the chest wall.

The pelycosaurs thus had a special synapsid solution to Carrier's Constraint: they evolved adaptations that went some way toward reducing its consequences. But pelycosaurs could not solve Carrier's Constraint. There is no way that they were running freely, or breathing while they ran.

Fishes can swim in water with sustained energy because Carrier's Constraint does not apply to gill breathing. The same is probably true for the lung breathing of turtles, because their shell does not allow the lungs to be distorted as they swim and come to the surface to breathe.

Many living land vertebrates have evolved a beautiful answer to Carrier's Constraint. They have freed the mechanics of respiration from the mechanics of locomotion by evolving an **upright** stance. With upright limbs, the body is suspended more freely from the shoulders. The thorax does not twist much as the animal walks or runs, allowing it to make its breathing movements with hardly any distortion of the lungs.

The evolutionary solution to Carrier's Constraint that resulted from erect stance is shown best today in mammals. Mammals evolved the **diaphragm**, a set of muscles to pump air in and out of the chest cavity. Air is sucked in as the diaphragm contracts, and forced out by the reaction of the elastic tissues of the lung. At the same time, the locomotion in most mammals has evolved to encourage breathing on the run. The backbone flexes and straightens in an up-and-down direction with each stride, alternately expanding and compressing the rib cage evenly (Fig. 11.11). This rhythmic pumping of the chest cavity in the running action

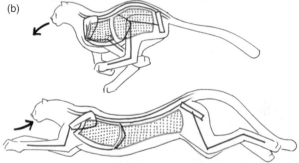

Figure 11.11 a) a running cheetah. Photography by Malene Thyssen (http://commons.wikipedia.org/wiki/User:Malene). b) a cheetah breathing while it runs, aided by the diaphragm. Diagram by Coluberssymbol, and placed into Wikimedia.

can be synchronized with the action of the diaphragm to move air in and out of the lungs with little effort. Thus quadrupeds running at full speed—gerbils, jackrabbits, dogs, horses, and rhinoceroses—take one breath per stride, and wallabies take one breath per hop (Fig. 11.12). Trotting is far more complex, but that doesn't harm the line of argument presented here. Human runners usually take a breath every other stride. It is such a natural action that we don't notice it. (Runners should try to breathe out of phase to get some idea how automatic it is!)

Animal locomotion often involves cyclic movements such as the strides and strokes of running or swimming limbs, or wingbeats in flight. Breathing may be made more efficient if it is synchronized with certain phases of limb movement. This is particularly important in human swimming, but it is a general principle. Flying insects synchronize their respiration with their wingbeats: the same muscular actions that raise and lower the wings also act to expand and compress the body, forcing air in and out of the spiracles. Birds do much the same thing (Chapter 13).

These principles are pieces of basic animal physiology, and they should be as true for extinct animals as they are for modern ones. Therefore, erect stance might be necessary for sustained running in any land animal, and its

(a)

(b)

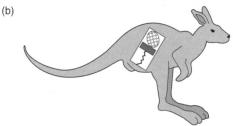

Figure 11.12 A leaping kangaroo. Photograph by PanBK and placed in Wikimedia. Diagram showing how the oscillation of the guts during the bounce can aid rhythmic breathing at high speed.

Figure 11.13 Life reconstruction of a running *Euparkeria*. An analog might be the living basilisk lizard, which can run across a river if it has to, even though it has an anatomy like other lizards. Image by Taenadoman and placed into Wikimedia.

evolution should represent a great breakthrough in any tetrapod lineage, giving the basis for greatly improved running speed and stamina. Living reptiles are successful, but they are limited in the ecological roles they can perform because they have a sprawling gait and cannot sustain fast movement for very long.

Diapsids living today, such as lizards, don't have erect stance or sustained energy output, but we must not be fooled into thinking that all diapsids always lacked those capabilities. David Carrier suggested that Triassic diapsids, and the archosauromorphs in particular, were the first amniotes to make the breakthrough to erect gait and rapid, sustained locomotion. That breakthrough is preserved in the fossil record in the structure of the limbs and shoulder girdles of the early archosauromorphs (Gauthier et al. 2011). Erect gait and sustained locomotion was most likely the key innovation that made possible the diapsid, and in particular the archosaur, takeover of the Late Triassic.

An accident of history may have played an important role in forming the differences between Triassic diapsids and synapsids. Therapsids evolved largely in cool climates of the late Permian, in northern Laurasia and southern Gondwana, while Permian diapsids evolved in warmer cli-

mates. Part of a heat-retaining syndrome in cool climates is to have stocky, compact bodies and short limbs and appendages, and therapsids are characteristically built that way (for example, *Keratocephalus*, Figure 10.16). Diapsids, on the other hand, had long, strong tails, and much of their body weight was on the hind limbs. It was relatively easier for diapsids to evolve to become partly or totally bipedal, and therefore to evolve erect limbs from a bipedal stance. Therapsids, with short tails, did not have that option: all of them were quadrupeds with a good deal of weight on the front feet. It may have been difficult to escape from the wheelbarrow locomotion that the therapsids inherited from the pelycosaurs, especially at larger body size. Truly erect gait, the solution to Carrier's Constraint, did not evolve among synapsids until the tiny mammals of the Early Jurassic.

The poster child for evolving upright stance and fast running in the archosaur lineage is *Euparkeria*, from the Late Permian of South Africa. It is a (perhaps the) basal archosaur (Fig. 11.1). It was about a meter long, very lightly built, and had a long, strong tail to give balance in running. Its skull was long and light, with many long, sharp stabbing teeth (Fig. 11.7). *Euparkeria* is usually drawn as a fast bipedal runner (Fig. 11.13) but in low gear it probably walked slowly on all four feet. Its speed and agility may have promoted its success in comparison to contemporary therapsids. Paleontologists are (usually) more skeptical than artists, but *Euparkeria* was at least semi-upright and capable of fast running.

Rhynchosaurs

Rhynchosaurs evolved in the Middle and Late Triassic with the decline of most large vegetarian therapsids and the disappearance of some. They were all herbivores, pig sized animals with hooked snouts bearing a powerful cutting beak and hind limbs that look as if they might have been used for digging. Strong jaws bore batteries of slicing teeth,

Figure 11.14 The Triassic rhynchosaur *Hyperodapedon*. About 1.3 meters (4 feet) long. Art by Nobu Tamura, and placed into Wikimedia.

Figure 11.15 One of the first archosauromorphs: *Archosaurus* from the Late Permian of Russia. It's one of the wonderful legal quirks of cladistics that Archosaurus is NOT an archosaurian, but merely an archosauromorph! Art by Dmitry Bogdanov, and placed into Wikimedia.

which are unusual among reptiles in that they were fused to bone at the base, not set into normal sockets. The teeth were ever-growing and were not replaced during life. As rhynchosaurs grew, they simply added more bone and more teeth at the back of the growing jaw as the teeth at the front became worn out. This style of tooth addition allowed rhynchosaurs great precision in tooth emplacement, so their bite was very effective for slicing vegetation with a scissor like action (Fig. 11.14).

Rhynchosaurs have been difficult to classify because of their peculiar features. They are probably a basal archosauromorph group (Fig. 11.1). They were abundant and widespread in the Middle and Late Triassic and may have replaced therapsid groups because they too evolved an erect gait. However, rhynchosaurs rapidly became extinct at the end of the Triassic.

Figure 11.16 A typical large powerful carnivorous Triassic archosaur, *Ornithosuchus*. About 4 meters (13 feet) long. Art by Nobu Tamura, and placed into Wikimedia.

Triassic Archosauromorphs

There was a repeated evolution of advanced locomotion among different early archosauromorphs. The early archosauromorphs were impressive carnivores, but they were large and dominantly quadrupedal, like *Archosaurus* from the late Permian of Russia (Fig. 11.15).

Once we get to true archosaurs, we find a surprising number of varied Triassic groups: but they all had long upright hind limbs (which is therefore a shared ancestral character of archosaurs). That means that Carrier's Constraint no longer applied, and implies that all archosaurs had an improved ability to breathe as they ran, and therefore to be active energetic animals compared with contemporary therapsids.

To walk (and run) bipedally, legs have to move dominantly forward, and that means that the ankle joint should not only be hinged in a forward-and-backward direction, but should be well braced so that it does not flop around sideways. There are many bones in the ankle region, and

possibly because of this skeletal legacy, there are alternative joint layouts that all can make an efficient ankle for bipedal running. Two lineages of archosaurs can be distinguished on this basis (and other characters): **Crurotarsi**, which are represented today only by crocodilians, and **Ornithodira**, which evolved into pterosaurs and dinosaurs, and are represented today by birds. Each lineage exploited the ankle to achieve more erect gait, culminating not only in erect-limbed dinosaurs, but erect-limbed crurotarsians too.

For most of the Middle and Late Triassic, the largest carnivorous groups were rauisuchians and ornithosuchians, two groups of basal cruritarsians typified by *Ornithosuchus* (Fig. 11.16). *Postosuchus*, from the Late Triassic of Texas, was about 4 meters long including the tail, and stood 2 meters high. It was lightly built, and could have walked on four feet (Fig. 11.17), but it surely ran bipedally. It was a hunter, with a heavy killing head, impressive wide-opening jaws, and serrated stabbing and cutting teeth. The eyes were large, set for forward stereoscopic vision, with bony eyebrows to shade them. *Postosuchus* is uncannily like

Figure 11.17 *Postosuchus*, a 4-meter long rauisuchian from the Triassic of Texas. Photograph by Dallas Krentzel, and placed into Wikimedia.

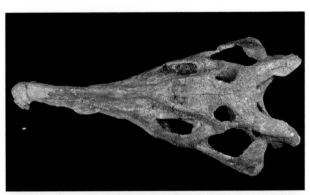

Figure 11.19 A phytosaur skull from the Triassic of Petrified Forest National Park, Arizona. National Park Service photograph.

Figure 11.18 *Poposaurus*, from the Late Triassic of Utah, reconstructed in the light of a new analysis by Gauthier et al. (2011). Art by smokybjb, and placed into Wikimedia.

Figure 11.20 The terrestrial crocodilian *Saltoposuchus* from the Late Triassic of Scotland, presumably a small fast agile predator. Art by Nobu Tamura, and placed into Wikimedia.

a small version of the much larger and later carnivorous dinosaurs *Allosaurus* and *Tyrannosaurus* in overall body plan and presumably in ecology.

These carnivores were perhaps close in ecology to living monitor lizards, though the monitors have only a semi-erect gait. Even so, in some ecosystems today they are active predators with a preferred body temperature close to 37°C (98°F). The Komodo dragon of Indonesia is the top predator in its ecosystem, weighing over 100 kg (200 pounds). Many Triassic archosaurs were mostly about the same size and, if anything, were more active, because they had erect gait and could probably run faster and further. *Poposaurus*, found in the Late Triassic of Utah, is well enough known that we can confidently interpret it as a powerful running carnivore (Fig. 11.18).

One group of cruritarsians explored a way of life that we now associate with living crocodiles: ambush hunting at the water's edge. **Phytosaurs** were large, long-snouted carnivores from the tropical belt of the Late Triassic. They evolved toward a crocodilian appearance and ecology (Fig. 11.19). Two phytosaurs from India more than 2 meters (7 feet) long had stomach contents that included small bipedal archosaurs, and one had eaten a rhynchosaur!

Living crocodiles may well be some guide to the physiology, locomotion, and ecology of phytosaurs. Crocodiles have a good circulatory system, with more advanced heart and lung modifications than other living reptiles. Although they normally walk slowly on land, in a sprawling stance, they are also capable of a faster run in which the limbs are nearly vertical. The little freshwater crocodile of Australia can gallop (briefly) at 16 kph (10 mph), but some phytosaurs could probably have done even better.

Phytosaurs disappeared at the end of the Triassic, with many other archosaur groups. They were replaced in that ecological niche by true crocodilians. Earlier crocodilians had been small, long-legged, terrestrial predators, like *Saltoposuchus*, from the Late Triassic of Western Europe (Fig. 11.20). Jurassic crocodilians adapted to water, replacing the phytosaurs, and only then did they become much larger. They evolved a secondary palate so that they could bite and chew under water without flooding their nostrils, and they also lost some of the feature of their terrestrial gait, becoming secondarily sprawling.

Dinosaur Ancestors

The first dinosaurs appeared in the Late Triassic, around 225 Ma. They were small, agile, and bipedal at first (their large to enormous sizes evolved later). Although the first

Figure 11.21 *Silesaurus*, which is technically a "dinosauromorph", from the Late Triassic of Poland. It differs from true dinosaurs by only a few characters. Art by Nobu Tamura, and placed into Wikimedia.

dinosaurs were almost certainly small bipedal running carnivores, they quickly evolved into different feeding styles. *Silesaurus* (Fig. 11.21), from the late Triassic of Poland, was almost a dinosaur, lacking only a couple of skull features: it was a small herbivore with a beak for cropping vegetation, like many later dinosaurs.

Herrerasaurus and *Eoraptor*, the best-known of the earliest dinosaurs, were carnivores living in Argentina alongside a fauna dominated by rhynchosaurs, with synapsids present too. The community seems to have been stable for at least 10 m.y. or so, with the dinosaurs forming perhaps one third of the carnivores. After that, the ecology seems to have changed rapidly, and dinosaurs became dominant, as we shall see in Chapter 12.

Further Reading

Carrier, D. R. 1987. The evolution of locomotor stamina in tetrapods: circumventing a mechanical constraint. *Paleobiology* 13: 326–341. [A breakthrough paper.]

Carrier, D. R. and C. G. Farmer 2000. The evolution of pelvic aspiration in archosaurs. *Paleobiology* 26: 271–293. Another astounding paper from David Carrier, this time with Colleen Farmer. A novel and convincing reconstruction of respiration in archosaurs, with major application to early crocodilians. Available at http://biologylabs.utah.edu/farmer/publications%20pdf/2000%20Paleobiology26.pdf

Dzik, J. 2003. A beaked herbivorous archosaur with dinosaur affinities from the early Late Triassic of Poland. *Journal of Vertebrate Paleontology* 23: 556–574. [*Silesaurus*.]

Farmer, C. G. and W. J. Hicks 2000. Circulatory impairment induced by exercise in the lizard *Iguana iguana*. *Journal of Experimental Biology* 203: 2691–2697. More proof of, and more amazing implications of, Carrier's Constraint. Available at http://jeb.biologists.org/content/203/17/2691.full.pdf

Gauthier, J. A. et al. 2011. The bipedal stem crocodilian *Poposaurus gracilis*: inferring function in fossils and innovation in archosaur locomotion. *Bulletin of the Peabody Museum of Natural History* 52: 107–126. Available at http://www.geo.uni-tuebingen.de/fileadmin/website/arbeitsbereich/palaeo/biogeologie/Images/Joyce_Publications/32__Gauthier_et_al.__Poposaurus__2011.pdf

Nesbitt, S. J. 2011. The early evolution of archosaurs: relationships and the origin of major clades. *Bulletin of the American Museum of Natural History* 352.

Owerkowicz, T. et al. 1999. Contribution of gular pumping to lung ventilation in monitor lizards. *Science* 284: 1661–1663. Monitor lizards avoid Carrier's Constraint (a little) by pumping air from a throat pouch. Available at http://biologylabs.utah.edu/farmer/publications%20pdf/1999%20Science284.pdf

Question for Thought, Study, and Discussion

Explain the advantages that the upright locomotion of archosaurs seems to have given them. Then summarize the evidence that the origin of archosaurs was deep in the Triassic, perhaps even in the early Triassic. Then think of reasons why the archosaurs did not take over the terrestrial world as soon as they evolved.

TWELVE
Dinosaurs

In This Chapter

The whole chapter is devoted to dinosaurs. I outline the history of the four main groups: theropods, prosauropods, sauropods, and ornithischians, and briefly describe their probable biology and ecology. There has been so much fine research over the past 20 years on dinosaur paleobiology that I focus on the main evidence and conclusions. All dinosaurs laid eggs, and probably all of them had some sort of parental care for the nests and the hatchlings. Ironically some of the best finds are of nests and hatchlings that did not survive, but died and were buried as fossils for us to study. Dinosaur bones often have growth lines, so we now know how they grew and how long it took. (Dinosaurs had a "teenage" growth spurt just as we humans do!) The old arguments about dinosaur body temperatures are over: dinosaurs all had good temperature regulation, and were warm-blooded, though we do not know exact temperatures.

Vegetarian dinosaurs include many 5- to 7-ton animals that are comparable ecologically with rhinos, but nothing living is like the giant sauropods that weighed several tens of tons. Increasingly, we are discovering dinosaurs with feathers, usually small or young animals. This nails down the warm-blooded nature of dinosaurs, but raises the question of why feathers evolved. Certainly the number of non-flying dinosaurs with feathers means that feathers originally had nothing to do with flight. They may have been for thermoregulation, though their placement on the ends of arms and tails suggests that they also played a role in display. Dinosaur stampedes are known from their footprints, and some bizarre offshoots of the nostril system suggest that they made sounds for communicating. The amazing array of dinosaur adaptations makes their extinction at the end of the Cretaceous all the more puzzling.

History of Life, Fifth Edition. Richard Cowen.
© 2013 by Richard Cowen. Published 2013 by Blackwell Publishing Ltd.

Dinosaurs

We are familiar with dinosaurs in many ways: they have been with us since kindergarten or before, in comic strips, toys, stories, movies, nature books, TV cartoons, and advertising. Yet it's still not easy to understand them as animals. The largest dinosaurs were more than ten times the weight of elephants, the largest land animals alive today. Dinosaurs dominated land communities for 100 million years, and it was only after dinosaurs disappeared that mammals became dominant. It's difficult to avoid the suspicion that dinosaurs were in some way competitively superior to mammals and confined them to small body size and ecological insignificance.

We are in a golden age of dinosaur paleontology. Dinosaurs are being discovered, described, and analyzed faster than ever, and new techniques are giving us better insight into their biology and ecology. Fortunately, the basic outline of dinosaur history has been stable for decades, and the iconic dinosaurs we all know and love remain iconic. We know that dinosaurs were all warm-blooded, with high metabolic rates. They lived from pole to equator, and on all continents.

The earliest dinosaurs were small bipedal carnivores, which appeared in the Late Triassic of Gondwana at about the same time as the first mammals. All the spectacular variations on the dinosaur theme came later, but all four major dinosaur groups (Fig. 12.1) had evolved by the end of the Triassic.

Most people know now that birds are highly evolved dinosaurs. That means that in cladistic terms, the clade Dinosauria includes birds as well as those dinosaurs that are not birds (the "non-avian dinosaurs"). That is a clumsy phrase. I shall use dinosaurs with a small d to refer to non-avian dinosaurs, and Dinosauria to mean dinosaurs plus birds.

Theropods

The earliest known dinosaurs were theropods, and almost all later theropods retained their body plan as bipedal runners and their ecological character of a carnivorous way of life. *Eoraptor* and *Herrerasaurus*, both from the Late Triassic of Argentina, are right at the point where theropods diverge from other saurischians. *Eoraptor* was a very small animal with a skull only about 8 cm (3 inches) long (Fig. 12.2). Even so, it was a fast-running carnivore, with sharp teeth and grasping claws on its fore limbs. *Herrerasaurus* was very like *Eoraptor* but much larger, between 3 and 6 meters (10–20 feet) long.

Small bipedal theropods formed the only dinosaur clade that still survives (as birds). However, at least four lineages of theropods evolved to giant size: the Jurassic allosaurs and three Cretaceous groups: the North American tyrannosaurs, *Argentinosaurus* from Argentina, and *Carcharodontosaurus* from North Africa. These varied theropods were the largest land carnivores of all time, each weighing 6 or 7 tons or more (more than an elephant), and standing about 6 meters (20 feet) high, with a total length around 12 meters (40 feet). Weight and mass estimates change all the time, but the very large *Tyrannosaurus rex* called "Sue" is possibly the largest of them all, with an estimated mass of 9 tons (Hutchinson et al. 2011). All these giant theropods must have relied on massive impact from the head for killing, aided by huge stabbing teeth that would have caused severe bleeding, usually lethal, in a prey animal (Fig. 12.3). *Tyrannosaurus* had the most powerful bite of any land carnivore that has ever lived (Bates and Falkingham 2012).

Early theropods included *Coelophysis* from the Late Triassic of North America. At 2.5 meters (8 feet) long, and lightly built (perhaps only 20 kg or 45 pounds), it was

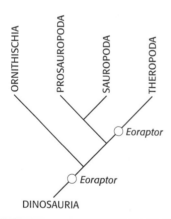

Figure 12.1 A phylogram showing the major groups of dinosaurs. The earliest well-known dinosaur, and the earliest well-known theropod, is *Eoraptor*. It is simple enough to be the ancestor of one, or both, so I've shown it in both places. Prosauropods are known from the Late Triassic, so the phylogram predicts that all four groups of dinosaurs had diverged by that time, and that there are "ghost ornithischians" and "ghost sauropods" still to be found somewhere in Late Triassic rocks. New fossils will clean up these uncertainties (or make them worse!).

Figure 12.2 The small early dinosaur *Eoraptor* from Argentina, the earliest theropod. Photograph by Kentaro Ohno and placed into Wikimedia.

clearly adapted for fast running. The bones of its skeleton were more extensively fused into stronger units than in the earliest theropods, so *Coelophysis* is placed into the first of the derived theropod groups, the **coelophysoids**. Jurassic **ceratosaurs** (Fig. 12.4b) and **allosaurs** included large powerful predators up to 6 meters long.

The large clade of **coelurosaurs** includes more advanced theropods of all sizes. *Compsognathus* is a basal coelurosaur that was small but an active predator with long arms and clawed fingers (Fig. 12.5). **Ornithomimids** are the so-called *ostrich dinosaurs*. Their body plan is much like that of a living ostrich, except that they had long arms and slim, dexterous fingers instead of wings. Ornithomimids had long legs and necks, large eyes, and rather large brains, but

no teeth. They could have been formidable carnivores, of course, but perhaps they specialized on smaller prey animals. Many of them lacked large claws and had long fingers that could have been used to manipulate objects. **Tyrannosaurs** are well known, of course (Fig. 12.3). They had enormous heads and tiny arms, a combination of characters that is still not understood. It is still a debate whether tyrannosaurs were giant predators or giant scavengers (the easiest and most likely answer is both, based on analogy with modern hyenas).

The third major lineage of coelurosaurs is made up of small- to medium-sized, agile carnivores, the **maniraptorans**. **Oviraptors** are nest-building dinosaurs from Mongolia, to be discussed later. **Therizinosaurs** are large, superficially bird-like theropods. *Mononykus*, from the Late Cretaceous of Mongolia, is a small therizinosaur with a rather long tail, but it has a breastbone like a bird. Its arms were much modified, so the hand had only one strong, blunt, clawed finger (Fig. 12.6). The scientists who described *Mononykus*

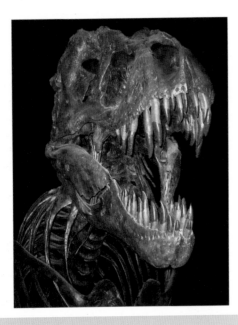

Figure 12.3   The skull of *Tyrannosaurus rex*. Photograph by Quadell and placed into Wikimedia.

Figure 12.5   *Compsognathus*, a small basal coelurosaur from the Late Jurassic of Germany. Photograph by Zach Tirrell and placed into Wikimedia.

(a)

(b)

Figure 12.4   a) *Coelophysis*, a small Triassic theropod from Arizona. Art by John Conway and placed into Wikimedia. b) the skull of *Ceratosaurus*, a large and early theropod from the Jurassic of North America. Photograph by Tremaster and placed into Wikimedia.

wondered whether it dug with these strange hands, but they recognized that burrowing would not suit a rather tall, long legged theropod. I suggest that it used the hands for digging out its prey (small mammals?) or for molding its nest. Ecologically, I would compare it with the big, flightless, megapod birds of Australia and New Guinea (Fig. 12.7a), which build a huge mound of leaves that they collect and shape into a fermenting incubator (Fig. 12.7b) by kicking backwards with their very big feet. This is a truly comical sight, especially as the bird has to keep looking over its shoulder to see what it is doing. *Mononykus*, I suspect, had a much easier time making its nest. The arms of *Mononykus* seem especially well designed for adduction (moving together under a load) so I envisage *Mononykus* digging a shallow nest, then sweeping together vegetation or sand to cover its eggs.

Figure 12.6 *Mononykus*, a therizinosaur from the Cretaceous of Mongolia, had strangely stubby clawed hands. Art by Nobu Tamura and placed into Wikimedia.

The other major clade of maniraptors is **deinonychosaurs**. They are mostly small, fast, and agile. **Troodonts** are the least specialized group. **Dromaeosaurs** include *Velociraptor*, supposed star of the movie *Jurassic Park*, and *Deinonychus*, from the Early Cretaceous, which actually was the dinosaur on which the movie was based. *Deinonychus* was one of the most impressive carnivores that ever evolved (Fig. 12.8). It was about 3.5 meters long, it was clearly fast and agile, and had murderous slashing claws on both hands and feet, and a most impressive set of teeth. And dromaeosaurs include *Archaeopteryx*, which is also the earliest bird (a derived dromaeosaur). Birds are therefore also deinonychosaurs, but I shall deal with them and their evolution in Chapter 13.

Ornithischians

The earliest ornithischians were the first herbivorous dinosaurs, and they gave rise to a spectacular radiation of dinosaurs, all of which were also herbivorous. Judging by their teeth, most ornithischians ate rather coarse, low-calorie vegetation, so many of them tended to be at least medium-sized. They were the most varied and successful herbivorous animals of the Mesozoic, and they were abundant in terrestrial ecosystems right to the end of the Cretaceous. Small bipedal basal ornithischians gave rise to several derived groups that were much heavier, with some animals weighing 5 tons or more (Fig. 12.9). The armored dinosaurs (**Thyreophora**) form one derived clade that includes **stegosaurs** and **ankylosaurs**, and another major clade is the **Marginocephalia**, which includes the horned dinosaurs or **ceratopsians** and the heavy-skulled **pachycephalosaurs**. However, most of the larger ornithischians were **ornithopods,** which include **iguanodonts** and the so-called "duckbill" dinosaurs, the **hadrosaurs**.

Figure 12.7 a) a brush turkey in Lamington National Park, Queensland, Australia, one of the strange birds called megapods. Photograph by P. Pouliquin and placed into Wikimedia. b) the mound of vegetation built by a male brush turkey to attract a mate who will lay her eggs in the mound. Photograph by Marissa Rose and placed into Wikimedia.

Figure 12.8   a) the skull of *Deinonychus*. Photograph by Didier Descouens and placed into Wikimedia. b, most of the skeleton of *Deinonychus*, posed in active running. Photograph by dinoguy2, modified by Conty, and placed into Wikimedia.

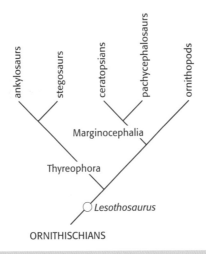

Figure 12.9   Cladogram of ornithischian dinosaurs.

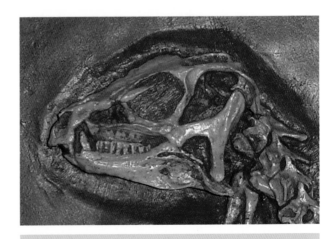

Figure 12.10   *Heterodontosaurus* had teeth that varied greatly along the jaw in size, shape, and presumed function. Skull about 10 cm (4 inches) long. Photograph by Sheep81 and placed into Wikimedia.

The best-known early ornithischians are small bipedal dinosaurs from the Early Jurassic of Gondwana. *Lesothosaurus* was small, agile, and fast-running, but it clearly had vegetarian teeth. *Heterodontosaurus* had teeth that were even more specialized for a vegetarian diet (Fig. 12.10). Small teeth at the front of the upper jaw bit off vegetation against a horny pad on the lower jaw. The back teeth evolved into shearing blades for cutting vegetation. (The sharp incisors were for display or fighting.) The cheek teeth were set far inward, with large pouches outside them to hold half-chewed food for efficient processing.

The earliest ornithopods were less than a meter long, but they soon increased significantly in size. A general theme of ornithopod evolution was the successive appearance of groups that in different ways evolved toward the 5- to 6-ton size that seems to have been a weight limit for most terrestrial herbivores. Even at this size, many ornithopods remained bipedal. Others probably walked most of the time on all fours but raised themselves up on two limbs for running, or browsing on high vegetation (like goats and gerenuks today).

Ornithopods evolved large batteries of teeth, and newly evolved modifications of the jaws and jaw supports allowed complex chewing motions. Iguanodonts were particularly abundant in the Early Cretaceous: they reached 9 meters (30 feet) in length and stood perhaps 5 meters (16 feet)

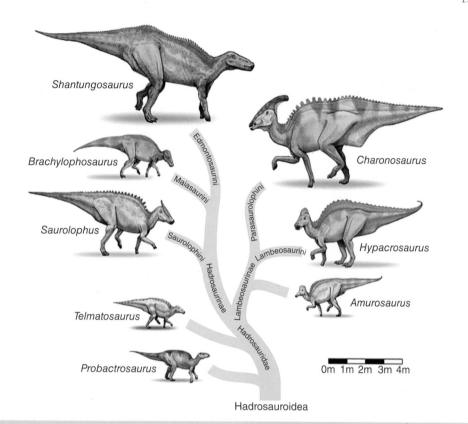

Shantungosaurus

Brachylophosaurus

Saurolophus

Telmatosaurus

Probactrosaurus

Charonosaurus

Hypacrosaurus

Amurosaurus

Edmontosaurini

Maiasaurini

Saurolophini

Hadrosaurinae

Parasaurolophini

Lambeosaurini

Lambeosaurinae

Hadrosauridae

Hadrosauroidea

0m 1m 2m 3m 4m

Figure 12.11 A variety of hadrosaurs. The names don't matter. Body size may have varied, but basic body plan did not. Image by Debivort and placed into Wikimedia.

high. They cropped off vegetation with powerful beaks before grinding it. Most iguanodonts were replaced ecologically by a variety of hadrosaurs (duckbilled dinosaurs) in the Middle and Late Cretaceous (Fig. 12.11). Hadrosaurs were about the same in size and body plan as iguanodonts, but had tremendous tooth batteries, with several hundred teeth in use at any time.

The other ornithischians were dominantly quadrupeds, but they betrayed their bipedal ornithopod ancestry with hind limbs that were usually longer and stronger than the fore limbs. Stegosaurs (Fig. 12.12), with their characteristic plates set along the spine, were the major quadrupedal ornithischians in the Jurassic, but they were replaced in the Early and Middle Cretaceous by the armored ankylosaurs (Fig. 12.13, Fig. 12.14). Later in the Cretaceous, the ornithischians were particularly abundant and varied in their body styles. Many quadrupedal forms lived alongside the hadrosaurs, including the ceratopsians, or horned dinosaurs (Fig. 12.15).

Figure 12.12 *Stegosaurus.* Art by Nobu Tamura, and placed into Wikimedia.

Sauropodomorphs

Some early dinosaurs evolved to become very large, heavy quadrupedal vegetarians with broad feet and strong pillar-like limbs. The sauropodomorphs had an early radiation as prosauropods and a later radiation as the famous sauropods with which we are all familiar.

Prosauropods were abundant, medium to large dinosaurs of the Late Triassic. They were typically about 6

Figure 12.13 The ankylosaur dinosaur *Euplocephalus*. Art by Nobu Tamura, and placed into Wikimedia.

Figure 12.14 The ankylosaur dinosaur *Minotaurosaurus*. Art by Nobu Tamura, and placed into Wikimedia.

Figure 12.15 The ceratopsian dinosaur *Nedoceratops*. Art by Nobu Tamura, and placed into Wikimedia.

meters (20 feet) long, but *Riojasaurus* was unusually large at 10 meters (over 30 feet). Prosauropods lived on all continents except Antarctica, with rich faunas known from Europe, Africa, South America, and Asia. They ranged into the Early Jurassic, when they were replaced by sauropods.

Prosauropods were all browsing herbivores. The teeth were generally good for cutting vegetation but not for pulping it. Opposing teeth did not contact one another, and all the grinding must have been done in a gizzard. (Masses of small stones have been found inside the skeletons of several prosauropods.) Prosauropods have particularly long, lightly built necks and heads, and light forequarters. They were clearly adapted to browse high in vegetation, perhaps reaching up from the tripod formed by the hind limbs and heavy tail. Only *Riojasaurus*, the largest, was always quadrupedal because of its weight, but it had a very long neck to compensate. Prosauropods were the first animals to browse on vegetation high above the ground, and they represent a completely new ecological group of herbivores exploiting an important new resource in the zone up to perhaps 4 meters (13 feet) above ground. The same adaptation was re-evolved later in sauropods, and again in mammals such as the giraffe.

Prosauropods began small, like other dinosaur groups, but were soon the largest and heaviest members of their communities, and they were abundant. *Plateosaurus* (Fig. 12.16) accounts for 75% of the total individuals in a well-collected site in Germany, and probably over 90% of the animal mass in its community.

Sauropods (Fig. 12.17) include the largest land animals that ever evolved. Remember that there is a natural human ambition to discover the largest or the oldest of anything; we must be cautious in assessing claims about the size and weight of dinosaurs without also surveying the evidence. But even a cautious person must admit that well-documented sauropod body weights are at least 50 tons;

Figure 12.16 The large prosauropod *Plateosaurus*, GPIT/RE/7288 from Tübingen University, reconstructed digitally by Heinrich Mallison. It is not a trivial matter to reconstruct a dinosaur, whether one is dealing with a complete skeleton made of many separate bones, or with many virtually digitized bones. The process is carefully detailed in Mallison 2010a, 2010b. Image © Heinrich Mallison, and used by permission.

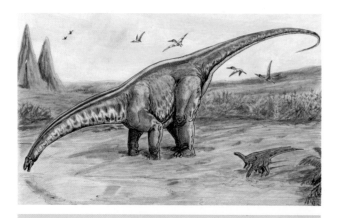

Figure 12.17   The huge sauropod *Apatosaurus*, once known as *Brontosaurus*. Life reconstruction by Dmitry Bogdanov and placed into Wikimedia, with corrections by Funkmonk and Dinoguy2 to take account of new research.

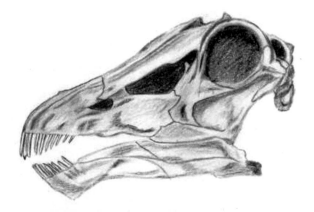

Figure 12.18   The skull of the huge sauropod *Diplodocus*, showing the peg-like teeth. Drawing by Nobu Tamura and placed into Wikimedia.

Argentinosaurus (from Argentina of course) has been estimated at between 60 and 88 tons, with 73 tons as a best guess. This is the best documented most massive dinosaur as I write. Famous names and enormous numbers are associated with sauropod anatomy: "*Seismosaurus*", from New Mexico, was at least 28 meters (90 feet) long, and maybe much more (it is probably a huge individual of *Diplodocus*). *Brachiosaurus* had long fore limbs carrying it over 12 meters (40 feet) high, as tall as a four-story building, and with a weight estimated at 50 tons.

However, despite these favorites from children's books, North Americans may welcome a new find of some poorly preserved but huge neck vertebrae from the Late Cretaceous of New Mexico, probably belonging to the titanosaur *Alamosaurus*. Judging from these bones, *Alamosaurus* was in the same size range as Argentinosaurus and a few other South American titanosaurs (Fowler and Sullivan 2011).

Sauropods were all herbivores, of course; no land animals that size could have been carnivorous. They had curiously small heads and very long necks that allowed them to browse on anything within 10 meters (33 feet) of ground level. The tails were long also, but the body was massive, with powerful load-bearing limb bones and pelvis. All sauropods were quadrupedal. The major body mass was centered close to the pelvis, which was accordingly more massive than the shoulder girdle. Within sauropods, there were two major lineages. One, the Diplodocoidea, obviously includes *Diplodocus* itself, and other sauropods with fairly long skulls and peg-like teeth (Fig. 12.18). The other is the Macronaria, named for the fact that the compact skull has huge gaps for nostrils (Fig. 12.19). This group includes the camarasaurs and titanosaurs. The nostrils may have been associated with sound production, but it is difficult to see how to test that.

Figure 12.19   The skull of the huge sauropod *Camarasaurus*. Photograph by Quadell and placed into Wikimedia.

Dinosaur Paleobiology

Now that we have surveyed the dinosaurs, it's time to try to reconstruct their biology. Fortunately, new discoveries over the past 20 years have given us much more detail about their daily lives.

Dinosaur Eggs and Nests

All dinosaurs laid eggs, and as far as we know they laid them in carefully constructed nests, usually scooped out of

Figure 12.20   a) a *Troodon* nest. I have no other data on this fine specimen. It came from "Central Asia", probably under dubious circumstances, and was to be auctioned off. b) part of a Cretaceous dinosaur nest in Indroda Fossil Park in India (the rest was broken off and fell down the ravine). The eggs are carefully spaced. Photograph by SBallal and placed into Wikimedia.

the ground. Major finds of fossilized dinosaur eggs and nests have been made in Jurassic and Cretaceous rocks, world-wide, and we have eggs from all major dinosaur groups.

Where we can reconstruct the nests, they were rounded hollows. In each nest the eggs were laid or arranged (by the mother) in a single layer and in a neat pattern so they would not roll around (Fig. 12.20a, b). Sometimes many nests are clustered together at regular close intervals, suggesting that they were in communal breeding grounds: nesting colonies, if you like. The nests are sometimes remodeled and reused, perhaps in successive seasons. Many large (long-lived) birds do this today, including the red-tailed hawks in my pine tree.

Late-stage embryos are preserved inside some dinosaur eggs. Once they hatched, very young dinosaurs seem to have stayed in or around the nest, sometimes until they grew to twice their hatching size. This is good evidence for long-term parental care by dinosaurs. One nest from the Cretaceous of Montana contained 15 baby duckbilled dinosaurs. They were not new hatchlings because they were about twice as large as the eggshells found nearby, and because their teeth had been used long enough to have wear marks. But they were together in the nest when they died and were buried and fossilized, with an adult close by—named *Maiasaura*, the "good mother" (see Horner and Gorman 1988).

Dinosaur eggs and nests were found in Mongolia in the 1920s, associated with the most abundant dinosaur in the area, *Protoceratops*. To everyone's surprise, when an embryo was finally discovered inside one of the eggs, it was well enough preserved to be identified not as *Protoceratops*, but

as a little oviraptor theropod called *Citipati*. The irony here is that the genus *Oviraptor* and the family of oviraptors had originally been identified (and named) as nest-robbing egg eaters, preying on innocent *Protoceratops*! In the iconic specimen, the adult *Citipati* was crouched in a sheltering position over its nest (for description and images see Norell et al. 1995). Several oviraptors have now been found in or near their nests.

Most recently, a nest containing 15 young *Protoceratops* huddled together was found in the same area under layers of wind-blown sand. The direction of the wind can be inferred from the layering in the sand. The baby dinosaurs had crowded into the downwind side of the nest, with their faces pointing away from the wind (and blowing sand), but they were eventually overcome and buried. There were no eggshells, and the babies were all the same size (Fig. 12.21). The simplest explanation is that this was a single clutch that had stayed in or around the nest, receiving parental care, until the disaster of the sandstorm.

A vivid, and gruesome, insight into the dangers faced by dinosaur hatchlings comes from a nest found in Late Cretaceous rocks of India. Here a nest of titanosaur eggs had been invaded by a giant snake which was in the process of catching and killing baby dinosaurs when the entire nest was covered by a flood of sediment and the complete scene was buried and fossilized (Fig. 12.22, Wilson et al. 2010; Figure 12.23, Benton 2010).

For those hatchling dinosaurs that survived their first few weeks, it is reasonable to reconstruct post-hatchling dinosaurs in the company of, and being cared for by, adults (Fig. 12.24a). At a very different body size, partridges hatch a large number of chicks, which are taken by a parent on

Figure 12.21 A nest of the little ornithischian *Protoceratops*, from the Cretaceous of Mongolia. Fifteen very young individuals, larger than hatchlings, were overcome and buried by blowing sand in a windstorm. Scale in cm. The nest was described by Fastovsky et al. (2011). Photograph courtesy of Kh. Tsogtbataar of the Paleontological Center, Mongolian Academy of Sciences, Ulaanbataar.

Figure 12.23 Reconstruction of the giant Cretaceous snake *Sanajeh* with unattended titanosaur nest. Sculpture by Tyler Keillor; photography by Ximena Erickson, modified by Bonnie Miljour. This is Figure 1 from Benton (2010), published in PLoS Biology at http://www.plosbiology.org/article/info%3Adoi%2F10.1371%2Fjournal.pbio.1000321. © Michael J. Benton. Publication in PLoS places the image into Wikimedia.

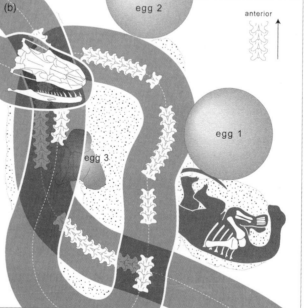

Figure 12.22 These blocks of rock, from the Late Cretaceous of India, fossilize a drama that began when a giant snake *Sanajeh* found an unattended titanosaur nest. For details and more images, see Wilson et al. 2010. These images are Figures 2 and 3 of their paper, published in PLoS Biology at http://www.plosbiology.org/article/info%3Adoi%2F10.1371%2Fjournal.pbio.1000322. © Wilson et al. 2010. Publication in PLoS places the images into Wikimedia.

Figure 12.24 a) maiasaur adults and babies on the Cretaceous plains of Montana. Art by Debivort, and placed into Wikimedia. b) an ostrich on the plains of the Serengeti, with many chicks in her care. Photograph by Caelio, and placed into Wikimedia.

trips from the nest to forage very soon after hatching. Ostriches run a crèche system for the care of foraging young (Fig. 12.24b).

Thus, in terms of their reconstructed behavior, including parental care, complex social structure, and intelligence (needed to run a complex society), dinosaurs should be compared not with living reptiles, but with living mammals and birds.

Large Size and Growth Rates

There are small dinosaurs: *Compsognathus* was only the size of a chicken, for example. But the dominant feature of dinosaurs, and the dominant aspect of their paleobiology, is the enormous size of the largest ones. Ornithischian dinosaurs are easier to understand than the others because they were vegetarians in the 5-ton range, comparable with living elephants or rhinos, perhaps. On the other hand, there are no 5-ton carnivores alive today on land that we can compare with dinosaurs such as *Tyrannosaurus*, and there are no 50-ton vegetarians that we can compare with the sauropods. Despite this, we can make some reasonable inferences about dinosaur biology.

Dinosaur Growth

A few dinosaur species are known from enough specimens to get an idea of their life history. But we need a clock to find out how fast they grew. Fortunately, the long bones (the femur, for example) of dinosaurs have growth rings built into the bone, and they are spaced regularly enough to be annual growth rings, recording seasonal fluctuations of the environment during growth (Fig. 12.25). This has revolutionized our understanding of dinosaur life history.

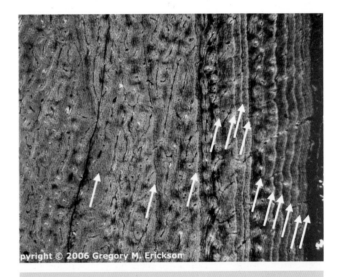

Figure 12.25 LAGs (Lines of Annual Growth) in a long bone of *Tyrannosaurus rex*. You can see rapid growth (well-spaced growth lines) over much of this specimen, during its growth spurt, followed by slow growth as it reached its full adult size. © Gregory M. Erickson of Florida State University, used by permission.

Juvenile dinosaurs grew moderately slowly, but adolescents went through a growth spurt that is uncannily like that of humans. At this stage in their lives, they are large enough to be a lot safer from predators, and the growth spurt takes them quickly to sexual maturity. After that, growth slows down as more energy is put into reproduction (Fig. 12.25). This pattern is found commonly among larger animals today, especially those for whom adult size makes them less vulnerable to predators.

The growth rates required for enormous body size are not outrageous compared to those of large living mammals: but obviously they have to be maintained for longer. They do imply that dinosaurs had a high metabolic rate, and therefore that they had a relatively high body temperature.

Metabolic Rate and Temperature in Dinosaurs

Any vertebrate cells are capable of high-energy output if they are kept fueled with oxygen and food. Thus, the secret of evolving thermoregulation at high levels and at high resting metabolic rates lies in the engineering around the cells rather than in their biochemistry. Respiration and circulation systems, which transport oxygen and food, are the crucial factors. For example, the hearts of living mammals and reptiles are very different, and David Carrier showed how and why their respiratory systems and locomotory systems are different too (Chapter 11).

David Carrier and Colleen Farmer (Carrier and Farmer 2000a, 2000b) have documented astonishing similarities between the respiration systems of birds and crocodiles (the surviving archosaurs). Both groups, in different ways, use bone and muscle systems, and a system of air spaces set into the body, to aid breathing by moving the pelvis and guts to generate a pumping action that in turn affects the lungs. Air is moved through the lungs in a one-way system, rather than the two-way system we have, increasing its efficiency. Furthermore, the pumping system is almost inextricably linked with locomotion. All these observations allow this same pumping system to be reconstructed in dinosaurs, linking their respiration with their active motion, in an analogous way to that seen in mammals and birds (Chapter 10).

Given that this is a general archosaur system, found in birds and crocodiles, it certainly applied to all dinosaurs and to pterosaurs as well. The oxygen supply system of archosaurs would certainly have been capable of generating high energy flow in a high-level metabolism, as long as the muscle/bone/air sac supporting infrastructure was present (Fig. 12.26, Wedel 2003, 2006; O'Connor and Claessens 2005, Codd et al. 2008).

There is also an ecological factor to consider. Whatever the advantages of high metabolic rate, it has a cost: more food must be eaten. The higher the metabolic rate, the greater the cost. We would expect dinosaurs to have excellent adaptations for food processing.

Vegetarian Dinosaurs

Large animals on purely vegetarian diets almost always have bacteria in their guts to help them break down cellulose. Large animals have slower metabolic rates than small ones, and for vegetarians this means a slower passage of food through the gut and more time for fermentation. Alternatively, a large vegetarian can digest a smaller per-

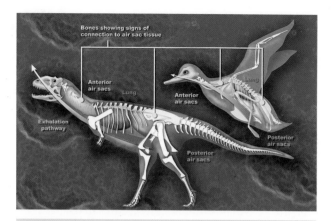

Figure 12.26 A comparison of the respiration system that is known in living birds, and reconstructed for dinosaurs. Drawn by Zina Deretsky of the National Science Foundation, in the public domain.

centage of its food and live on much poorer quality forage. Vegetarians usually grind their food well so that it can be digested faster, and this was accomplished in two different ways in the two major groups of vegetarian dinosaurs, the ornithischians and the sauropods.

Sauropods had very small heads for their size. This has sometimes been thought to indicate a soft diet that did not need much chewing. However, sauropods probably had a small head for the same mechanical reason that giraffes do: the head sits on the end of a very long neck. In both sets of animals the food is gathered by the mouth and teeth, then swallowed and macerated later.

Giraffes are ruminant mammals, and boluses of food are regurgitated and chewed at leisure with powerful batteries of molar teeth. Sauropods probably used a different system for grinding food. They probably had large stomachs in which food was ground up between stones that the dinosaurs swallowed. Wild birds seek and swallow grit to help them grind food, and poultry farmers can increase egg production by feeding grit to their chickens. [A fossil moa of New Zealand was found with 2.5 kg (over 5 pounds) of stones in its stomach area.] There are literally millions of dinosaur gastroliths (Fig. 12.27) in Cretaceous rocks in the western interior of the United States. A high proportion of these are made of very hard rocks, often colored cherts. William Stokes made the irresistible suggestion that dinosaurs specifically sought brightly colored, rounded pebbles in stream beds for swallowing, providing themselves with perfectly shaped and very hard grinding stones (Stokes 1987). (Dinosaurs as the first rock hounds!)

Ornithischians generally had impressive batteries of teeth, especially in advanced hadrosaurs and ceratopsians, and they would have chewed up their food thoroughly, as living mammals do. Even so, there are gizzard stones associated with fossils of the little ceratopsian *Psittacosaurus*.

Overall, food gathering and processing do not seem to have posed problems difficult enough to prevent vegetarian

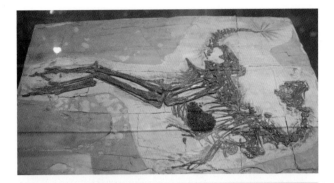

Figure 12.27 Gastroliths preserved within the ribcage of the little theropod dinosaur *Caudipteryx* from China. There is no reason why gastroliths should not have been useful for carnivores too: any dinosaur that swallowed before chewing would have benefited from them. (Notice also the tuft of feathery structures at the end of the tail, and shorter tufts elsewhere.) Photograph by Kabacchi and placed into Wikimedia.

Figure 12.28 Left foot of *Deinonychus*, showing the way the second claw is reflexed upward and backward for a powerful strike. Photograph by Didier Descouens and placed into Wikimedia.

dinosaurs from reaching enormous size. Dinosaurs evolved to be much larger than other land vertebrates for reasons not connected with diet.

Carnivorous Dinosaurs

Carnivorous dinosaurs typically have impressive teeth (Fig. 12.3, Fig. 12.4b). The big theropods in particular (*Tyrannosaurus* for example) have teeth with serrations that are beautifully shaped to break through membranes and muscle fibers in meat (Abler 1999). While tyrannosaurs must have killed prey with their skulls, the more lightly built maniraptorans had long arms and vicious claws, and many deinonychosaurs had huge foot claws that would have made fearsome ripping weapons (Fig. 12.28).

Dinosaur Metabolism and Feathers

Feathers have always been regarded as structures unique to birds: in fact, for 200 years they were used as one of the most important characters that define birds. Recently discoveries have shown that dinosaurs had feathers too.

Theropods from Early Cretaceous beds in China are of great interest because they are so well preserved. *Sinosauropteryx* and *Beipiaosaurus* have a halo of very fine structures on the body surface that look like down. *Protarchaeopteryx* has down feathers on its body, tail, and legs, and a fan-shaped bunch of long feathers, several inches long, at the very end of its tail. *Caudipteryx* also has down and strong tail feathers, but it has feathers on its arms as well. They are shorter toward the fingers, and longest toward the elbow, in contrast to the feathers on the wings

Figure 12.29 The feathered dinosaur *Caudipteryx* from the Cretaceous of China. Compare with Figure 12.27. Art by M. Martyniuk and placed into Wikimedia.

of flying birds (Fig. 12.27, Fig. 12.29). Finally, and most important, *Microraptor* (Fig. 12.30) and *Sinornithosaurus* have true branching feathers on all four limbs and on the tail.

Feathers have now been found in many clades of theropods, and in one ornithischian (the small genus *Psittacosaurus*) (Mayr et al. 2002). Feathers, then, were likely evolved in early theropods, were widely spread within that clade, and were inherited by birds. We do not yet know

Figure 12.30 The feathered dinosaur *Microraptor* from the Cretaceous of China. Feathers are indicated by white arrows. Figure 1 from Hone et al. 2010, http://www.plosone.org/article/info%3Adoi%2F10.1371%2Fjournal.pone.0009223. Publication in PLoS ONE automatically places the image into Wikimedia.

whether feathers evolved independently in *Psittacosaurus*, or whether they were a basal feature of all dinosaurs, along with high metabolism and warm body temperatures.

Not all dinosaurs were feathered. Direct evidence from skin impressions shows that some adult ornithischians and sauropods had scales. But baby elephants have hair, so the evidence about scaly adult dinosaurs may not tell the whole story. *Psittacosaurus* had long bristles around its tail, for example, while the rest of its skin had scales.

The Chinese theropods confirm that birds evolved from small ground-running predatory theropods, and that feathers evolved before birds and before flight. So how do we now distinguish a bird from a small theropod? With difficulty, and certainly not by its feathers! (By now there is no major feature of the skeleton that can be used with confidence.)

This is perhaps a good opportunity to illustrate how insignificant the transition can be from one group to another. The first bird hatched out of an egg laid by a theropod dinosaur, but unless there were many other hatchlings at almost exactly the same evolutionary stage to form a breeding population, the lineage would have gone extinct. Even if you had been there, watching those feathery chicks hatch and grow up, you would not have been aware that you were seeing the transition from theropod to bird in that generation of that evolving population. (And, of course, that is true of any other evolutionary transition that has ever occurred.) That means that if the fossil record is relatively complete, the change that defines the transition will necessarily be one so trivial that it will look artificial—and, of course, it is trivial and it is artificial!

The Origin of Feathers

The proteins that make feathers in living birds are completely unlike the proteins that make reptilian scales today.

Feathers originate in a skin layer deep under the outer layer that forms scales. Feathers probably arose as new structures under and between reptile scales, not as modified scales. Many birds have scales on their lower legs and feet where feathers are not developed, and penguins have such short feathers on parts of their wings that the skin there is scaly for all practical purposes. So feathers evolved in theropods as completely new structures, and any reasonable explanation of their origin has to take this into account. Obviously, feathers did not evolve for flight. They evolved for some other function and were later modified for flight.

Feathers may have evolved to aid thermoregulation. The feathered Chinese theropods all have down, probably as insulation to keep their bodies at an even temperature. It doesn't matter whether they used their feathers to conserve heat in cold periods, or to keep heat out in hot periods, or both. Insulation would have been useful in either case.

The thermoregulatory theory for the origin of feathers is probably the most widely accepted one today, but it does have problems. Why feathers? Feathers are more complex to grow, more difficult to maintain in good condition, more liable to damage, and more difficult to replace than fur. Every other creature that has evolved a thermoregulatory coat—from bats to bees and from caterpillars to pterosaurs—has some kind of furry cover. There is no apparent reason for evolving feathers rather than fur even for heat shielding.

Furthermore, thermoregulation cannot account for the length or the distribution of the long feathers on the Chinese theropods. Short feathers (down) can provide good thermoregulation, but thermoregulation does not require long feathers, and it would not help thermoregulation very much to evolve long strong feathers on the arms and tail. So it is difficult to suggest that feathers evolved for thermoregulation alone. It would be better to think of another equally simple explanation.

I naturally prefer an idea that I developed years ago, with my colleague Jere Lipps (Cowen and Lipps 1982). In living birds, feathers are for flying, for insulation, but also for camouflage and/or display. Lipps and I suggested that feathers evolved for display. The display may have been between females or between males for dominance in mating systems (sexual selection), or between individuals for territory or food (social selection), or directed toward members of other species in defense.

Living reptiles and birds often display for one or all of these reasons, using color, motion, and posture as visual signals to an opponent. Display is often used to increase apparent body size; the smaller the animal, the more effectively a slight addition to its outline would increase its apparent size. Lipps and I therefore proposed that erectile, colored feathers would give such a selective advantage to a small displaying theropod that it would encourage a rapid transition from a scaly skin to a feathery coat.

Display would have been advantageous as soon as any short feathers appeared, and it would have been most effective on movable appendages, such as forearms and tail. (Display on the legs would not be so visible or effective.)

Figure 12.31 Forearm display would have drawn attention to the powerful claws of the first bird (and derived theropod), Archaeopteryx. From Heilmann.

Figure 12.32 A dinosaur stampede: trackways at the Lark Quarry Conservation Park, near Winton, Queensland, Australia (see text). Courtesy of the Queensland Museum.

Forearm display by a small theropod would also have drawn particular attention to the powerful weapons the theropod carried there, its front claws (Fig. 12.31). *Caudipteryx* carried long feathers on its middle finger, between the two outside claws, and it could fold that middle finger away, with the feathers out of harm's way.

The display hypothesis explains more features of feathered theropods, including *Archaeopteryx*, the first bird, than other hypotheses, with fewer assumptions. It explains completely the feather pattern: the evolution of long strong feathers on arms and tail.

Once they evolved, feathers could quickly have been co-opted for thermoregulation, and the down coat on the Chinese theropods may show that process. Down can only be for thermoregulation. Although down is not proof of warm blood, it is very strong evidence in favor of it. In living birds, down feathers are associated with solving the problem of heat loss for hatchlings. However, the discovery of extensive feathers on the giant Early Cretaceous tyrannosaur *Yutyrannus* opens up a different set of questions (Xu et al. 2012).

Dinosaur Behavior

Dinosaur behavior can be judged by footprints; for example, a dinosaur stampede has been discovered (Fig. 12.32). Rocks laid down about 90 Ma as sediments near a Cretaceous lakeshore in Queensland, Australia, bear the track of a large dinosaur heading down toward the lake with a 2-meter (6-foot) stride. Superimposed on this track are thousands of small footprints made by small, bipedal, lightly built dinosaurs, running back up the creek bed away from the water. More than 3000 footprints have now been uncovered, showing all the signs of a panicked stampede. At least 200 animals belonging to two species were stampeding. One of the species, probably a coelurosaur, ranged up to about 40 kg (90 pounds), and the other, probably an ornithopod, ranged up to twice as large. Juveniles and

Figure 12.33 The skull of *Pachycephalosaurus*, showing the dome of bone that leads to the idea of head-butting in this dinosaur. Photograph by Ballista and placed into Wikimedia.

adults of both species were digging in their toes as they tried to accelerate: 99% of the footprints lack heel marks. The footprints show slipping, scrabbling, and sliding, and the smaller species usually avoided the tracks of the larger one. They may have felt hemmed against the lakeshore, breaking away in a terrified group.

The stampede sheds light on ecology as well as behavior, telling us that at least some dinosaurs gathered in herds and behaved just as African plains animals do today at water-holes on the savanna, responding immediately and instinctively to the approach of larger animals.

Figure 12.34 The skull of *Saurolophus*, showing the crest that extends upward and backward on the skull. Photograph by Didier Descouens and placed into Wikimedia.

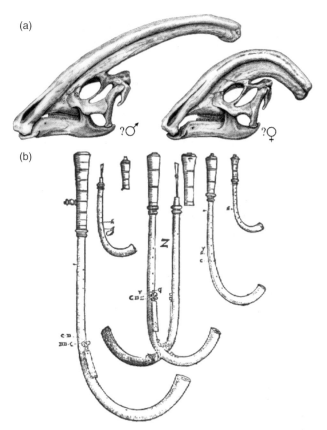

Figure 12.35 a, two skulls of the hadrosaur *Parasaurolophus*, from two different fossil localities. These skulls show the very long bone tunes that lead from the nostrils in a recurved tube leading to the back of the mouth. They could be adult and juvenile, or they could be male and female, or they could be two different species. More collecting and better age dating could resolve the issue. Whatever the answer, it is obvious that the two dinosaurs would have made very different sounds by blowing through their tubes! Illustration © Dr. Paul E. Olsen, used by permission. b, a collection of Renaissance wind instruments called crumhorns. David Weishampel (1981) made this striking analogy with *Parasaurolophus*. Drawings from the book *Syntagma musicum*, published in 1620.

Other indicators of behavior are preserved in dinosaur skeletons. The ornithischian dinosaur *Pachycephalosaurus* had a dome-shaped area of bone on the top of its skull (Fig. 12.33). It is not solid bone but has air cavities in it, with an internal structure very much like that found in the skulls of sheep and goats that fight by ramming their heads together. Did *Pachycephalosaurus* do that? It weighed eleven times as much as a bighorn sheep, so it may have butted opponents from the side rather than head-to-head. *Triceratops* and other ceratopsians, however, also had air spaces between the horns and the brain case, which may indicate that they competed by direct head-to-head impact and/or wrestling.

Some hadrosaurs had huge crests on the head (Fig. 12.34). The crests were not solid but contained tubes running upward from the nostrils and back down into the roof of the mouth. Only large males had large crests; females had smaller ones, and juveniles had none at all. The tubes are unlikely to have evolved for additional respiration or thermoregulation. (If so, adults would have needed large tubes whether they were female or male.) In 1981 David Weishampel suggested that the tubes were evolved for sound production. In *Parasaurolophus*, for example, they look uncannily like medieval pipes (Fig. 12.35). (Reconstructed tubes can be blown to give a note.) The varying sizes of crests allow us to infer differences between the sounds produced by young, by adult females, and by adult males, to go with the different visual signals provided by the crests (Fig. 12.36). These hadrosaurs may have had a sophisticated social system, as complex as those we take for granted in mammals and birds.

With advances in technology, CAT scans have become powerful ways to look inside solid objects such as dinosaur skulls. It turns out that the same sound-producing reso-nant tubes occur in the skulls of other hadrosaurs (Fig. 12.37a), and once we have seen that, then one can look for (and find) anomalous (and analogous) nostril pathways in other dinosaurs such as ankylosaurs (Fig. 12.37b).

In all aspects of their biology, therefore, dinosaurs seem as modern as mammals and birds. That makes their extinction at the end of the Cretaceous even more puzzling (except for the survival of the smallest dinosaurs, which we call birds). I will discuss that extinction in Chapter 16.

"Procheneosaurus praeceps"
(Holotype 3577 Ct. R.O.M.P.) Juvenile

"Lambeosaurus clavinitialis"
(351 G.S.C.) Adult female

"Procheneosaurus cranibrevis"
(Holotype 8633 G.S.C.) Juvenile

Lambeosaurus lambei
(5131 Ct. R.O.M.P.) Adult male

"Corythosaurus frontalis"
(Holotype 5853 Ct. R.O.M.P.) Juvenile

300 mm

"Lambeosaurus clavinitialis"
(Holotype 8703 G.S.C.) Adult female

Figure 12.36 Reconstructed heads of several "species" of lambeosaurine dinosaurs, drawn to imply that all of them could be different age and gender members of only one species, *Lambeosaurus lambei*. Art by Nobu Tamura, and placed into Wikimedia.

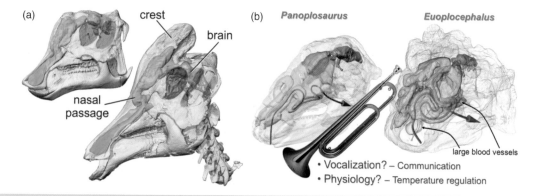

Figure 12.37 a) images from CT scans of the skulls of a young and a sub-adult hadrosaur *Corythosaurus*. The older individual has a nostril pathway that was more complex, and was divided into resonating lengths. Research details in Evans et al. 2009. b) images from CT scans of two ankylosaurs, with trumpet for visual effect. The nasal passages could have been use to emit sounds. Of course, nostrils can also be used in thermoregulation (cooling), but both functions would have worked here. Look back at Figure 12.13, which shows the ankylosaur *Euoplocephalus*, and compose a signature trill for it. Research details in Miyashita et al. 2011. Both images courtesy Witmer Lab at Ohio University, http://www.oucom.ohiou.edu/dbms-witmer/lab.htm

Further Reading

Abler, W. L. 1999. The teeth of the tyrannosaurs. *Scientific American* 281 (3): 40–41.

Bates, K. T. and P. L. Falkingham 2012. Estimating maximum bite performance in *Tyrannosaurus rex* using multi-body dynamics. *Biology Letters*, in press.

Carrier, D. R. and C. G. Farmer 2000a. The evolution of pelvic aspiration in archosaurs. *Paleobiology* 26: 271–293. [A novel and convincing reconstruction of respiration in archosaurs, with major application to dinosaurs.] Available at http://biologylabs.utah.edu/farmer/publications%20pdf/2000%20Paleobiology26.pdf

Carrier, D. R. and C. G. Farmer 2000b. The integration of ventilation and locomotion in archosaurs. *American Zoologist* 40: 87–100. [Some overlap with the previous paper.] Available at http://icb.oxfordjournals.org/content/40/1/87.full

Chiappe, L. 1998. Dinosaur embryos. *National Geographic* 194 (6): 34–41.

Chin, K. et al. 1998. A king-sized theropod coprolite. *Nature* 393: 680–682.

Codd, J. R. et al. 2008. Avian-like breathing mechanics in maniraptoran dinosaurs. *Proceedings of the Royal Society B* 275: 157–161. Available at http://rspb.royalsocietypublishing.org/content/275/1631/157.full

Cowen, R. and J. H. Lipps 1982. An adaptive scenario for the origin of birds and of flight in birds. *Proceedings of the 3rd North American Paleontological Convention, Montréal*, 109–112. Available at http://mygeologypage.ucdavis.edu/cowen/HistoryofLife/Montreal.html, with successive updates during previous editions of this book at http://mygeologypage.ucdavis.edu/cowen/HistoryofLife/feathersandflight.html

Erickson, G. M. et al. 2001. Dinosaurian growth patterns and rapid avian growth rates. *Nature* 412: 429–433. [See also Padian et al. 2001.]

Evans, D. et al. 2009. Endocranial anatomy of lambeosaurine dinosaurs: a sensorineural perspective on cranial crest function. *Anatomical Record* 292: 1315–1337. Available at http://www.oucom.ohiou.edu/dbms-witmer/Downloads/2009_Evans_et_al._lambeosaurine_brains_&_crests.pdf

Farlow, J. O. and M. K. Brett-Surman (eds.) 1997. *The Complete Dinosaur*. Bloomington: Indiana University Press. Many fine essays on all aspects. See especially Part 4: Biology of the Dinosaurs.

Fastovsky, D. E. and D. B. Weishampel 1996. *The Evolution and Extinction of the Dinosaurs*. New York: Cambridge University Press.

Fowler, D.W. and Sullivan, R.M. 2011. The first giant titanosaurian sauropod from the Upper Cretaceous of North America. *Acta Palaeontologica Polonica* 56: 685–690. Available at http://app.pan.pl/archive/published/app56/app20100105.pdf

Gillette, D. D. and M. G. Lockley (eds.) 1989. *Dinosaur Tracks and Traces*. Cambridge: Cambridge University Press. Paperback, 1991.

Hone D. W. E. et al. 2010. The extent of the preserved feathers on the four-winged dinosaur *Microraptor gui* under ultraviolet light. *PLoS ONE* 5(2): e9223. Available at http://www.plosone.org/article/info%3Adoi%2F10.1371%2Fjournal.pone.0009223

Hone, D. W. E. et al. 2012. Does mutual sexual selection explain the evolution of head crests in pterosaurs and dinosaurs? *Lethaia* 45: 139–156. [Yes]

Horner, J. R. and J. Gorman 1988. *Digging Dinosaurs*. New York: Workman. [An excellent read and a persuasive insight into the scientific investigation of dinosaurs. Paperback, 1990.]

Horner, J. R., and D. Lessem 1994. *The Complete T. rex*. New York: Simon and Schuster. [Paperback. Light, bright and breezy.]

Horner, J. R. 2000. Dinosaur reproduction and parenting. *Annual Reviews of Earth & Planetary Sciences* 28: 19–45. Available at http://www.lowellcarhart.net/ebay/papers/parenting.pdf

Hutchinson, J. R. and S. M. Gatesy 2000. Adductors, abductors, and the evolution of archosaur locomotion. *Paleobiology* 26:

734–751. Available at http://www.rvc.ac.uk/aboutus/staff/jhutchinson/files/JRH1.pdf

Hutchinson J. R. et al. 2011. A computational analysis of limb and body dimensions in *Tyrannosaurus rex* with implications for locomotion, ontogeny, and growth. *PLoS ONE* 6(10): e26037. Available at http://www.plosone.org/article/info%3Adoi%2F10.1371%2Fjournal.pone.0026037

Ji, Q. et al. 1998. Two feathered dinosaurs from northeastern China. *Nature* 393: 753–761, and comment, pp. 729–730. [*Caudipteryx*.] Available at http://dinosaurideas.com/Dinosaur_Ideas/pdf_files/1998Caudipteryx.pdf

Ji, Q. et al. 2001. The distribution of integumentary structures in a feathered dinosaur. *Nature* 410: 1084–1088.

Mallison, H. 2010a. The digital *Plateosaurus* I: body mass, mass distribution, and posture assessed using CAD and CAE on a digitally mounted complete skeleton. *Palaeoelectronica* 13.2 8A. Available at http://www.uv.es/~pardomv/pe/2010_2/198/index.html

Mallison, H. 2010b. The digital *Plateosaurus* II: An assessment of the range of motion of the links and vertebral column and of previous reconstructions using a digital skeletal mount. *Acta Palaeontologica Polonica* 55: 433–458. Available at http://www.app.pan.pl/archive/published/app55/app20090075.pdf

Mayr, G. et al. 2002. Bristle-like integumentary structures at the tail of the horned dinosaur *Psittacosaurus*. *Naturwissenschaften* 89: 361–365. Available at http://www.miketaylor.org.uk/tmp/papers/gmayr43.pdf

Miyashita, T. et al. 2011. The internal cranial morphology of an armoured dinosaur *Euoplocephalus* corroborated by X-ray computed tomographic reconstruction. *Journal of Anatomy* 219:661–675. Available at http://www.oucom.ohiou.edu/dbms-witmer/Downloads/2011_Miyashita_et_al._Euoplocephalus_head_anatomy.pdf

Norell, M. A. et al. 1994. A theropod dinosaur embryo and the affinities of the Flaming Cliffs dinosaur eggs. *Science* 266: 779–782, and comment, p. 731. See comment also in *Nature*, v. 372, p. 130. [*Oviraptor* embryo.]

Norell, M. A. et al. 1995. A nesting dinosaur. *Nature* 378: 774–776, and comment, pp. 764–765. Available at http://www.bi.ku.dk/dna/course/papers/I2.norell.pdf

Norell, M. A. et al. 2002. "Modern" feathers on a non-avian dinosaur. *Nature* 416: 36–37. [On a dromaeosaur.]

Norell, M. A. and X. Xu 2005. Feathered dinosaurs. *Annual Reviews of Earth and Planetary Sciences* 33: 277–299.

Norman, D. B. and P. Wellnhofer 2000. *The Illustrated Encyclopedia of Dinosaurs*. London: Salamander. [A fine "coffee-table" book, with good science. This edition includes pterosaurs.]

O'Connor, P. and L. Claessens 2005. Basic avian pulmonary design and flow-through ventilation in non-avian theropod dinosaurs. *Nature* 436: 253–256. Available at http://www.ohio.edu/people/ridgely/OconnorClaessensairsacs.pdf

Padian, K. et al. 2001. Dinosaurian growth rates and bird origins. *Nature* 412: 405–408. [See also Erickson et al. 2001.] Available at http://bigcat.fhsu.edu/biology/cbennett/bib-arch-pter/Padian-etal-2001.pdf

Prum, R. O. and A. H. Brush 2002. The evolutionary origin and development of feathers. *Quarterly Review of Biology* 77: 261–295.

Rayfield, E. J., et al. 2004. Cranial mechanics and feeding in *Tyrannosaurus rex*. *Proceedings of the Royal Society B* 271: 1451–1459. Available at http://rspb.royalsocietypublishing.org/content/271/1547/1451.full.pdf

Reisz, R.R. et al. 2012. Oldest known dinosaurian nesting site and reproductive biology of the Early Jurassic sauropodomorph *Massospondylus*. *PNAS* 109: 2428–2433.

Spicer, R.A. and A. B. Herman 2010. The Late Cretaceous environment of the Arctic: A quantitative reassessment based on plant fossils. *Palaeogeography, Palaeoclimatology, Palaeoecology* 295: 423–442. Available at http://oro.open.ac.uk/20876/1/Spicer_and_Herman_2010.pdf

Thulborn, R. A. and M. Wade 1979. Dinosaur stampede in the Cretaceous of Queensland. *Lethaia* 12: 275–279.

Wedel, M. J. 2003. Vertebral pneumaticity, air sacs, and the physiology of sauropod dinosaurs. *Paleobiology* 29: 243–255.

Wedel, M. J. 2006. Origin of postcranial skeletal pneumaticity in dinosaurs. *Integrative Zoology* 1: 80–85.

Weishampel, D. B. 1981. Acoustic analyses of potential vocalization in lambeosaurine dinosaurs (Reptilia: Ornithischia). *Paleobiology* 7: 252–261. Available at http://www.hopkinsmedicine.org/FAE/DBWpdf/R3_1981aWeishampel.pdf

Weishampel, D. B. 1997. Dinosaur cacophony. *Bioscience* 47: 150–159. Available at http://philoscience.unibe.ch/documents/TexteHS10/Weishampel1997.pdf

Weishampel, D. B. et al. (eds.). 2004. *The Dinosauria*, 2nd edition. Berkeley: University of California Press.

Wilson J. A. et al. 2010. Predation upon hatchling dinosaurs by a new snake from the Late Cretaceous of India. *PLoS Biology* 8(3): e1000322. Available at http://www.plosbiology.org/article/info%3Adoi%2F10.1371%2Fjournal.pbio.1000322

Wings, O. and P. M. Sander 2007. No gastric mill in sauropod dinosaurs: new evidence from analysis of gastrolith mass and function in ostriches. *Proceedings of the Royal Society B* 274: 635–640. Available at http://rspb.royalsocietypublishing.org/content/274/1610/635.full.pdf

Witmer, L. M. 2001. Nostril position in dinosaurs and other vertebrates and its significance. *Science* 293: 850–853. Available at http://bill.srnr.arizona.edu/classes/182h/vertebrate%20evolution/DinosaurNoses.pdf

Xu, X. et al. 1999. A therizinosauroid dinosaur with integumentary structures from China. *Nature* 399: 350–354. [*Beipiaosaurus*.]

Xu, X. et al. 1999. A dromaeosaurid dinosaur with a filamentous integument from the Yixian Formation of China. *Nature* 401: 262–266. [*Sinornithosaurus*.]

Xu, X. et al. 2000. The smallest known non-avian theropod dinosaur. *Nature* 408: 705–708. [*Microraptor*.]

Xu, X. et al. 2000. Branched integumental structures in *Sinornithosaurus* and the origin of feathers. *Nature* 410: 200–204. [Real feathers, not just "integumental structures".]

Xu, X. et al. 2003. Four-winged dinosaurs from China. *Nature* 421: 335–340, and comment, pp. 323–324; also comment in *Science* 299, p. 491. [*Microraptor gui*.] Available at http://www.sciteclibrary.ru/ris-stat/st597/dynapage-1.htm

Xu, X. et al. 2012. A gigantic feathered dinosaur from the Lower Cretaceous of China. *Nature* 484, 92–95. [*Yutyrannus*]

Question For Thought, Study, and Discussion

We know that dinosaurs laid large clutches of eggs, and clearly a female dinosaur would lay a LOT of eggs during her lifetime. Yet on average, only two of her eggs would survive to be full adults (or the world would have been overrun with dinosaurs). So do the best you can to describe the hazards of being a dinosaur egg (or hatchling, or adolescent), citing evidence where you can.

THIRTEEN
The Evolution of Flight

In This Chapter

Flight in animals began with insects in the Carboniferous coal forests, but there were gliding diapsids in the Permian and Triassic, all evolved independently. The earliest animals with powered flapping flight were the pterosaurs of the Late Triassic. They were mostly fish-eating, but came in a variety of sizes and shapes, consistent with living in different habitats and catching different prey. Some were filter-feeders with many very fine teeth, rather like a flamingo. Spectacular preservation in a few localities has given us beautiful pterosaur fossils, some with apparent visual display features. Pterosaurs were undoubtedly warm-blooded, and they laid eggs, though very few fossil eggs have been found. Pterosaurs included the largest flying animals at a wingspan of over 10 meters. The earliest bird is Late Jurassic in age, and birds clearly evolved from small theropod dinosaurs that already had feathers. The first bird *Archaeopteryx* has the skeleton of a dinosaur, but with strong feathers that may or may not have given it the ability to fly. I discuss the issue of the origin of bird flight at some length. More modern-looking birds were well evolved in the Cretaceous, and there is no question that many of them were strong fliers with ecologies that we would easily recognize today. Birds survived the extinction at the end of the Cretaceous that wiped out their dinosaur and pterosaur relatives, and they had a dramatic radiation in the Cenozoic that has given us birds ranging in size and biology from a humming-bird to an ostrich. Bats are flying mammals, but they evolved quickly after the Cretaceous extinction, with sonar already perfected by 50 Ma. They are now one of the most diverse mammal groups: fruit bats to vampires!

Styles of Flight

There are four kinds of flight: passive flight, parachuting, soaring, and powered flight. Passive flight can be used only by very tiny organisms light enough to be lifted and carried by natural winds and air currents, and light enough to suffer no damage on landing. Tiny insects, baby spiders, frogs' eggs, and many kinds of pollen, spores, and seeds can

History of Life, Fifth Edition. Richard Cowen.
© 2013 by Richard Cowen. Published 2013 by Blackwell Publishing Ltd.

be transported this way. But their "flight" duration, direction, and destination are entirely at the mercy of chance events.

The first land organisms and the first aerial organisms were microscopic. As reproduction adjusted to the problems of life in air, spores were evolved by fungi, plants, and other small organisms—their soft reproductive cells were protected by a dry, watertight coating, rather than the damp slime that is sufficient in water. Dry spores could then be spread as passive floaters on the wind—Earth's first fliers. Plant spores occur in Ordovician rocks (Chapter 8), and they are numerous and widespread enough in Devonian rocks to be used as guide fossils in relative age dating. Apart from anything else, this suggests that some plant species had such a large area of tolerable habitats open to them that a long-range dispersal method had become worthwhile.

Gliding flight includes **parachuting**, in which the flight structures slow a fall, and **soaring**, in which the flight structures allow an organism to gain height by exploiting natural air currents. Parachuting and soaring may seem to grade into one another, but their biology is very different, and the two flight modes probably have distinctly different origins. Parachuting organisms have simpler flight structures and much less control over the direction, speed, and height of flight than soarers (Fig. 13.1). They seek short-range travel from one point to another, and their landing point is reasonably predictable because they do not seek external air currents for lift. Parachuting is used in habitats where external air currents are minimal, especially in forests. Wind gusts and air currents are potentially disastrous to animal parachutists, just as they are to human paratroops.

Powered flight is usually accomplished by some sort of flapping motion with special structures (wings). It needs a lot of energy, but gives much greater independence from variations in air currents, and it is usually accompanied by a high level of control over flight movements. Because powered flight is achieved by controllable appendages, almost all powered fliers can glide to some extent, some very poorly (no better than parachutists) and some very well indeed. Raptors and soaring seabirds are examples of powered fliers that glide well.

Soaring is used by flying organisms that range widely over a broad habitat (Fig. 13.2). It is a low-energy flight style because the lift comes from external air currents rather than muscular expenditure by the flier. Energy costs are mainly related to the maintenance and adjustment of gliding surfaces in the air flow. Soarers may need occasional bursts of flapping flight if there are no up-currents, or in transferring from one up-current cell to another. Flapping is sometimes needed for takeoff, until airspeed exceeds stalling speed, or for final adjustments of attitude and speed in landing. Because flapping flight is needed occasionally by all soarers today (especially in emergencies), soaring probably cannot evolve from parachuting but only from powered flight.

Flight of all kinds demands a light, strong body. Soaring especially emphasizes lightness in muscle mass as well as

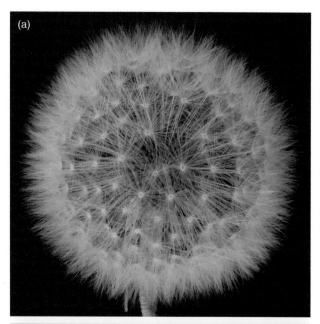

Figure 13.1 Parachuting by dandelion seeds. a) the seeds still attached to the plant. Photograph by Jdparker, and released into the public domain. b) seeds parachuting down to find a place to settle and germinate. Photo by PiccoloNamek and placed into Wikimedia.

overall structure. Powered flight has more requirements, including a significant output of energy and strength. Even the best soarers among living birds, albatrosses and condors, cannot flap for long before they are exhausted, because their flight muscles are small relative to their size and total weight. It might be difficult for a specialized soarer to re-evolve the ability to sustain flapping flight.

One would imagine that powered flight could evolve easily from gliding. Any evolving wing should be a fail-safe device, allowing a gliding fall during flight training. But an animal that has already evolved efficient gliding would not easily improve its flight by flapping in mid-glide, because that would disturb the smooth airflow over the gliding

Figure 13.2   Soaring by the turkey vulture *Cathartes aura* at Bodega Head, California. Photograph by VivaVictoria and placed into Wikimedia.

surfaces. Aerodynamic analysis shows that an evolutionary transition is possible from gliding to flapping, but only in very special circumstances. The glider must add fairly large, rapid wing beats, not little flutters, and because wing beats require considerable expenditure of energy, there must be a corresponding payoff in energy saved (for example, the animal must save some walking or climbing, or must reach a larger food supply).

The evolution of flight demands lightening and strengthening of the whole body structure, and the evolution of a flight organ from a pre-existing structure (a limb, for example) that could otherwise perform some other function. Flight may have strong advantages in locomotion—for food-gathering, escape, rapid travel between base and food supply, or migration—but it also has costs, not just in energy but also in constraints on body form and function that may have accompanying drawbacks. Flight has evolved many times in spite of all these problems.

Flight usually involves relatively large lifting structures; in almost every case a small lifting structure is no better than none at all. Lifting structures must already be present before flight can evolve, and they must therefore have evolved for some different function. This theme dominates the discussion of flight origins in this chapter.

Flight in Insects

Primitive insects are known from Devonian rocks, but flying insects are not found until the Late Carboniferous, as insects radiated in the Carboniferous forest canopy.

Many insects of all sizes are known from the coal beds of the Carboniferous, and half of all known Paleozoic insects had piercing and sucking mouthparts for eating plant juices. In turn, these smaller insects were a food source for giant predatory dragonflies (Fig. 9.14) and for early amniotes (Chapter 9).

In living insects (except mayflies), only the last molt stage, the adult, has wings, and there is a drastic metamorphosis between the last juvenile stage (the nymph) and the flying adult. Wings have to be as light and strong as possible; in living insects this is achieved by withdrawing as much live tissue as possible. Most of the wing is left as a light mass of dead tissue that cannot be repaired. This gives great flying efficiency, though it usually means a short adult lifespan. The automatically short life expectancy of flying insects has played a strong part in the evolution of social behavior among some insects, in which the genes of a comparatively few breeding but nonflying adults are passed on with the aid of a great number of cheap, throwaway, sterile flying individuals (worker bees, for example). Some insects shed their wings. In many ants, for example, the wings are functional only for a brief but vital period during the mating flight. Insects do not have a long enough life expectancy to have the luxury of learning, so they carry with them a "read-only memory" that seems to govern their behavior entirely by "instinct."

But these are characters of living insects, and they would not necessarily apply to early insects that had not yet evolved flight. However, there is now good evidence from Carboniferous insects on the evolution of flight in this group, the first animals to take to the air under control. Jarmila Kukalová-Peck has pointed out that insects (and angels) are the only flying creatures that evolved flapping flight without sacrificing limbs to form the wings. Insects have thus lost little of their ability to move on the ground.

Many living insects are good gliders. Dragonflies, which were among the earliest insects to evolve flight, have wings arranged so that they are very stable in a gliding attitude, but dragonflies use flapping flight to chase rapidly and expertly after their prey. We still do not understand dragonfly flight. Somehow, complex eddies are produced between the two sets of wings, which beat out of phase. At some phases in the wing cycle, dragonfly wings produce lift forces that are 15 to 20 times the body weight.

Other insects have complex locking devices to hold their wings in a gliding position without energy expenditure. They need these locking devices for gliding because in powered flight their wings flap freely during complex movements. This line of reasoning suggests that insects evolved flight as flappers and later adjusted in a complex way to gliding.

The critical fact about the evolution of insect wings is that arthropod limbs consisted originally of two branches: a walking leg and another jointed unit—the exite—that was used as a filtering device or a gill. These structures are still found in most marine arthropods, but at first sight they seem to have been lost in insects, which have only walking legs. They were not lost: exites disappeared because they

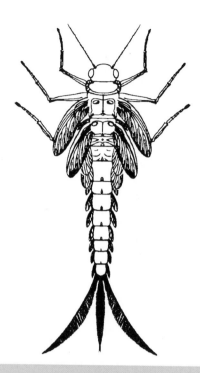

Figure 13.3 A mayfly nymph from the Early Permian of Oklahoma. The wings on the thorax are big and look functional, but they are for underwater rowing rather than flight. There are also smaller winglets on all the segments of the abdomen. Redrawn after a reconstruction by Jarmila Kukalová-Peck.

Figure 13.4 *Draco sumatranus*, a species of flying lizard, with its wing membranes stretched out to show the ribs built into them. You can also see that the wings can be used in display. Photograph by Biophilia curiosus, and placed into Wikimedia.

evolved into wings. We can see some of the stages in this evolution. In the young water-dwelling stages—nymphs—of living mayflies, the exites along the abdomen are shaped into platelike gills. The same structure is found on the thoracic exites of larval dragonflies, some beetles, and several other groups of insects.

There are fossilized nymphs of Late Carboniferous insects, and many of them had exites modified into plate-like gills (Fig. 13.3). The plates were probably also used for swimming (living mayfly nymphs use their plated gills for swimming in the same way). However, the plates were also pre-adapted to flight (short at first, of course), and this pathway to flapping flight is the leading hypothesis for the origin of insect flight.

An intermediate stage in the evolution of insect flight may still exist in some primitive living insects, mayflies and stoneflies. James Marden and Melissa Kramer showed that these primitive insects "skim" across water surfaces, using a wing action that is exactly like flying, but they also receive some lift from the legs, which remain in contact with the water surface (Marden and Kramer 1994).

Parachuting Vertebrates

Several living forest-dwelling vertebrates have evolved parachuting flight, using skin flaps as flight surfaces. They include flying squirrels, three different lineages of Australian gliding marsupials (greater gliders, squirrel gliders, and feathertail gliders), *Draco* the flying lizard (Fig. 13.4), flying geckos, flying frogs, and even flying snakes (Socha 2011)! This suggests that parachuting adaptations evolve in animals of the forest canopy that habitually jump from branch to branch, from tree to tree, or from trees to the ground. Any method of breaking the landing impact or of leaping longer distances would be advantageous and might evolve rapidly.

None of these parachuting animals has powered flight, however. The energy for gliding flight is gravitational, generated as the animal climbs in the tree and released as it parachutes off the branch. Parachuting can evolve in animals with rather low metabolic rates. It does not require the high metabolic rate of birds and bats, which have powered flight. Most parachuting vertebrates have short limbs, long trunks, flexible spines, and quadrupedal stance.

Early Gliding Vertebrates

The earliest known gliding vertebrate is the Late Permian reptile *Coelurosauravus*. Its fossils have been found in Germany, Britain, and Madagascar, so it was widespread across Pangea. All these areas were near tropical shores at the time. *Coelurosauravus* is an ordinary, small diapsid reptile in the structure of its skull and body, about the size of a small squirrel. But the trunk is dominated by 20 or more long, curving, lightly-constructed rod-shaped bones that extended outward and sideways from the body. They supported a skin membrane that was close to an ideal airfoil in shape, 30 cm (1 foot) across, and could only have been used for gliding (Fig. 13.5). More impressive still, the bones are jointed so that the airfoil could have been folded back along the body when it was not in use. Extra-long vertebrae allowed space for this folding along the spinal column. These bones are not ribs, but must have evolved specifically under the skin as a gliding structure. Because of this unique character, *Coelurosauravus* is placed in a major basal diapsid group of its own, the Weigeltisauria.

We can judge how well *Coelurosauravus* was adapted for gliding flight by comparing it with Triassic and living reptilian gliders. Kuehneosaurids are a family consisting of two gliding reptiles from the Late Triassic, *Kuehneosaurus* from Britain and *Icarosaurus* from New Jersey (Fig. 13.6). They too had effective airfoils, but since they were stretched out on elongated ribs, gliding must have evolved independently in kuehneosaurids and *Coelurosauravus*. Astonishingly, *Kuehneosaurus* is found in two forms, one with a broader airfoil than the other, yet in all other features of the skeleton they are identical. The most vivid (and convincing) interpretation is that these are male and female of the same species, with the male having broader wings for better gliding (and displaying) (Stein et al. 2008).

The Cretaceous lizard *Xianglong* from China (Fig. 13.7) (Li et al. 2007), and the living lizard *Draco*, which also use long ribs to support an airfoil, both evolved gliding independently. Ligaments and muscles between the ribs of *Draco* give precise control of the gliding surface, and this

Figure 13.5 The Late Permian gliding reptile *Coelurosauravus*. The supports for its airfoil are clearly not ribs: they do not begin at the vertebrae. Art by Nobu Tamura, and placed into Wikimedia.

Figure 13.7 The Cretaceous gliding lizard *Xianglong* from China. Art by Nobu Tamura, and placed into Wikimedia.

Figure 13.6 Reconstructions of Late Triassic gliding reptiles. a) male and female *Kuehneosaurus* from Britain (see text). b) *Icarosaurus* from New Jersey. In these and later gliding lizards, the airfoils were supported by long ribs. Both pieces of art by Nobu Tamura, and placed into Wikimedia.

Figure 13.8   The Triassic gliding lizard *Sharovipteryx* from Central Asia. Art by Dmitry Bogdanov, and placed into Wikimedia.

Figure 13.9   Body plan of a pterosaur. The leading edge of the wing membrane is supported for most of its length by the fourth finger, and its trailing edge is fixed to the hind limb. This pterosaur is *Anhanguera*, from the Cretaceous of Brazil. Scale bar, 10 cm. Image by Leon Claessens, Patrick O'Connor, and David Unwin, Figure 1d in (http://www.plosone.org/article/info%3Adoi%2F10.1371%2Fjournal.pone.0004497) © Classens et al. 2009. Placed into the public domain by publication in PLoS ONE.

was probably true also in all the fossil gliders. In all of them, all four limbs remain free for walking, grasping, and climbing. All of them can fold up the airfoil when it is not in use. *Icarosaurus* may have been the best of the fossil lizards in terms of gliding performance (McGuire and Dudley 2011).

In *Draco* and in the Triassic gliders, the ribs are single, unjointed bones. When *Draco* folds its airfoil, the spinal ends of the ribs have to be moved between the back muscles, which means that the ribs cannot be very big or very strong. The Triassic gliders had long levers mounted on the spinal ends of their ribs to get around this problem, and of course *Coelurosauravus* avoided the problem altogether by having a separate jointed airfoil that was not made from ribs.

Sharovipteryx was discovered by accident in a search for fossil insects in Late Triassic rocks in Central Asia. It was a small reptile, and preserved skin clearly indicates that it had a gliding membrane. *Sharovipteryx*, however, was unique in that the membrane was stretched between very long, strong hind limbs and a long tail, so a large, broad wing surface was set well behind the trunk and head (Fig. 13.8), rather like some modern aircraft—the supersonic Concorde, for example.

Several studies agree that *Sharovipteryx* glided very well, most likely with the aid of a canard, or accessory membrane, set in front of the short, normal-looking fore limbs (Fig. 13.8). Sensitive control over flight could have been maintained by slight backward-and-forward motion of the hind limbs, as in many gliding birds today. A similar but cruder system is used in swing-wing aircraft.

Longisquama is a strange reptile from the same Triassic rocks as *Sharovipteryx*. Its remains include a series of long, flattened bones with flared, curved tips. Susan Evans suggested that these were ribs from a gliding airfoil. The lightness and flattening of the bones and the curvature of their tips would all make sense if that were true. *Longisquama* was probably a glider, very much like the kuehneosaurids.

An airfoil does not appear by magic, especially a folding one. Robert Carroll points out that there may have been

other good reasons for evolving a folding, extended rib structure. The great area of exposed skin could have been used in thermoregulation, for example. If the extinct reptiles behaved like *Draco*, they may have used their airfoils for display as well as for flight, and may even have evolved them first for display.

We recognize these fossil reptiles as gliders because they had specialized skeletons. By comparison with small, insect-eating vertebrates in forests today, there were probably many other jumping and gliding reptiles in Permian and Triassic forest canopies, with skin flaps unsupported by bones. The forest canopy was probably rich in many species of small insectivorous amphibians and reptiles.

Pterosaurs

Pterosaurs are the most famous flying reptiles. The earliest pterosaurs known are Late Triassic, when they were already well evolved for flight. The earliest well-preserved pterosaur, *Austriadactylus*, already had a bony crest on its skull and a very long tail.

Most analyses place pterosaurs within archosaurs, and in particular into the bird-like archosaurs, the Ornithodira. If so, then you remember from Chapter 11 that pterosaurs must have evolved in the Early Triassic, with a very long ghost lineage that we have not yet found, probably consisting mostly of their terrestrial ancestors.

Pterosaurs have very lightly built skeletons, with air spaces in many of the bones. Their forelimbs were extended into long struts that supported a wing, as in birds and bats. Pterosaurs were unique, however, in that most of the wing membrane was supported on one extraordinarily long finger, while three other fingers were normal and bore claws (Fig. 13.9). The fourth finger was about 3 meters (10 feet) long in the largest pterosaurs. In contrast, birds support the wing with the whole arm, and bats use all their fingers as

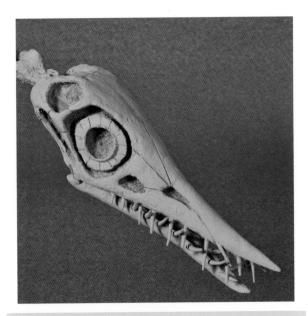

Figure 13.10 *Rhamphorhynchus*, a pterosaur from the Jurassic of Germany, with fish-eating teeth. Photograph by Amy Martiny, courtesy Witmer Lab at Ohio University, http://www.oucom.ohiou.edu/dbms-witmer/lab.htm.

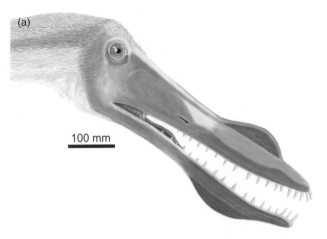

Figure 13.11 a) *Anhanguera*, from the Cretaceous of Brazil, reconstructed by Larry Witmer. (Art by C. McQuilkin, courtesy Witmer Lab at Ohio University, http://www.oucom.ohiou.edu/dbms-witmer/lab.htm). b) *Pterodaustro* may have been a filter-feeder, perhaps like a flamingo. Art by Nobu Tamura, and placed into Wikimedia. c) *Ctenochasma* may also have been a filter-feeder. Image by Ghedoghedo, and placed into Wikimedia.

bony supports through their wing membranes. Pterosaurs thus have a unique wing anatomy, but as the largest flying creatures ever to evolve and as a group that flourished for more than 140 m.y., they can't be dismissed as primitive or poorly adapted.

Most pterosaurs had large eyes sighting right along the length of long, narrow, lightly-built jaws. The teeth were usually thin and pointed, often projecting slightly outward and forward, as in *Rhamphorhynchus* (Fig. 13.10). This is most likely an adaptation for catching fish. Almost all pterosaur fossils are preserved in sediments laid down on shallow seafloors, and where stomach contents have been preserved with pterosaur skeletons, they always contain fish remains such as spines and scales. Some pterosaurs may have fished on the wing, like living birds such as gadfly petrels or skimmers, which fly along just above the water surface and dip in their beaks to scoop up fish or crustaceans. One can imagine *Anhanguera* doing this (Fig. 13.11a). Other pterosaurs may have fed like terns, which dive slowly so that only the head, neck, and front of the thorax reach under the water, while the wings remain above the surface. Some pterosaurs with long sharp beaks may have fished standing in the water, or slowly patrolling, like herons, or sitting on the water. It seems unlikely that pterosaurs crash-dived into water like pelicans or gannets, or swam underwater like penguins: pterosaur wings were too long and too fragile. At least one pterosaur, *Pterodaustro* from Argentina, had teeth that were so fine, long, and numerous that it must have been a filter feeder, perhaps like a flamingo (Fig. 13.11b); and *Ctenochasma* looks like a

filter-feeder too (Fig. 13.11c). Some short-jawed pterosaurs may have eaten shore crustaceans or insects.

A slab from the Late Jurassic Solnhofen Limestone of Germany preserves dramatic evidence of a very bad day in a tropical lagoon. A *Rhamphorhynchus* had caught a little

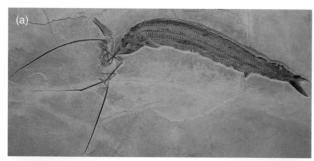

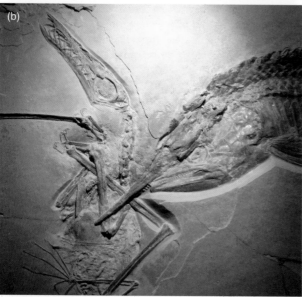

Figure 13.13 Some *Nyctosaurus* specimens have been preserved with a strange crest on the head. It is difficult to see this as anything other than a display structure. Art by Matt Martyniuk and placed into Wikimedia.

Figure 13.12 A bad day in the Solnhofen lagoon. A pterosaur had just caught a small fish right at the water surface when a larger fish struck at its wing. The pterosaur was too large for the fish to swallow, but its teeth were stuck so far into the elastic membranes of the wing that the fish could not get free, and both animals died. This tragic accident was beautifully preserved for paleontological detectives to explain (Frey and Tischlinger 2012). Their images are published in a paper in the open access journalPLoS ONE (http://www.plosone.org/article/info:doi/10.1371/journal.pone.0031945), and are thereby placed into the public domain.

fish, presumably by dipping its beak into the water. The fish was still in the throat pouch of the pterosaur as it flapped strongly to regain height. The pterosaur may have touched the water with its wingtip, or came so close to the water that it caused a strong shadow, because a large fish struck at it and seized the wing. The pterosaur was pulled into the water, but it was too big for the fish to swallow. The teeth of the fish were firmly fixed in the elastic membranes of the pterosaur wing, and although the fish struggled enough to severely damage the pterosaur wing, it could not get free, and all three animals died (Fig. 13.12a, b) (Frey and Tischlinger 2012).

There are two main groups of pterosaurs. **Rhamphorhynchoids** (Late Triassic to Late Jurassic) are the stem group of early pterosaurs, rather than a clade. Most of them had wingspans under 2 meters (6 feet), and some were as small as sparrows. At least some of them had extravagant crests on the head (*Nyctosaurus*, for example, Fig. 13.13), and others had long, thin, stiff tails that carried a vertical vane on the tip.

Pterodactyloids are a clade of advanced pterosaurs that replaced rhamphorhynchoids in the Late Jurassic and flourished until the end of the Cretaceous. Pterodactyloids had no tails, and many were much larger than rhamphorhynchoids. The large forms were adapted for soaring rather than continuous flapping flight, although they all flapped for takeoff. *Pterodactylus* itself was sparrow-sized, but *Pteranodon*, from the Cretaceous of North America, had a wingspan of about 7 meters (22 feet); and the gigantic pterosaur from Texas, *Quetzalcoatlus*, was 10–11 meters (35–35 feet) in wingspan, the largest flying creature ever to evolve. (An incomplete set of pterosaur fossils from Romania may be pieces of an even larger form: guesses vary around 12 meters.)

Although pterosaur bones were light and fragile, several examples of outstanding preservation have shown us many details of their structure. Black shales in Lower Jurassic rocks of Germany have shown details of rhamphorhynchoids; Late Jurassic members of both pterosaur groups

Figure 13.14 Lyrical reconstruction of two ornithocheirid pterosaurs in the Cretaceous skies above England. However, the artist has allowed some brutal reality: you will notice that a fish is being stolen. (This happens often among seabirds today.) Art by Dmitri Bogdanov and placed into Wikimedia.

Figure 13.15 A white-tailed kite hovering, with its head held still, looking downward to fix on a prey. Photograph by Zoipes and released into the public domain.

have been found exquisitely preserved in the Solnhofen Limestone of Germany and in lake deposits in Kazakhstan in Central Asia. From the Lower Cretaceous of Brazil we have partial skeletons preserved without crushing, and the Upper Cretaceous chalk beds of Kansas have yielded huge specimens of *Pteranodon*. Discoveries of skin, wing membranes, and stomach contents allow biological interpretations of these exciting animals.

However, those interpretations vary widely. Pterosaurs have no living descendants that we can study, and we have not found their ancestors. It is difficult to choose a living analog: some people look at bats for anatomical and functional guidance, others look at birds. A few things are very clear: all pterosaurs, including the giant forms, were capable of powered, flapping flight (Fig. 13.14), and that includes the giant forms.

The pterosaur wing was attached low on the hind limb, in a "broad-wing" reconstruction (Fig. 13.8) (Elgin et al. 2011). The wing itself was not simply a giant skin membrane: that would have been too weak to power flapping flight. Furthermore, with bones, joints, and ligaments only on the leading edge of the wing, a pterosaur needed a way to control the aerodynamic surface of the wing. Beautifully preserved specimens show that the wing had special adaptations. It was stiffened by many small, cylindrical fibers, which were probably tied together by small muscles. The combination of structural stiffeners and muscles allowed fine control over the surface, and at the same time made the wing reasonably strong, not easily damaged or warped, and not likely to billow in flight like the fabric of a hang glider.

A research team led by Larry Witmer of Ohio University made CT scans of two uncrushed pterosaur skulls. The scans revealed the size and shape of the pterosaur brain. In both brains, the lobes associated with balance were very large, and this allowed the researchers to reconstruct the head to be arranged in the usual, or preferred, attitude it had in life. While the little early Jurassic pterosaur *Rhamphorhynchus* apparently held its head horizontally (as birds do in normal flight), the later and larger Cretaceous pterosaur *Anhanguera* seems to have held its head angled downward, perhaps in fishing position (Fig. 13.11a). This is not unreasonable. Herons spend hours in this kind of attitude as they stand waiting for fish, even though they fly with their heads horizontal. In completely different ways of life, kites and pelicans (at least the white-tailed kite and the white pelican of California) hold their heads "normally" as they fly from place to place, but kites hover over potential prey sites, and pelicans go into slow searching flight mode, both with their heads tilted dramatically downward (Fig. 13.15).

All the small rhamphorhynchoids and many of the pterodactyloids had active, flapping flight. Naturally, the gigantic pterosaurs could not have flapped for long, and they probably spent most of their time soaring, as does the living albatross. Aerodynamic analysis shows that pterosaurs were the best slow-speed soaring fliers ever to evolve.

Flapping flight involves very high energy expenditure. Birds are warm-blooded, as are bats and many large insects when they are in flight: dragonflies, moths, and bees are examples. Thus one might guess that pterosaurs too were warm-blooded. Several Jurassic pterosaurs have fur preserved on the skin: if pterosaurs had fur, they were probably warm-blooded. Flapping flight has evolved only three times among vertebrates (in pterosaurs, in birds, and in bats) and in each case the animal was apparently warm-blooded before or just as it achieved flight. Pterosaur bones had air spaces running through them in the same way that living bird bones do. In birds, this system helps to provide air cooling, and it is reasonable to interpret pterosaur bone structure in the same way.

But the air spaces are more than that: we now realize, thanks to Claessens et al. (2009), that pterosaurs had much the same respiration system as dinosaurs and birds, including the one-way flow through the lungs that is much more efficient than our mammalian in-and-out system (Chapter 12). There are other potential implications to this inference, particularly the question of how early in archosaur history it evolved. But that is still to be worked out (or argued, at least!)

If most pterosaurs ranged widely over the ocean searching for fish, it would have been impossible for pterosaur nestlings to feed themselves until they had reached a fairly advanced stage of growth and flight capability. Nesting behavior and care of the young would therefore have been mandatory. Kevin Padian has described a "pterosaur nursery" preserved in Cretaceous rocks in Chile.

The social behavior of pterosaurs may have been complex. Many pterosaurs were dimorphic. Males were larger, with long crests on the back of the head and with relatively narrow pelvic openings. Females were smaller, with smaller crests but larger pelvic openings. New discoveries show that the soft tissues associated with some crests were extravagantly large, and were much more likely to have been display structures than aids to flight (Fig. 13.13).

The largest pterosaurs, *Quetzalcoatlus* and related forms (together called azhdarchids) lived right at the end of the Cretaceous. The fossils of *Quetzalcoatlus* were found in nonmarine beds in Texas, deposited perhaps 400 km (250 miles) inland from the Cretaceous shoreline. Perhaps it was the ecological equivalent of a vulture, soaring above the Cretaceous plains and scavenging on carcasses of dinosaurs. *Quetzalcoatlus* did have a strangely long, strong neck, but its beak seems too lightly built for this method of feeding. Witton and Naish (2008) have argued that it was more like a gigantic heron (Fig. 13.16), standing and fishing in inland lakes and swamps, or picking up frogs, turtles, baby dinosaurs, or arthropods such as crayfish from shallow water.

We do not know why pterosaurs became extinct. As we have seen, they were most likely active, warm-blooded animals with flapping flight much like that of birds. Yet pterosaurs became extinct at the end of the Cretaceous, at the same time as the dinosaurs disappeared, while birds did not. We shall return to that question in Chapter 18.

Birds

Living birds are warm-blooded, with efficient thermoregulation that maintains body temperatures higher than our own. Birds breathe more efficiently than mammals, pumping air through their lungs rather than in and out. They have better vision than any other animals. Birds build extraordinarily sophisticated nests: bowerbirds are second only to humans in their ability to create art objects. New Caledonian crows learn to make tools faster than chimpanzees do. And above all, birds can fly better, farther, and faster than any other animals, an ability that demands

Figure 13.16 Dawn patrol. A group of *Quetzalcoatlus* forages across a Late Cretaceous wetland in Texas. Note the baby sauropod! There is no specific evidence that *Quetzalcoatlus* ate them, though it certainly could have done so. Image © 2008 Mark Witton and Darren Naish: Figure 9 in http://www.plosone.org/article/info%3 Adoi%2F10.1371%2Fjournal.pone.0002271 Publication in PLoS One automatically places the image in Wikimedia.

complex energy supply systems, sensing devices, and control systems.

Birds include ostriches and penguins, which cannot fly, and hummingbirds, which can hardly walk. But birds share enough characters for us to be sure that they form a single clade, descended from (that is, part of) archosaurs. The skull, pelvis, feet, and eggs of birds are so clearly archosaurian that Darwin's friend T. H. Huxley called birds "glorified reptiles."

Archaeopteryx

Archaeopteryx, from Upper Jurassic rocks in Germany, is perhaps the most famous fossil in the world. It is a feathered dinosaur that looks remarkably like a bird until it is examined carefully (Fig. 13.17). Only eleven specimens have been found, plus a single feather. The first complete *Archaeopteryx* was immediately seen as a fossil bird, because it had feathers on its wings and tail. But without feathers, it looks very much like a small theropod dinosaur. Two of the specimens lay unrecognized for a long time, labeled as small theropods.

Archaeopteryx has a theropod pelvis, not the tight, boxlike structure of living birds. It has a long, bony tail, clawed fingers, and a jaw full of savage little teeth. These are all theropod features. *Archaeopteryx* lacks many features of living birds. The only birdlike features on the entire bony skeleton of *Archaeopteryx* are a few characters of the skull,

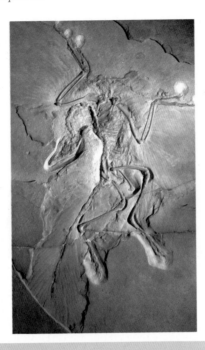

Figure 13.18 The supracoracoideus muscle in living birds attaches to the breastbone, then passes through the shoulder joint to insert on the upper side of the humerus. This is the muscle system that raises the wing. *Archaeopteryx* does not have it.

Figure 13.17 *Archaeopteryx lithographica* from the Late Jurassic of Germany, about the size of a large crow. Look carefully for feather impressions. Photograph by H. Raab and placed into Wikimedia.

but a CT scan of its braincase shows that the brain was very bird-like (Alonso et al. 2004).

Archaeopteryx is always preserved in an unusual body attitude, with the neck severely ricked back over the body (Fig. 13.17). We know why this happens. If an animal dies today on or near the beach or on a desert salt pan, it may be mummified by wind and salt spray before it rots or is eaten by predators. The muscles slacken and the tendons dry out. The long tendons that support the head contract severely, dragging the skull backwards over the spine. At the same time, any body feathers on a bird usually drop off, but the stronger wing and tail feathers stay fixed in position.

Occasionally, birds mummified on a beach may be washed out to sea on a high tide, or blown into the sea by a gale. They may float for several weeks before becoming waterlogged, and even when they finally sink, they retain their peculiar body attitude. There is no need to suggest that *Archaeopteryx* could fly because it sank and was buried at sea.

There are good reasons to argue that *Archaeopteryx* could not fly. It could not have sustained flapping flight. There is no breastbone, and no hole through the shoulder joint through which to pass the large tendon that gives the rapid, powerful, twisting wing upstroke in living birds. This tendon passes *through* the shoulder joint, and as well as raising the wing, it twists it. On the upstroke, the twist arranges the wing and feathers so that they slip easily through the air, with little drag. At the top of the upstroke,

the wing is in exactly the right position to give a powerful downbeat. Without this **supracoracoideus system** (Fig. 13.18), which is easily identified in fossils because it leaves a strong trace on the shoulder joint, a bird cannot fly by wing flapping. In fact, it cannot even take off and land, because the greatest power from rapidly beating wings is required during slow flight.

In small flying birds today, the wishbone or **furcula** is flexible and acts as a spring that repositions the shoulder joints after the stresses of each wing stroke. It is needed to give the rapid flaps necessary for flight (a starling flies with 14 complete wing beats per second). The wishbone also helps to pump air in respiration, and to recover some of the muscular energy put into the downstroke. But the wishbone in *Archaeopteryx* and the wishbones of theropods are U-shaped and strong and solid; they could not have acted as effective springs. Furthermore, *Archaeopteryx* did not have the long primary feathers on the wing tips, or the breastbone anchoring the muscles, that are needed for routine takeoff and landing. It could not have raised its arms high above its body for an effective downstroke. In fact, *Archaeopteryx* evolved structures that were active deterrents to flight. Its tail was long and bony, with long feathers. Among living birds with display feathers, this sort of tail is aerodynamically the worst of all possible tail styles, adding a lot of drag and little lift (Balmford et al. 1993).

Archaeopteryx, then, was a fierce little fast-running, displaying theropod (specifically a dromaeosaur), which probably spent its life scurrying around the Solnhofen shore, hunting for small prey such as crustaceans, reptiles, and mammals. It probably foraged much like the roadrunner of the American Southwest. If so, *Archaeopteryx* did not compete in the air with the pterosaurs that are also found in the Solnhofen Limestone, and it did not fly.

Feathers have been found on a number of dinosaurs, but they are particularly evident in dromaeosaurs. These are small-bodied agile running little theropods, known best from the Early Cretaceous of China. *Archaeopteryx* fits neatly into the dromaeosaurs, but since it also has some characters linking it directly to birds, its position as a particularly bird-like dromaeosaur, or as a basal bird, has not been seriously questioned.

The Origin of Flight in Birds

Almost all paleontologists are now convinced that birds evolved from theropods, and in particular from dromaeosaurs. The best hypothesis suggests that flight evolved in a ground-running dromaeosaur very much like *Archaeopteryx*. With long, erect limbs, a comparatively short trunk, and bipedal locomotion, *Archaeopteryx* and its feathered relatives are exactly the opposite in body plan of all living mammals and reptiles that jump and glide from tree to tree. There is nothing in the ancestry of birds as we now know it to suggest any arboreal adaptations at all (Dececchi and Larsson 2011).

Since ground-running theropods had feathers, the origin of flight in birds has nothing to do with the appearance of feathers. I argued in Chapter 12 that display and thermoregulation may have been involved in the origin of feathers, but not flight. (Flight evolved in bats and pterosaurs without feathers.)

Perhaps some features of a ground-running theropod could provide some of the anatomy and behavior necessary for flight, such as lengthening the forearms, especially the hands, placing long, strong feathers in those areas, and evolving powerful arm movements. Baron Nopsca, perhaps impressed by the apparent speed and dexterity of *Compsognathus* (Fig. 12.5), suggested 100 years ago that flight could evolve in a fast-running reptile. His ideas do not work in detail, but Burgers and Padian (2001) offered the most plausible recent version of the "**running raptor**" hypothesis. They envisage flight evolving from running, with the early advantage of flight being greater speed over the ground. Achieving flight would replace thrust from the ground by foot traction by aerodynamic forward thrust from the wing. In the take-off run, energy expended by the fore limbs would replace energy expended by the hind limbs, after a transition period in which all limbs would be contributing to forward thrust.

Lift is not important at first. The first stages of this fast, low-level flight would be aided by the phenomenon of **ground effect**. Essentially, eddies generated by the wings interact with the ground immediately under the wings, providing enough lift at very low altitude to achieve take-off. Thus the wing stroke would not have to produce much lift as long as there was no advantage in acquiring height.

In the scenario, the bird is now capable of fast-flapping low-level flight, but its advantage ends if it ascends out of the shallow zone of ground effect. All the wing action is energetically expensive, especially in the early stages of lift-off. Rapid flapping is essential throughout the scenario. And finally, none of this scenario begins to work until (unless) wing thrust is powerful enough to replace the (powerful) leg thrust of a running theropod. (The earliest feathered wings would not have been very effective as thrust devices.)

The Display and Fighting Hypothesis

Flight in birds requires long feathers. Jere Lipps and I suggested 30 years ago that display was involved in the evolution of flight as well as feathers. Theropods had long, strong display feathers on arms and tail (Chapter 12). Successful display was increased by lengthening the arms, especially the hand, and by actively waving them, perhaps flapping them rapidly and vigorously. Flapping in display would have encouraged the evolution of powerful pectoral muscles, and the supracoracoideus system.

Display can be very effective, and not just for sexual ends. Frigate birds and bald eagles often try to rob other birds of food instead of catching prey themselves. Because the penalty for wing injury is high, many birds can be intimidated by display into giving up their catch rather than fighting to defend it.

But a threat display cannot always be an empty bluff. Fighting is the last resort. Living birds often fight on the ground, even those that fly well. The wings no longer have claws but are still used as weapons in forward and downward smashes (steamer ducks are particularly deadly at this). Beaks and feet can be used as weapons too, and are most effective when used in a downward or forward strike.

A strong wing flap, directed forward and downward, is also the power stroke that gives lift to a bird in takeoff. Lipps and I suggest that strong wing flapping is a simple extension of display flapping, honed in fighting behavior. Powerful flapping used to deliver forearm smashes could have lifted the bird off the ground, allowing it also to rake its opponent from above with its hind claws. The more rapidly the wings could be lifted for another blow, the more effective the fighting. This would rapidly encourage an effective wing-lifting motion that minimized air resistance, so the wing action would then be almost identical to a takeoff stroke.

Kevin Padian also sees the wing stroke evolving from the arm strike used by a theropod in predation (rather than competition). It is not clear (to me) how this could have led easily to whole-body takeoff, however. Living raptors such as hawks and eagles try to avoid situations in which they need a prolonged struggle to subdue a prey. (Sexual competition is another matter: you may be fighting for your posterity, not just for another meal.)

Archaeopteryx fits our display-and-fighting hypothesis well. It was well adapted for display. Like any small theropod, it was well equipped for fighting with its teeth and the strong claws on hands and feet. *Archaeopteryx* did not have

Figure 13.19 Reconstruction of the wing of *Archaeopteryx* by Gerhard Heilmann. There are no primary feathers on the fingers, because they would have interfered with the claws.

long primary feathers on its fingers (Fig. 13.19), probably because they would have hidden the claws in display and would most likely have broken in a fight.

From Fight to Flight

Display and fighting in birds takes a lot of energy, whether it is for territory, dominance, or food, but it provides an enormous payoff in survival and selection. Sexual display in most living birds must be done correctly, or no mating takes place. New behaviors can be evolved rapidly, and they are often evolutionarily cheap, because they usually don't require important morphological changes in their early stages.

Our scenario stresses lift as well as thrust. It suggests that the earliest birds evolved flight behavior, anatomy, and experience at low ground speed and low height: ideal pre-flight training. The selective payoff for successful mastery of the flight motions gave significant advantages, even before flight itself was possible. Short-lived but intense activity could provide major adaptive advantage, an advantage that would begin as soon as the feathered surfaces of the wings could generate any life at all. Rapid wingbeats would not be essential, as they are in the cursorial hypotheses: the wingbeats would simply have to be faster than those of the competition.

From the stage exemplified by *Archaeopteryx*, the many advantages of flight were added to those of social or sexual competition. I do not think it is a coincidence that the males of the Early Cretaceous Chinese birds *Confuciusornis* and *Changchengornis* had extravagantly long (display) feathers on the tail (Fig. 13.20)! In more advanced birds than *Archaeopteryx*, the supracoracoideus tendon system evolved in the shoulder, while the wishbone evolved into a spring. The breastbone evolved as the anchor for the flight muscles. The forearms became longer, lighter, and more fragile in bone structure, becoming specialized as wings, and losing the finger claws. Meanwhile, the feet and beak

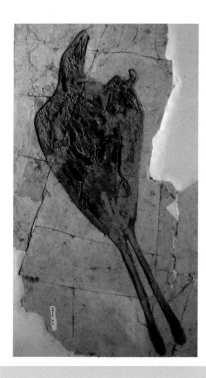

Figure 13.20 Specimens of the Cretaceous basal bird *Confuciusornis* often had long tail feathers that look as if they were for display. Photograph by Laikayui and placed into Wikimedia.

became the dominant fighting weapons, as in most living birds.

Cretaceous Birds

The radiation of birds was very rapid. Early Cretaceous rocks have yielded bird remains in all the northern continents and in Australia. *Sinornis*, a sparrow-sized bird from the Early Cretaceous of China, had many features directly related to much better flight and perching than was possible in *Archaeopteryx*. The body and tail were shorter, and the tail had fused vertebrae at its end that provided a firm but light base for strong tail feathers. The center of mass of the body was much farther forward, closer to the wings. *Sinornis* had a breastbone, a shoulder joint that allowed it to raise its wings well above the horizontal, and fingers that were adapted to support feathers rather than grasping and tearing claws. The wrist could fold much more tightly forward against the arm than the 90° seen in *Archaeopteryx*, so the wing could be folded away cleanly in the upstroke or on the ground, reducing drag. The foot was much better adapted for perching. Even so, *Sinornis* still had some very primitive features: the skull and pelvis were much like those of *Archaeopteryx*, and it had teeth (Fig. 13.21).

New Early Cretaceous fossils from China tell the same story. Rapid evolution among Early Cretaceous birds dramatically improved their flying and perching ability;

perhaps this is why most of them were small and light. *Confuciusornis* and *Changchengornis* had lighter bones than *Archaeopteryx*, and had genuine beaks rather than jaws with teeth.

Most Cretaceous bird fossils are from shoreline habitats, but that may reflect preservation bias rather than ecological reality. We have good fossils of Late Cretaceous diving birds such as *Hesperornis* (Fig. 13.22a), and *Ichthyornis* was tern-like in its adaptations (Fig. 13.22b).

Cenozoic Birds

When the dinosaurs died out at the end of the Cretaceous, there must have been a very interesting opportunity for surviving creatures to invade the ecological niches associated with larger body size on the ground. The two leading contenders were birds and mammals, and although mammals quickly became large herbivores, it was birds that became the dominant land predators in some regions in the Paleocene. These birds evolved to become flightless terrestrial bipeds once more.

Large, flightless birds called **diatrymas** lived across the Northern Hemisphere in the Paleocene and Eocene. They were close to 2 meters (6 feet) tall, and they had massive legs with vicious claws and huge, powerful beaks (Fig. 13.23). It is not clear whether they were powerful but slow predators, or whether the beaks were used for crushing nuts! Diatrymas became extinct at the end of the Eocene.

Truly carnivorous birds with very similar appearance, the **phorusrhacids**, dominated the plains ecosystem of South America from the Paleocene to the Pleistocene (Fig. 13.24). The skull and beak of phorusrhacids were much more rigid than they are in most birds, and they were particularly strong in resisting the stresses involved in a downward strike (Degrange et al. 2010). Some phorusrhacids were 2.5 meters (8 feet) tall, and a spectacular

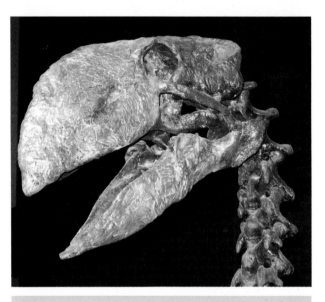

Figure 13.23 *Gastornis giganteus*, a diatryma from the Eocene of North America. The skull and beak of diatrymas were thought to indicate carnivory, but they may have been for cracking nuts! Photograph by Mitternacht90 and placed into Wikimedia.

Figure 13.21 The Cretaceous basal bird *Sinornis* had more advanced characters (see text). Art by Pavel Riha and placed into Wikimedia.

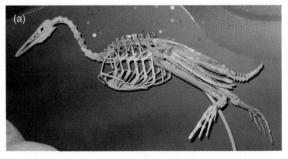

Figure 13.22 a) skeleton of the Cretaceous diving bird Hesperornis. Photograph by Quadell and placed into Wikimedia. b) the Cretaceous bird *Ichthyornis* seems to have been tern-like. Art by Nobu Tamura and placed into Wikimedia.

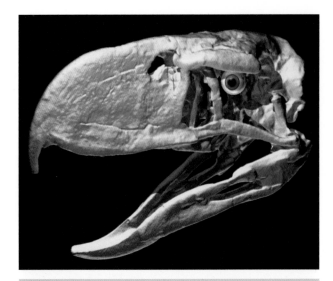

Figure 13.24 *Andalgalornis* from the Miocene of Argentina. (The eyeball is added for dramatic effect.) Courtesy Witmer Lab at Ohio University, http://www.oucom.ohiou.edu/dbms-witmer/lab.htm

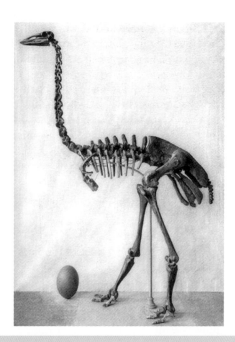

Figure 13.25 The recently extinct elephant bird *Aepyornis* from Madagascar. Image from Monnier 1913, in the public domain.

late phorusrhacid, *Titanis*, crossed to Florida from South America less than 3 m.y. ago. It was larger than an ostrich and no doubt caused at least temporary consternation among the Floridian mammals of the time.

The southern continents have a number of large flightless birds. Living forms such as the ostrich, cassowary, rhea, and emu are familiar enough, but even more interesting forms are now extinct. The moas of New Zealand reached well over 3 meters (10 feet) in height. *Aepyornis*, the "elephant bird" of Madagascar (Fig. 13.25), was living so recently that its eggshells are still found lying loose on the ground. The eggs are unmistakable because they had a volume of 11 liters (2 gallons). Early Muslim traders along the African coast certainly saw these eggs, and they may even have seen living elephant birds in Madagascar, giving rise to folktales about the fearsome roc that preyed on elephants and carried Sinbad the Sailor on its back (Fig. 13.26). *Aepyornis* and *Dromornis*, a giant extinct Australian bird related to ducks (Chapters 18 and 21), are close competitors for Heaviest Bird Ever to Evolve. The *Guinness Book of World Records* currently favors *Dromornis*, which was powerfully built and weighed perhaps 500 kg (1100 pounds).

The Largest Flying Birds

The largest flying birds so far discovered are **teratorns**, immense birds from South America, now extinct, who reached North America during the Pleistocene. Hundreds of specimens have been found in the tar pits of La Brea in Los Angeles, California, and from Florida and Mexico. But the largest teratorn was *Argentavis* from the Late Miocene

Figure 13.26 The roc, Smizurgh, a fearsome legendary bird that captured elephants three at a time, and gave Sinbad the Sailor a ride on its back. Art © Fred Lu 2011, based on a 19th century engraving, and used by permission.

of Argentina: it had a wingspan of 7.5 meters (24 feet). By contrast, the largest living bird is the royal albatross, just over 3 meters (10 feet) in wingspan.

The beak of *Argentavis* suggests that it was a predator, not a scavenger. It probably stalked prey on the ground. With a skull 55 cm (2 feet) long and 15 cm (6 inches) wide, it could have swallowed prey animals 15 cm across. Its bones are associated with other vertebrate fossils, but 64% of those are from *Paedotherium*, a little mammal about the size of a jackrabbit (in other words, an easy swallow for *Argentavis*).

In the same size range as teratorns were **pelagornithids**, gigantic marine birds that must have spent most of their time soaring over water. They ranged worldwide from the Eocene to the Late Miocene. They were lightly built, but the wingspan was close to 6 meters (nearly 20 feet) in the largest specimens. Their beaks were very long, with tooth-like projections built into their edges, presumably to help them hold squirming prey. More than any other living birds, pelagornithids were the ecological equivalents of pterosaurs, and it will be fascinating when further research allows us to reconstruct their mode of life accurately.

Bats

The latest evolution of flapping flight among vertebrates took place among bats. In all bats, the wing is stretched between arm, body, and leg, with the fingers of the hand splayed out in a fan toward the wingtip (Fig. 13.27). The wing membrane has little strength of its own, but it is elastic, and tension has to be maintained in it by muscles and tendons. The hind leg is used as an anchor for the trailing edge of the membrane, which means that the limb is not free for effective walking and running. Bats therefore are forced into unusual habits, which include roosting in inaccessible places where they hang upside down. Because bats are placental mammals, they have evolved special adaptations to maintain flight during pregnancy and nursing. For example, the pelvis has features that allow the body to be streamlined yet still have a rather large birth canal. Baby bats have needle-sharp milk teeth that allow them to hold tightly to the mother's fur in flight (ouch at feeding time!).

The earliest bats, *Onychonycteris* and *Icaronycteris*, are known from a few extremely well-preserved fossils from Early Eocene lake beds in Wyoming (Fig. 13.28) (Simmons et al. 2008). A little later in time, the Messel Oil Shale, in Middle Eocene rocks of Germany, has yielded dozens of bat skeletons. Some of them still contain the bats' last meals (primitive moths). Even the smallest ear bones are preserved, and they tell us that these bats were already equipped with the echo-locating sonar that all insect-hunting, fishing, and frog-eating bats have today. The baby bats at Messel already had sharp milk teeth.

Bat sonar presumably evolved from the acute hearing of little, nocturnal, insect-hunting mammals in the forest canopy of the Late Cretaceous (fruit bats have lost their sonar). Obviously, bats must already have had an eventful

Figure 13.27 The bat *Corynorhinus townsendii*, showing the arm and hand bones supporting the wings. Note also the tail membrane or uropatagium, extending between the hind limbs and supported by a long tail. By coordinating movements of the hind limbs and tail, the membrane can be contracted or expanded, and raised or lowered. In other words, it is an active component of the flight system. Image from the Bureau of Land Management; US government image, in the public domain.

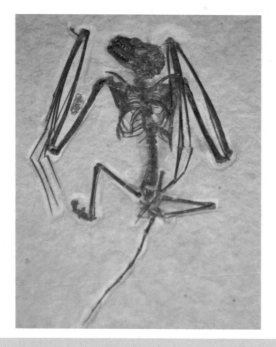

Figure 13.28 *Onychonycteris*, the earliest known bat, from the Eocene of Wyoming. Photograph of a replica of the only specimen, taken by Arvid Aase for the U.S. National Park Service.

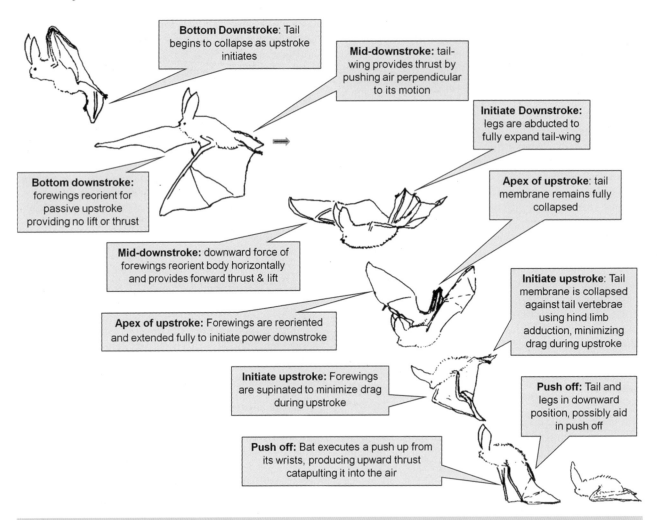

Figure 13.29 The way the wings and tail membrane are coordinated during take-off in a small vespertilionid bat. This system may also have operated during the first evolution of flight in bats. It may be important that the earliest bat, *Onychonycteris*, also had a long bony tail that could have supported and operated a relatively large uropatagium. The diagram is part of Figure 1 in Adams et al. 2012, published in PLoS ONE 7(2): e32074. Publication in PLoSONE automatically places this diagram into Wikimedia.

evolutionary history before the Eocene, but we still have to find the fossils that will tell us precisely who their ancestors were, and how they evolved flight.

A recent breakthrough in the latter problem came with a report that a number of small bats today begin flight by flapping the tail membrane as well as the wings as they take off (Adams et al. 2012). The coordination of the two systems is exact, and at very slow speed the tail is a vital component of bat flight (Fig. 13.29). If anything, this suggests that bats never glided and never parachuted. Instead they used the legs to jump into the air, gradually evolving longer and faster flights under better and better control from the actions of the skin membranes of the wings and uropatagium. Clearly we will learn much more about this aspect of bat flight in a year or two. Meanwhile it may be important that the earliest bat, *Onychonycteris*, also had a long bony tail that could have supported and operated a relatively large uropatagium (Simmons et al. 2008).

Further Reading

Flight (but not Birds)

Adams, R. A. et al. 2012. Flapping tail membrane in bats produces potentially important thrust during horizontal takeoffs and very slow flight. *PLoS ONE* 7(2): e32074. Available at http://www.plosone.org/article/info%3Adoi%2F10.1371%2Fjournal.pone.0032074

Averof, M. and S. M. Cohen 1997. Evolutionary origin of insect wings from ancestral gills. *Nature* 385: 627. [Genetic evidence from living arthropods.] Available at http://lepdata.org/monteiro/Evo-devo%20pdfs/Averof_Cohen_1997.pdf

Carrier, D. R. and C. G. Farmer 2000. The evolution of pelvic aspiration in archosaurs. *Paleobiology* 26: 271–293. [Relevant to the evolution of pterosaur and bird respiration.] Available at http://biologylabs.utah.edu/farmer/publications%20pdf/2000%20Paleobiology26.pdf

Claessens, L. P. A. M. et al. 2009. Respiratory evolution facilitated the origin of pterosaur fight and aerial gigantism. *PLoS ONE* 4(2): e4497. Available at http://www.plosone.org/article/info%3Adoi%2F10.1371%2Fjournal.pone.0004497

Clark, J. M. et al. 1998. Foot posture in a primitive pterosaur. *Nature* 391: 886–889. Dimorphodon could not have run, in this interpretation.

Dudley, R. 2000. The evolutionary physiology of animal flight: paleobiological and present perspectives. *Annual Reviews of Physiology* 62: 135–155. [Good discussion of the evolution of insect flight (by jumping, says Dudley). He tries to make all other instances of flight evolution fit the same model, but I don't believe that (yet). As with everything Dudley does, this is packed with good ideas.] Available at http://student.biology.arizona.edu/honors2007/group06/smithreview.pdf

Dyke, G. J. et al. 2006. Flight of *Sharovipteryx mirabilis*: the world's first delta-winged glider. *Journal of Evolutionary Biology* 19: 1040–1043.

Elgin, R. A. et al. 2011. The extent of the pterosaur flight membrane. *Acta Palaeontologica Polonica* 56: 99–111. Available at http://app.pan.pl/archive/published/app56/app20090145.pdf

Frey, E. et al. 1997. Gliding mechanism in the Late Permian reptile *Coelurosauravus*. *Science* 275: 1450–1452, and comment, p. 1419.

Kellner, A. W. A. and D. de A. Campos 2002. The function of the cranial crest and jaws of a unique pterosaur from the Early Cretaceous of Brazil. *Science* 297: 389–392. [*Thalassodromeus* seems to have been functionally and ecologically like the living skimmer.] Available at http://www.bios.niu.edu/davis/bios458/Kellner2002.pdf

Kingsolver, J. G. and M. A. R. Koehl 1994. Selective factors in the evolution of insect wings. *Annual Review of Entomology* 39: 425–451. Available at http://www.famu.org/mayfly/pubs/pub_k/pubkingsolverj1994p425.pdf

Kramer, M. G. and J. H. Marden 1996. Almost airborne. *Nature* 385: 403–404.

Kukalová-Peck, J. 1987. New Carboniferous Diplura, Monura, and Thysanura, the hexapod ground plan, and the role of thoracic side lobes in the origin of wings (Insecta). *Canadian Journal of Zoology* 65: 2327–2345.

Li, P-P. et al. 2007. A gliding lizard from the Early Cretaceous of China. *PNAS* 104: 5507–5509. Available at http://www.pnas.org/content/104/13/5507.full

Marden, J. H. and M. G. Kramer 1994. Surface-skimming stoneflies: a possible intermediate stage in insect flight evolution. *Science* 266: 427–430, and comment, v. 270, p. 1685. Available at https://courses.washington.edu/danielab/labwiki/images/8/8c/Marden_%26_Kramer_1994.pdf. See also Marden's non-technical 1995 article in *Natural History* 104 (2): 4–8.

McGuire, J. A. and R. Dudley 2011. The biology of gliding in flying lizards (genus *Draco*) and their fossil and extant analogs. *Integrative and Comparative Biology* 51: 983–990.

Simmons, N. B. et al. 2008. Primitive Early Eocene bat from Wyoming and the evolution of flight and echolocation. *Nature* 451: 818–821. [*Onychonycteris*] Available at http://deepblue.lib.umich.edu/bitstream/2027.42/62816/1/nature06549.pdf

Socha, J. J. 2011. Gliding flight in *Chrysopelea*: turning a snake into a wing. *Integrative and Comparative Biology* 51: 969–982.

Springer, M. S. et al. 2001. Integrated fossil and molecular data reconstruct bat echolocation. *PNAS* 98: 6241–6246. [Echolocation evolved with the earliest bats, and was lost later on in the "megabats", the fruit bats.] Available at http://www.pnas.org/content/98/11/6241.full

Tintori, A. and D. Sassi 1992. *Thoracopterus* Bronn (Osteichthyes: Actinopterygii): a gliding fish from the Upper Triassic of Europe. *Journal of Vertebrate Paleontology* 12: 265–283.

Weimerskirch, H. et al. 2003. Frigatebirds ride high on thermals. *Nature* 421: 333–334. [A wonderful analog for pterosaurs, in my opinion.]

Witmer, L. M. et al. 2003. Neuroanatomy of flying reptiles and implications for flight, posture and behaviour. *Nature* 425: 950–953, and comment by Unwin, pp. 910–911. Available at http://129.116.97.2/specimens/Anhanguera_santanae/Witmer_02048.pdf

Witton, M. P. and D. Naish 2008. A reappraisal of azhdarchid pterosaur functional morphology and paleoecology. *PLoS ONE* 3(5): e2271. Available at http://www.plosone.org/article/info%3Adoi%2F10.1371%2Fjournal.pone.0002271

Birds

Alonso, P. D. et al. 2004. The avian nature of the brain and inner ear of *Archaeopteryx*. *Nature* 430: 666–669. [Results of a CT scan]. Available at http://www.geo.utexas.edu/faculty/rowe/Publications/pdf/045%20archaeop1.pdf

Balmford, A. et al. 1993. Aerodynamics and the evolution of long tails in birds. *Nature* 361: 628–631. [*Archaeopteryx*-type tails make flight worse, not better.]

Chiappe, L. M. and L. M. Witmer (eds.) 2002. *Mesozoic Birds: Above the Heads of Dinosaurs*. Berkeley, California: University of California Press. [Massive overview in 20 chapters, with full references. See especially chapters by Witmer, Clark et al., Chiappe, and Gatesy.]

Chiappe, L. M., and G. J. Dyke 2002. The Mesozoic radiation of birds. *Annual Review of Ecology and Systematics* 33: 91–124. Available at http://biology-web.nmsu.edu/houde/Chiappe%26Dyke_2002.pdf

Cowen, R. and J. H. Lipps 1982. An adaptive scenario for the origin of birds and of flight in birds. *Proceedings of the 3rd North American Paleontological Convention, Montréal*, 109–112. Available at http://mygeologypage.ucdavis.edu/cowen/HistoryofLife/Montreal.html, with successive updates during previous editions of this book at http://mygeologypage.ucdavis.edu/cowen/HistoryofLife/feathersandflight.html

Dececchi T. A. and H. C. E. Larsson 2011. Assessing arboreal adaptations of bird antecedents: testing the ecological setting of the origin of the avian flight stroke. *PLoS ONE* 6(8): e22292. [Birds evolved flight from the ground.] Available at http://www.plosone.org/article/info%3Adoi%2F10.1371%2Fjournal.pone.0022292

Degrange F. J. et al. 2010. Mechanical analysis of feeding behavior in the extinct "terror bird" Andalgalornis steulleti (Gruiformes: Phorusrhacidae). *PLoS ONE* 5(8): e11856. Available at http://www.plosone.org/article/info%3Adoi%2F10.1371%2Fjournal.pone.0011856

Dial, K. P. 2003. Wing-assisted incline running and the evolution of flight. *Science* 299: 402–403, and comment, p. 329. [I and others don't believe it.] Available at https://bcrc.bio.umass.edu/courses/spring2012/biol/biol544/sites/default/files/dial.pdf

Earls, K. D. 2000. Kinematics and mechanics of ground take-off in the starling *Sturnus vulgaris* and the quail *Coturnix coturnix*. *Journal of Experimental Biology* 203: 725–739. [Both these small birds take off with a powerful jump: this is fully compatible with our display hypothesis.] Available at http://jeb.biologists.org/content/203/4/725.full.pdf

Hone, D. W. E. et al. 2012. Does mutual sexual selection explain the evolution of head crests in pterosaurs and dinosaurs? *Lethaia* 45: 139–156. [Yes]

Ji, Q. et al. 1999. A new Late Mesozoic confuciusornithid bird from China. *Journal of Vertebrate Paleontology* 19: 1–7. [*Changchengornis*, with display tail feathers just like *Confuciusornis*.] Available at http://dinosaurs.nhm.org/staff/pdf/1999Qiang_et_al.pdf

Murray, P. F. and P. Vickers-Rich 2004. *Magnificent Mihirungs: the Colossal Flightless Birds of the Australian Dreamtime.* Bloomington: Indiana University Press.

Nudds, R. L. and Bryant, D. M. 2000. The energetic cost of short flights in birds. *Journal of Experimental Biology* 203: 1561–1572. Short flights are even more expensive in terms of energy than we had thought. This supports a display-and-fighting idea, because it places the origin of flight as a short-flight, intense-effort situation. Available at http://jeb.biologists.org/content/203/10/1561.full.pdf

Shipman, P. 1998. *Taking Wing: Archaeopteryx and the Evolution of Bird Flight.* New York: Simon and Schuster.

Xu, X. et al. 2003. Four-winged dinosaurs from China. *Nature* 421: 335–340, and comment, pp. 323–324. Available at http://www.sciteclibrary.ru/ris-stat/st597/dynapage-1.htm; see also comment in Science 299, p. 491, and also Padian 2003. [*Microraptor gui*.]

Zhou, Z. and F. Zhang 2002. A long-tailed, seed-eating bird from the Early Cretaceous of China. *Nature* 418: 405–409. [*Jeholornis*: early, fairly primitive, seed-eating bird.]

Zhou, Z. et al. 2003. An exceptionally preserved Lower Cretaceous ecosystem. *Nature* 421: 807–814. [Convenient review of the Jehol Biota from China that includes feathered dinosaurs and early birds. No new science.] Available at http://webpages.fc.ul.pt/~maloucao/Zhou.pdf

Questions for Thought, Study, and Discussion

1. Pterosaurs seem to have shared many adaptations for flying, feeding, nesting, and so on, with birds. Try to find a reasonable suggestion for the fact that birds survived the Cretaceous extinction but pterosaurs did not. (I do not know of one, but you would imagine that there must have been some reason.)

2. In The Lord of the Rings (the book!), J. R. R. Tolkien made a big deal out of the observation that the evil flying Ringwraiths were gliding against the wind! (Gasp!!) This is the only mistake I can find in all three volumes. What's wrong with it?

3. Why would a bird lose the ability to fly? We know from their bone structure that ostriches, penguins, and many island birds lost the ability they once had to fly. This loss has to make evolutionary sense by giving an advantage to flightless birds over their flying relatives. Think of some reasons.

4. There are many insects and birds that have lost the ability to fly. But I do not know of a flightless bat, out of 1240 or so living species. Why no flightless bats? Again, you would imagine that there must have been some reason.

FOURTEEN

The Modernization of Land and Sea

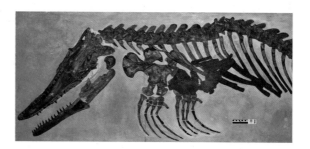

In This Chapter

In the Mesozoic, a number of diapsid lineages invaded the sea to become powerful air-breathing carnivores, apparently competing on equal terms with the sharks and other fishes that filled the same ecological niche. Turtles were in the sea by Late Triassic times, and they include the 10-foot-long giant turtle *Archelon* in the Late Cretaceous. Crocodiles are mostly freshwater and estuarine today, but there were marine crocodiles in the Jurassic. Ichthyosaurs were streamlined powerful swimmers with long jaws that carried fish-eating teeth. The largest specimens, from the Triassic were up to 50 feet long. Placodonts seem to have crushed mollusks between big tooth plates. A very large clade, the sauropterygians, began at fairly small size but eventually evolved into plesiosaurs. These reptiles had large strong limbs formed into paddles that probably moved in an up-and-down motion, generating an underwater "flight" pattern. Again, some plesiosaurs were over 50 feet long. Mosasaurs are simply water-going lizards, but they evolved long powerful streamlined bodies and limbs that made hydrofoils for efficient swimming. Amazingly, rare finds of embryos inside fossils show that mosasaurs, ichthyosaurs and plesiosaurs all had lineages that had live birth at sea, just like living whales and dolphins.

Finally, I discuss the way in which the land plants evolved in the Mesozoic. Most important, the flowering plants or angiosperms evolved dramatically in the Cretaceous, until they were worldwide and successful in the Late Cretaceous. Much of their success seems to be linked to their ability to attract insects or other visitors to them for pollination, using flowers that give visual or scent clues.

Mesozoic Ocean Ecosystems

The world's biology was decimated at the end of the Permian. The Mesozoic is the time (era, if you like) when that biology was not only reconstituted as a diverse global fauna and flora, but took on many of the characteristics of the modern world. A SCUBA diver in Permian seas would not have seen many familiar creatures, even if they were playing familiar ecological roles. However, a SCUBA diver in the Late Cretaceous would have found a much more familiar world.

In this chapter we will look at some of the marine and terrestrial organisms that display this major change. For the ocean, I will concentrate on the top predators of

History of Life, Fifth Edition. Richard Cowen.
© 2013 by Richard Cowen. Published 2013 by Blackwell Publishing Ltd.

the Mesozoic oceans, chiefly marine reptiles, and on land I will concentrate on the engine that drove the change: the transition from a land flora dominated by conifers to a land flora dominated by flowering plants. They gave not only a different structure to land ecosystems, but filled them with beautiful blossoms and fragrant scents.

After the demise of the giant placoderms at the end of the Devonian (Chapter 7), the larger carnivores in open water in the late Paleozoic were various lineages of cephalopods, ammonoids that were essentially squids with shells. They were relatively slow-moving and clumsy.

But after the P–Tr extinction, fishes became the mid-sized predators of the ocean. Ammonites were still abundant, but the major additions to the global oceans in terms of large-bodied predators were not fishes, but fish-eaters, and they were dominated by marine reptiles. This says (to me) that Mesozoic oceans were productive enough that the ecosystem could sustain a level of large predators that could not have succeeded in Paleozoic oceans. Mesozoic ecosystems differ dramatically from Paleozoic ones because large-bodied animals on land (dinosaurs) and in the air (pterosaurs) had their oceanic counterparts in large marine reptiles. Most of these reptile groups evolved in Triassic times, but reached their greatest abundance in the Jurassic and Cretaceous. Several different clades of reptiles evolved spectacular adaptations to life at sea.

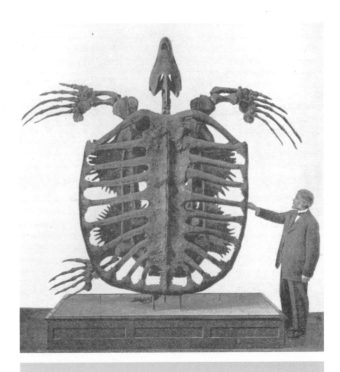

Figure 14.1 The giant Cretaceous marine turtle *Archelon* evolved a carapace that was lightened so that the turtle could maintain buoyancy in the water. This 1902 image is in the public domain.

Turtles

We have already seen that turtles are diapsids, though it is still debated whether they are basal diapsids or basal archosauromorphs (Chapter 11). The first well-known turtle *Proganochelys* is from the Late Triassic of Europe, and it already had bony plates on its surface, though it had not yet accomplished the turtle trick of having the shoulder blades inside the ribs.

Turtles were widespread and successful in Jurassic and Cretaceous seas and estuaries. Perhaps the most famous is the giant Cretaceous turtle *Archelon*, which was 3 meters (10 feet) long and nearly 4 meters (13 feet) in flipper span. It was so large that it couldn't have swum with a complete solid carapace, so it had only a bony framework (Fig. 14.1). Large marine turtles are anything but primitive in their biology. Their limbs are modified into hydrofoils, and they "fly" underwater. Marine turtles can navigate precisely over thousands of kilometers and they are warm-blooded, maintaining their body temperatures at levels significantly higher than the water around them.

Crocodiles

Crocodiles are archosaurs, and their ancestry is clearly terrestrial (Chapter 11). All crocodiles were terrestrial predators in the Late Triassic. There were large, powerful crocodile-like aquatic phytosaurs in the Triassic, and true crocodiles did not become aquatic until these others

became extinct. *Saltoposuchus*, a crocodile from the Triassic of Britain, had long, slim, erect limbs, and probably ran quite fast on land (Fig. 11.21). Terrestrial crocodiles lived on well into the Jurassic, but in the end may have been outcompeted on land by bipedal theropod dinosaurs.

Ever since the Early Jurassic, most crocodiles have been amphibious. Many of them are predators at or near the water's edge. Some became almost entirely aquatic, and others returned yet again to land in the Cenozoic to become powerful terrestrial predators in South America. Crocodiles that became amphibious or aquatic evolved to large size and were reasonably common in Mesozoic seas and rivers. There are several candidates for the largest crocodile that ever lived: four separate lineages had species over 10 meters (33 feet) long and perhaps 5 tons in weight (Fig. 14.2). *Deinosuchus* from the Late Cretaceous of Texas may have taken duckbilled dinosaurs as prey (they are found in the same rock formations) in the same way that living Nile crocodiles take hippos. But *Sarcosuchus* from the Early Cretaceous of Africa (Paul Sereno's "Supercroc") has received more publicity recently. All of these crocodiles were much larger than the largest living crocodiles, which reach 6 meters at most.

Crocodiles today are not equipped to kill large prey quickly. They usually kill large prey by holding them under the water until they drown. There's no reason to suppose that *Deinosuchus* did anything more sophisticated as it hunted large dinosaurs.

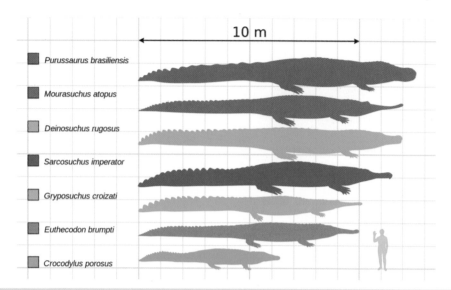

Figure 14.2 Huge fossil crocodiles were much larger than those living today. The silhouettes are placed on a meter grid. Four different genera reached over 10 meters long. From top to bottom, the crocodiles are *Purussaurus* and *Mourasuchus* from the Miocene of South America; *Deinosuchus* from the Late Cretaceous of Texas; *Sarcosuchus* ("Supercroc") from the Early Cretaceous of Africa; *Gryposuchus* from the Miocene of South America; *Euthecodon* from the Miocene of East Africa; and the largest living species, the salt-water crocodile of Australia, *Crocodylus porosus*. Diagram by Smokeybjb and placed into Wikimedia.

Figure 14.3 This ichthyosaur from the Jurassic of Britain is beautifully preserved. The bend in the tail is not a distortion of the fossil, but in life served as a strengthening structure for the lower part of the vertical tail fin. Photograph by Ballista, and placed into Wikimedia. The specimen is displayed at Dinosaurland, Lyme Regis, England, www.dinosaurland.co.uk

Figure 14.4 An ichthyosaur front limb, with the series of bones humerus, radius + ulna, wrist, hand, all modified to become part of a paddle or fin to act as a hydrofoil. Scale in cm.

Ichthyosaurs

Ichthyosaurs are not easily related to any other reptile groups, but the best guess is that they are highly derived basal diapsids. They were shaped much like dolphins, except that the tail flukes are horizontal in dolphins and vertical in ichthyosaurs. Advanced ichthyosaurs had a continuation of the spine running into the lower tail fin (Fig. 14.3). The main propulsion would then have been a side-to-side body motion, like a fish rather than a dolphin. The limbs were modified into small, stiff fins for steering and attitude control (Fig. 14.4), again like dolphins, so that

Figure 14.5 Many ichthyosaurs had big eyes that looked right along the jaw: a beautiful adaptation for sighting on a target fish. Photograph © Kevin Walsh, and used by permission.

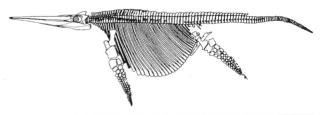

Figure 14.6 The huge Triassic ichthyosaur *Shastasaurus*. Note the powerful pectoral structure and front limbs. After Merriam.

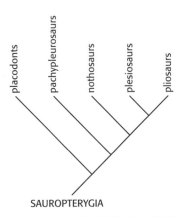

Figure 14.7 A cladogram of the sauropterygians, which included some of more spectacular marine reptiles.

ichthyosaurs would have been very maneuverable up and down in the water as well as sideways. The tail fin was usually very deep, which is characteristic of swimmers that use fast acceleration in hunting prey. Ichthyosaurs were beautifully streamlined, but would have been unable to move on land.

Beautiful ichthyosaur fossils have been known for 200 years, and they figured in many early discussions of evolutionary theory because everyone could recognize their exquisite adaptations for life in water. Ichthyosaurs all had good vision, with large eyes sighting right along the line of the jaw (Fig. 14.3, Fig. 14.5). In advanced ichthyosaurs the jaw was long and thin, with many piercing conical teeth that were well designed for catching fish. Preserved stomach contents include fish scales and hooklets from the arms of cephalopods, possibly soft-bodied squids. One spectacular Jurassic ichthyosaur, *Eurhinosaurus*, had a swordlike upper jaw projecting far beyond the lower, with teeth all along its length. It was probably an ecological equivalent of the swordfish, using its upper jaw to slash its way through a school of fish, then spinning around to catch its crippled victims.

Most early ichthyosaurs had blunt, shell-crushing teeth and may have hunted and crushed ammonites and other shelled cephalopods in a way of life that did not demand high levels of hydrodynamic performance. The earliest ichthyosaurs found to date, from the Early Triassic of the Northern Hemisphere, were small, about 1 meter (3 feet) long, but they were already specialized for marine life. *Mixosaurus* is a typical small, early ichthyosaur, from Middle Triassic rocks ranging from the Arctic to Nevada to Indonesia; but the best-preserved specimens come from the Alps. The spine had not yet turned down to form the lower tail fin, but almost all the other features show excellent adaptation to swimming, with the limbs totally modified into effective fins.

Shastasaurus from the Late Triassic, at 15 meters (50 feet) long, is one of the largest ichthyosaurs known. It was robust too, with a huge, strong deep body and long, powerful fins,

and nearly 200 vertebrae in the spine and tail (Fig. 14.6). It also had a very unusual mode of life. It had no teeth, and apparently fed by suction feeding, pulling in large quantities of water along with its prey of fishes and soft-bodied cephalopods (Sander et al. 2011). Some living whales (the beaked whales) feed in this way, using movements of a very large tongue to create the suction. The suction-feeding ichthyosaurs did not survive the end of the Triassic, but they formed an interesting evolutionary radiation, and they were the first vertebrates to evolve this unusual way of life.

Jurassic ichthyosaurs were abundant and varied, and are standard attractions at museums worldwide. But there was only one Cretaceous ichthyosaur, *Platypterygius*. It had lost the large tail for fast acceleration, and instead its limbs were modified into large fins. This suggests that it had more of a cruising style of hunting than most ichthyosaurs did. It may have used the limb fins as underwater wings for propulsion rather than steering, in the style used by sea turtles and penguins and reconstructed for some plesiosaurs.

Sauropterygians

Sauropterygians (Fig. 14.7) are a large clade of reptiles, probably descended from basal diapsids of the Permian

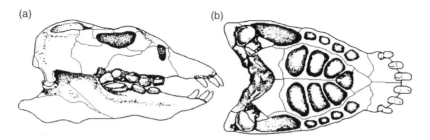

Figure 14.8 a) the skull of *Placodus*, a marine reptile from the Triassic of Germany that may have had an ecology much like the living walrus. b) the lower jaw of *Placodus*, showing the clam-crushing teeth of the palate. (What did it do with its tongue during this process?) After Broili.

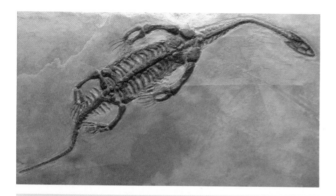

Figure 14.9 The pachypleurosaur *Keichousaurus* from the Middle Triassic of China. Note the powerful pectoral structure and front limbs. Photograph by Daderot, and placed into Wikimedia.

Figure 14.10 The nothosaur *Lariosaurus* from the Triassic of Switzerland. Photograph by Ghedoghedo and placed into Wikimedia.

(Chapter 11). Sauropterygians had unusually large limbs for land animals that had evolved toward life in water (compare crocodiles, seals, and whales). Most had small heads and comparatively long necks for their body size, so their prey (presumably fishes) must have been relatively small.

Placodonts are early but very specialized sauropterygians known only from the Triassic of Europe: in other words, they were an early offshoot from the basal part of the sauropterygian clade (Fig. 14.7). They had their own set of adaptations that may reflect a specialized ecology like the living walrus, which dives down to shallow seafloors to dig and crush clams. *Placodus* itself had unusual teeth that suited it for this way of life. The large comblike teeth at the front of the jaw (Fig. 14.8a, b) were probably used to dig into the seafloor to scoop up clams, and sediment could be washed off them by shaking the head with the mouth open. The clean clams were then crushed between flat molar teeth in the lower jaw and flat plates on the roof of the mouth (Fig. 14.8b). Placodonts did not need great maneuverability or speed, and many had heavy plated carapaces that covered them dorsally and ventrally, rather like a turtle.

Pachypleurosaurs were the simplest early sauropterygians (Fig. 14.7). They are small marine reptiles that are well known from Middle Triassic rocks of the Alps and in China. Pachypleurosaurs had thick ribs that presumably made the thorax quite stiff (Fig. 14.9). This adaptation is a solution to Carrier's Constraint (Chapter 11) in active, air-breathing swimmers. I suspect that some pachypleurosaurs swam like living monitor lizards, with propulsion from the tail, the front limbs tucked away against the rib cage, and the hind limbs used as rudders. *Keichousaurus*, however, had distinctly powerful forelimbs, set on a strong pectoral girdle (Fig. 14.9). Its hands probably made powerful paddles for swimming, but they might also have been useful for dragging the animal out onto land for breeding and for egg-laying (or even live birth).

Nothosaurs were more advanced Late Triassic sauropterygians (Fig. 14.7). All nothosaurs were large compared with their pachypleurosaur ancestors. They extended the rigid thorax of pachypleurosaurs by evolving ribs far back along the body (Fig. 14.10). With their bodies stiffened in this way, and a short tail, nothosaurs probably used their strong fore limbs for swimming power, and probably used a rowing action for propulsion. The hind limbs were not very well adapted for a swimming stroke either.

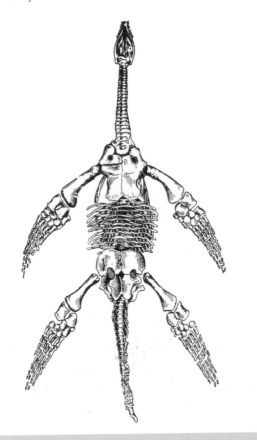

Figure 14.11 The plesiosaur *Rhomaleosaurus*, seen from underneath the skeleton. The limbs clearly dominated the swimming: but how? (After Fraas).

One nothosaur group evolved in Early Jurassic times into the largest and best-known sauropterygian clade, the **Plesiosauria**. The Plesiosauria had large bodies, and limbs that were very strong, equally well developed front and back, and highly modified for swimming. They swam with all four limbs that used the stiffened body as a solid mechanical base, in a further extension of the swimming style of nothosaurs. The limbs were strengthened and further modified for efficient swimming strokes (Fig. 14.11). The jaws have modifications that look very well evolved for fish eating.

The Plesiosauria flourished worldwide in marine ecosystems from the Early Jurassic until the end of the Cretaceous. They came in two versions, **pliosaurs** and **plesiosaurs** (Fig. 14.7). Pliosaurs had short necks and long, large heads, and they looked rather like powerful, long-headed ichthyosaurs. They swam mainly with the strong limbs, however, all four of which were large, paddle-shaped structures, shaped into effective hydrofoils. Some pliosaurs were huge: *Leiopleurodon* reached 20 meters (65 feet) long.

Plesiosaurs had the same limb structure but had very long necks and small heads. An average adult was about 3 meters (10 feet) long, with a neck that had 40 vertebrae. Some plesiosaurs were very large too. *Elasmosaurus* from

the Cretaceous of Kansas was 12 meters (40 feet) long, with 76 neck vertebrae.

Plesiosaurian limbs were jointed to massive pectoral and pelvic girdles (Fig. 14.11), presumably by very strong muscles and ligaments. Jane Robinson suggested that these structures could be explained if all four limbs were used in an up-and-down power stroke, in underwater "flying" like that of penguins—except, of course, that four limbs were involved instead of two (Robinson 1975). She realized that the plesiosaurian body had to be tightly strung with powerful ligaments to transmit the propulsion generated by the limbs to the body that they pulled through the water; and she found grooves in the skeleton where the ligaments had run (Robinson 1976).

But plesiosaurian limbs were not jointed strongly enough to the shoulder and pelvic girdles to allow strictly "flight" power strokes, and they could not have been lifted above the horizontal. They also show no sign of powerful muscle attachments. Steven Godfrey suggested instead that the propulsion stroke was downward and backward in a combination of "flying" and rowing (living sea lions swim this way). However it worked, plesiosaur swimming required precise coordination between the limb strokes.

But how did the limb strokes coordinate? Did all four limbs work in synchrony? Did the power stroke of both front limbs alternate with the power stroke of both back limbs? Or did right front and left back limb strokes coincide with left front and right back? Most people favor the first technique of synchronous strokes, which is also used by sea lions. The second option would involve a lot of stress on the trunk, which would be alternately extended and compressed if power strokes alternated between front and back limbs. The third option, however, would require only resistance to trunk twisting, which could easily be accomplished by the ligaments along the spine and those connecting the large bony masses along the underside (Fig. 14.11). This could potentially make a plesiosaurian much more maneuverable than the other techniques would.

It is difficult to envisage how plesiosaurians hunted. Perhaps, with their large heads, pliosaurs hunted large fishes at fairly high speed. But plesiosaurs are different. They have large bodies but small heads and long necks. Perhaps they stalked smaller prey and used sustained underwater "flight" mainly for migration or for cruising to feeding grounds.

Mosasaurs

Mosasaurs were essentially very large Late Cretaceous monitor lizards, up to 10 meters (30 feet) in length, the largest lizards that have ever evolved. Their evolution of aquatic adaptations in parallel with ichthyosaurs and plesiosaurs is astonishing.

Mosasaur bodies were long and powerful, with tails and limbs adapted for swimming. In early mosasaurs the main propulsion came from flexing the body and sculling with the tail, which was flat and deep, as it is in living crocodiles.

But, in addition, the limbs were modified into beautiful hydrofoils (Fig. 14.12). The elbow joint was rigid, and the shoulder joint was designed for up-and-down movement. Although the fore limbs could have given some lift, most mosasaurs probably used them as steering surfaces, as dolphins do.

Some forms like *Platecarpus*, however, had well-developed fore limbs (Fig. 14.13), and may have used them in a kind of underwater flying, like penguins. The hind limbs were like the fore limbs, though smaller, with the major muscle attachments also giving up-and-down movement. Because the pelvis was not strongly fixed to the backbone, the hind limb strokes cannot have delivered much power. The hind limbs could rotate, and probably worked like aircraft elevators to adjust pitch and roll. But the tail must have been large and powerful, because it had the backbone bent downward to strengthen the power stroke (Fig. 14.13).

Mosasaurs had long heads set on a flexible but powerful neck. The large jaws often had a hinge halfway along the lower jaw, which may have served as a shock absorber as the mosasaur hit a large fish at speed. This hinge and the powerful stabbing teeth (Fig. 14.14) suggest that most mosasaurs ate large fishes. Other mosasaurs had large, rounded, blunt teeth like those of *Placodus* (Fig. 14.8), and they probably crushed mollusc shells to reach the flesh inside.

It is rare to find fossilized skin, but the skin of the Cretaceous mosasaur *Ectenosaurus* (Fig. 14.15) gives special

Figure 14.12 The fore limb of a mosasaur evolved into an excellent swimming structure. Photograph by Dr. Mark A. Wilson of the College of Wooster, and placed into Wikimedia.

Figure 14.14 The skull of *Platecarpus*, showing its fish-stabbing teeth. Photograph by Yaakov and placed into Wikimedia.

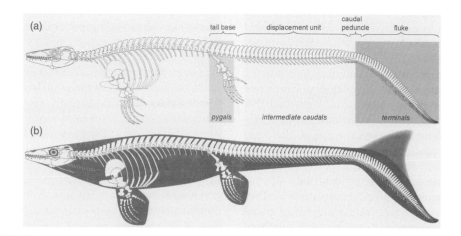

Figure 14.13 A, skeleton, and B, reconstruction of the body outline, in the Cretaceous mosasaur *Platecarpus*. From Lindgren et al. 2010, Figure 8 in PLoS One 5(8): e11998. Publication in PLoS ONE places the images into the public domain.

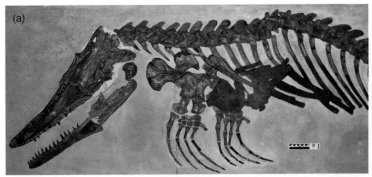

Figure 14.15 a) the Cretaceous mosasaur *Ectenosaurus* (scale in cm). b) a rare find of fossilized skin on this animal. It has diamond-shaped scales that were tightly bound to one another by strands of muscle fibers under then, helping to form a very stiff skin that probably helped fast swimming. The scales are about 2 mm × 2 mm in size. From Lindgren et al. 2011, Figure 1 and Figure 2 in PLoS ONE 6(11): e27343. Publication in PLoS ONE places the images into the public domain.

insight into its biology. The skin had diamond-shaped scales, tightly bound to one another by layers of muscle fibers. Thus *Ectenosaurus* (and possibly other mosasaurs) had a stiffened coat over the body, which would have allowed efficient water flow at speed (Lindgren et al. 2011). This is consistent with the teeth, jaws, and size of late mosasaurs, and makes them even more convincingly the ecological equivalent of the huge fish-chasing ichthyosaurs and plesiosaurs that preceded them in Cretaceous seas.

Air Breathers at Sea

All these Mesozoic reptiles were air breathers and therefore faced special problems for life in the sea. Precisely the same problems are faced today by marine mammals. The major one, of course, is the fact that air breathers must visit the surface for air, but there are also problems in introducing air-breathing young to a complex and dangerous world where they must be ready to use sophisticated skills immediately after birth.

Many marine reptiles, mammals, and birds return to the shore for reproduction. Turtles simply lay large clutches of eggs and leave them buried in the sand, a method that results in horrific mortality but has obviously worked successfully for 200 million years. Seals, sea lions, and penguins have their young on shore in safe nurseries, so that they can breathe air, be fed, and grow for a while before they take to a swimming and foraging life at sea.

But living dolphins and whales never come ashore. They have special adaptations for air breathing, breeding, giving birth and caring for the young at sea. The young are born tail first, and mothers and other related adults will push them to the surface until they learn to breathe properly. The young must be able to dive immediately to suckle, and whales feed their babies milk under high pressure. There is increasing evidence that all the major groups of Mesozoic

marine reptiles solved the same kinds of problems in spectacular fashion.

Several fossils of ichthyosaurs have been found with young preserved inside the rib cage of adults, evidence that ichthyosaurs had evolved live birth. The preserved fetuses have long, pointed jaws, showing that they would have been able to feed for themselves immediately after birth, and they were born tail first as whales are.

Mosasaurs do not look as if they would readily have come ashore to lay eggs, and no fetuses or even juveniles have been found associated with adults. However, the mosasaur pelvis is very unusual in being expanded. This may have resulted simply from adaptation to swimming, but perhaps the normal pelvis was expanded to give birth to live offspring much bigger than any normal egg. Embryos have been found inside an aigialosaur, which is a mosasauroid, a basal relative of mosasaurs proper. Therefore all mosasauroids may have had live birth, though it would be reassuring to have a lot more evidence.

The early sauropterygians had limbs that would have allowed them to haul themselves out onto a beach to lay eggs or to give birth, rather like sea lions. The fairly strong limbs of placodonts and the small body size of pachypleurosaurs make them particularly easy to imagine on the shore. However, embryos have been found inside the body of a pachypleurosaur, suggesting live birth.

Nothosaurs and plesiosaurs are usually much larger, and would have had to work much harder to drag themselves up a beach, so they were likely to have been totally seagoing, with live birth at sea as in ichthyosaurs, whales, and dolphins. This has been confirmed for one plesiosaur at least. A *Polycotylus* from the Late Cretaceous was fossilized with a large fetus inside it, certain evidence of live birth at sea (Switek 2011, commenting on research by O'Keefe and Chiappe).

All this implies that ichthyosaurs, mosasaurs, and plesiosaurs had special mechanisms for training their young to swim and feed, and it also suggests parental care on a scale comparable with that of dinosaurs (Chapter 12).

Air-breathers at sea are subject to Carrier's Constraint (Chapter 11): they cannot swim fast if they flex the body side-to-side. As in their terrestrial counterparts, marine mammals and birds do not have a problem: their bodies flex up and down as they swim. But mosasaurs, as lizards, certainly could not have swum at speed for long. As plesiosaurs evolved from nothosaurs, they also evolved stiffened trunks that avoided Carrier's Constraint, and their underwater flight is a reflection of that evolutionary breakthrough.

What about ichthyosaurs? They certainly look fast, yet their tail fin flexes sideways, and the body does not look stiff. My colleague Ryosuke Motani tells me that the size of the centers of the vertebrae imply considerable stiffness of the backbone, and that in turn implies that ichthyosaurs had solved Carrier's Constraint.

Many large and powerful swimming creatures today are warm-blooded to some extent: many sharks, tuna, and several turtles, as well as dolphins. The metabolic effort of swimming contributes to a warm body. Thus, one could guess that ichthyosaurs and plesiosaurs were warm-blooded. This and the likelihood that they had live birth does not make them mammals, but it does suggest that they were most impressive creatures.

Almost all these magnificent marine reptiles became extinct at the end of the Cretaceous, along with dinosaurs, pterosaurs, and a significant number of marine invertebrates. Only the egg-laying crocodiles and turtles have survived to give us some clues about the mode of life of large reptiles. Unfortunately, these survivors are far from being typical Mesozoic reptiles!

The Modernization of Land Plants

As plants invaded drier habitats from Devonian times onward, they evolved ways to retain water and protect their reproductive stages from drying out. The major advance was the perfection of seeds, which are fertilized embryos packed in a reasonably watertight container filled with food. The embryo can survive in suspended animation within the seed until the parent plant arranges for its dispersal. Germination can be delayed until after successful transport to a favorable location. The seedling then bursts its seed coat and grows, using the nutrition in the seed until its roots and leaves have grown large and strong enough to support and maintain the growing plant.

Seeds had evolved in Late Devonian times, and seed ferns were a successful component of Late Paleozoic floras, including the coal forests; they flourished into the Triassic. But Mesozoic gymnosperms perfected the seed system, making up 60% of Triassic and 80% of Jurassic species. Gymnosperms include conifers, cycads, and gingkos. Mesozoic forests had trees up to 60 meters (200 feet) high, forming famous fossil beds such as the Petrified Forest of Arizona. Conifers were the dominant land plants during the Jurassic and Early Cretaceous, and they are still by far the most successful of the gymnosperms. Finally, around the Jurassic-Cretaceous boundary, the flowering plants or angiosperms evolved and eventually came to dominate land floras.

Seed plant reproduction has two phases, fertilization and seed dispersal. The plant must be pollinated, and after the seed has formed it must be transported to a favorable site for germination. A major factor in the evolution of angiosperms is their manipulation of animals to do these two jobs for them.

Mesozoic Plants and Pollination

Conifers and many other plants are pollinated by wind. They produce enormous numbers of pollen grains, which are released to blow in the wind in the hope that a grain will reach the pollen receptor of a female plant of the same species. Wind pollination works, just as scattering sperm and eggs into the ocean works for many marine invertebrates. But the process looks very expensive. The pollen receptor in conifers is only about one square millimeter in area, so to achieve a reasonable probability of fertilization, the female cone must be saturated with pollen grains at a density close to one million grains per square meter.

Parent plants do some things to cut the costs of wind pollination. Male cones release pollen in dry weather in just the right wind conditions, for example (Fig. 14.16), and female cones are aerodynamically shaped to act as efficient pollen collectors. But for practical purposes, wind pollination is consistently successful only if many individuals of the same species live in closely packed groups: conifers in

Figure 14.16 Pine pollen, released in clouds when wind conditions are right. Photo by Dr. Beatriz Moisset, and placed in Wikimedia.

temperate forests or grasses in prairies and savannas. An ecological setting like a tropical rain forest, where many species have well-scattered individuals, is not the place for wind pollination.

We can imagine Jurassic floras that depended on wind pollination, with plants that produced large supplies of pollen. But insects then, as now, probably foraged for the food offered by plentiful pollen and soft, unripe female organs waiting for fertilization. We know that there were large clumsy beetles and scorpionflies in the Jurassic (Ren et al. 2009), and they probably visited plants looking for food. However, as they moved from plant to plant, they may have visited the same species frequently, collecting and transferring pollen by accident. Insects can help even by visiting one plant or one sex. In some living cycad gymnosperms, wind can only carry pollen to the surface of the female cone, but insects clustering around the cone carry it into the pollen receptors. It is therefore very likely that scorponflies and other insects were aiding in the pollination of gymnosperm plants before angiosperms even evolved (Fig. 14.17).

Over time, the plant structure may have evolved toward cooperation with insects in certain ways. Perhaps delicate

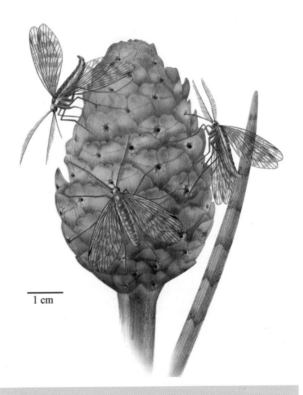

1 cm

Figure 14.17 Fossil scorpionflies have been found associated with gymnosperm cones in Jurassic sediments, suggesting that they were aiding pollination. Reconstruction by Mary Parrish under the direction of Conrad Labandeira. Courtesy of the Smithsonian Institution, and used with the permission of Conrad Labandeira and Mary Parrish.

structures were protected, but pollen was made easier to gather, and female pollen collectors were moved closer to the male pollen emitters. Such changes would have made pollen transfer by insects more likely, and less costly to the plant. Devices to attract insects—strong scents at first, then brightly colored flowers—perhaps evolved side by side with rewards such as nectar. Those plants that successfully attracted insects would have benefited by increasing their chances of fertilizing and being fertilized. Insects deliver pollen much more efficiently than wind.

An ideal pollinator should be able to exist largely on pollen and nectar, so that it can gather all its food requirements by visiting plants. It should visit as many (similar) plants as possible, so it should be small, fast-moving, and agile. A nocturnal pollinator should have a good sense of smell, and a daytime pollinator should have good vision or a good sense of smell, or both.

The only Jurassic candidates to fit this job description are insects. Birds and bats had not yet evolved, and small mammals were probably too sluggish and/or nocturnal. Insect pollinators had an increasing incentive to learn and remember certain smells and sights, and those that evolved rapid, error-free recognition of pollen sources, and clever search patterns to find them, would have become superior food gatherers and probably superior reproducers. Today, insects discriminate strongly between plant species, even between color varieties of particular species. Some insects congregate for mating around certain plant species.

Magnolias and Moths, Cycads and Beetles

The earliest known flowers, though small, had relatively large petals (Fig. 14.18), and the flowers could have produced many small seeds. Their closest living relatives are all very varied, but genetic evidence suggests that these include waterlilies, magnolias, and a family of tropical plants called Winteraceae.

Waterlilies have an intriguing pollination system. The plants bloom all summer, but there are never more than a few flowers open at once (usually only one per plant). Each flower lasts for three days in the giant water lily *Victoria amazonica*. On the first day it displays its female organs, which are white, and gives off an odor that attracts beetles (Fig. 14.19a). The odor is enhanced because the flower generates considerable heat. Beetles are trapped overnight as the flower closes, and usually the flower is pollinated. On the second morning it extends its male organs, which are pink, and any surviving beetles leave with a load of pollen. On the third day, the flower closes and its stalk bends to place it underwater. The flower develops its seeds there, and they are released when they are ripe into the water, where they float away or are spread by creatures that eat them.

No-one would seriously argue that this complex system was already present in the first waterlilies. But the **style** of pollination (tricking or trapping unwary beetles to achieve pollination) may be very ancient indeed, and perhaps a clue to the success of early angiosperms. Many living magnolias,

Figure 14.18 *Archaeanthus*, a very early angiosperm from the Early Cretaceous of China, is reconstructed with large flowers. Courtesy Professor David A. Dilcher of the University of Indiana.

Figure 14.19b Beetles congregating on a magnolia flower. Photograph by Dr. Beatriz Moisset and placed into Wikimedia.

Figure 14.19a *Victoria amazonica*, the giant waterlily. Photograph by Bilby, and placed into Wikimedia.

female angiosperm more mate choice than in other plants (remember Chapter 3). Pollen grains are haploid, so they cannot carry hidden recessive genes (as we do). A female plant with an abundant supply of pollen could in theory select certain pollen grains over others by placing chemical or physical barriers around the ovule; pollen grains that can cross the barrier are selected over others for fertilization. Experimentally, plants that are allowed to exercise pollen choice in this way have stronger offspring than others. Pollen choice may have been one of the most important factors in angiosperm success.

Of course, pollination encouraged tremendous diversity among the pollinators as they came increasingly to specialize on particular plants. The astounding rise in diversity of beetles and bees began in Cretaceous times, and there are now tens of thousands of species of each. The bees and beetles associated with angiosperms are many times more diverse than those associated with gymnosperms.

Pollination cannot be the whole story, however. Insects help to pollinate cycads too (Fig. 14.17), yet angiosperms are enormously successful while cycads have always been a relatively small group of plants. Several other Mesozoic plants experimented with ways of persuading organisms to transport pollen, and flowerlike structures evolved more than once.

Furthermore, if pollination were the key to angiosperm success, flowers could have evolved as soon as flying insects became abundant in the Late Carboniferous. There are some signs that insect pollination began then as something of a rarity. But angiosperms appeared much later and rather suddenly in the Early Cretaceous. Therefore, angiosperm success is not related simply to their evolution of flowers.

also primitive flowering plants, also have large, fragrant flowers where insects congregate to feed and mate (and pollinate) (Fig. 14.19b).

Animal pollination can deliver a large mass of pollen, rather than a few wind-blown grains. Competition between individual pollen grains to fertilize the ovule allows the

Mesozoic Plants and Seed Dispersal

If seeds fall close under the parent plant, they may be shaded so that they cannot grow, or they may be eaten by animals or birds that have learned that tasty seeds are often found under trees. Many plants rely on wind to disperse their seeds. Sometimes seeds are provided with little parachutes or airfoils to help them travel far away from the parent; winged seeds evolved almost as soon as seeds themselves, in the Late Devonian.

But seeds dispersed by wind will often fall into places that are disastrous for them. Although wind dispersal works, it seems very wasteful: it can work only in plants that produce great numbers of seeds. Wind-dispersed seeds must be light, so cannot carry much energy for seedling growth. They have to germinate in relatively well-lit areas, where the seedling can photosynthesize soon after emerging above ground.

Alternatively, a plant could have its seeds carried away by an animal and dropped into a good place for growth. Many animals can carry larger seeds than wind can, and larger seeds can successfully germinate in darker places. As with pollination, animals must be persuaded, tricked, or bribed to help in seed dispersal.

Some animals visit plants to feed on pollen or nectar, and others browse on parts of the plants. Others simply walk by the plant, brushing it as they pass. Small seeds may be picked up accidentally during such visits, especially if the seed has special hooks, burrs, or glues to help to attach it to a hairy or feathery visitor. Such seeds may be carried some distance before they fall off. Small seeds may be eaten by a visiting herbivore, but some may pass unharmed through the battery of gnawing or grinding teeth, through the gut and its digestive juices, to be automatically deposited in a pile of fertilizer.

Plants face two different problems in persuading animals to disperse seeds and in persuading them to pollinate. In pollination there is often a payment on delivery: the pollinator collects nectar or another reward as it picks up, and again as it delivers the pollen. There is no such payment on delivery of a seed. Any payment is made by the plant in advance, so that seed dispersers have no built-in payment for actual delivery of the seed. It would be better for them to cheat and to eat every seed. Thus plants often rely on tricks (burrs, for example) to fix seeds to dispersers. Velcro was evolved by plants long before the idea was copied by an astute human. Alternatively, plants may pack many small seeds into a fruit so that the disperser will concentrate on the fruit and swallow the seeds without crushing them (in strawberries, for example).

Some plants actually invite seed swallowing. They have evolved a tasty covering around the seed (a berry or fruit), and if the animal or bird eats the seed along with the fruit, every surviving seed is automatically planted in fertilizer. Tiny seeds are likely to be swallowed without being chewed, but can carry little food for the developing embryo. Large seeds loaded with nutrition are often protected by a strong seed coat or packed inside a nut.

Seed dispersal by animals is not cost free. Many dispersers eat the seeds, passing only a few unscathed through their gut. So there is a significant wastage of seeds, depending on a delicate balance between the seed coat and the teeth and stomach of the disperser. Too strong a seed coat, and the disperser will turn to easier food or germination will be too difficult; too weak a seed coat, and too many seeds will be destroyed. Some plants are so delicately adjusted to a particular disperser that the seeds germinate well only if they are eaten by that disperser.

Angiosperms evolved **carpels** as a new and unique protection for their ovules, and eventually for the developing seeds. Carpels probably evolved to protect against large, hungry insects. Soon, however, the angiosperm seed coat began to protect seeds as they passed through vertebrate guts. A seed with a strong coat was proof against many possible predators, but perhaps at the same time came to be desired food for one or a few animals that could break the seed coat. A plant could evolve to a stable relationship with a few such seed predators: the predators would receive enough food from the seeds to keep them visiting the plant regularly, but would pass enough seeds unscathed through the gut that the plant benefited too.

Seed dispersal by animals surely evolved after insect pollination. Jurassic insects may have become good pollinators, but they were too small to have been large-scale seed transporters. Jurassic reptiles were large enough, but often had low metabolic rates, so any seeds they swallowed were exposed to digestive juices for a long time. Reptiles do not even have fur in which seeds can be entangled (though feathered theropod dinosaurs might have done!).

Seeds were undoubtedly dispersed by dinosaurs to some extent, since the huge vegetarian ornithischians and sauropods ate great quantities of vegetation. But in spite of the size of the deposit of fertilizer that must have surrounded seeds passing through a dinosaur, browsing dinosaurs probably damaged and trampled plants more than they helped them. It's unlikely that any Mesozoic plant would have encouraged dinosaur browsing.

Effective transport over a long distance can take a seed beyond the range of its normal predators and diseases, and can allow a plant to become very widespread provided that there are pollinators in its new habitat. As angiosperms adapted to seed dispersal by animals, they probably dispersed into new habitats much faster than other plants. Other things being equal, we might expect a dramatic increase in the angiosperm fossil record as they adapted toward seed dispersal by animals rather than wind. (Some living angiosperms are pollinated by wind but have their seeds dispersed by animals. These include grasses, which did not evolve until well into the Cenozoic.)

There were few effective animal seed-transporters in the Jurassic, and dinosaurs are unlikely candidates in the Cretaceous. Philip Regal suggested that birds and mammals triggered the radiation of angiosperms by aiding them in seed dispersal. Birds and mammals have feathers and fur in which seeds easily become entangled; seeds pass quickly through their small bodies with their high metabolic rates and are likely to be unharmed unless they have been deliberately chewed. Angiosperm seeds would have been especially suited to vertebrate transport because of their extra protective coating. Conifer seeds are usually small and light, designed to blow in the wind, and conifers depend on close clusters for pollination. Isolated conifers are likely to be unsuccessful reproducers, and additional transport would make little difference to their long-term success.

However, the early angiosperm radiation took place in the Early and Middle Cretaceous, when mammals and birds were still minor members of the ecosystem. This early success of angiosperms may be explained better by the "fast seedling" hypothesis. This idea is based on the fact that angiosperm seeds germinate sooner, and the seedlings grow faster and photosynthesize better, than those of gymnosperms. Angiosperms may simply have outcompeted gymnosperms in the race for open spaces.

Regal's idea applies better to the later radiation of mammals and land birds in the Cenozoic, when angiosperms increased greatly in diversity, size, and abundance, and came to dominate most land floras. Today ferns are characteristic only of damp environments, conifers dominate mainly in temperate forests, and other ancient plants such as cycads and gingkos are rare.

Bruce Tiffney documented that Cenozoic angiosperm seeds are much larger than Cretaceous ones. The ability of angiosperms to become dominant forest trees in ecosystems, and their successful evolution of large-seed dispersal aided by animals and birds, were Early Cenozoic events.

The rise to dominance of the angiosperms provided a food bonanza for seed dispersers. Birds and small mammals, especially early primates and bats, all joined the seed- and fruit-eating guilds in Early Cenozoic times. Some tropical flowers today still rely for pollination on bats, small marsupials, or lemurs.

We can imagine a whole set of pollinators and seed dispersers evolving together with the plants on which they specialized. For most plants, it would be best not only to be conspicuous, but also to be different from other plant species, to encourage pollinators and seed seekers to be faithful visitors.

Suppose that particular techniques are needed to extract seeds or pollen from a plant. A visitor that learns the secret has an advantage over others and will tend to visit that species rather than foraging at random, which might require learning several collecting techniques. Fewer strangers are likely to visit the plant and rob its regular visitors of their rewards. The plant is much more likely to be fertilized or dispersed by faithful visitors than random browsers. An insect, which has only a short adult life, a limited memory, and a limited learning capacity, is more likely to be a faithful pollinator to the first plant it learns to forage from, or to the species for which it is genetically programmed. It's easy to imagine the evolution of a great variety of bright and highly scented flowers and fruits, together with a great variety of their specialized pollinators and seed seekers. Again, the evolution of the faithful visitor may have been much later than the evolution of angiosperms

themselves. Only in the Early Cenozoic do angiosperm flowers show evidence of pollination by faithful visitors such as bees, wasps, bats, and other small animals, and seed dispersal by birds, mammals, and large insects.

Living angiosperms have extraordinary devices for pollination as well as seed dispersal. One Arctic flower provides its insect pollinators with a bowl of petals that forms a perfect parabolic sun-bathing enclosure. Orchids have petals shaped and colored like female insects (Fig. 14.20), and they are pollinated by undiscriminating and optimistic males.

It's quite by accident that we happen to sense and appreciate the scents and colors of the flowers around us, because most of them were selected for the eyes and senses of insects. (We probably have color vision to help us choose between ripe and unripe fruit.) But we can gain a scientific as well as an aesthetic kick from looking at flowers if we admire their efficiency as well as their beauty.

Angiosperms and Mesozoic Ecology

The first angiosperms appeared around the Jurassic-Cretaceous boundary, but we see only their pollen. The earliest well-preserved angiosperm plants are from the famous sediments in northern China that have also yielded feathery dinosaurs and early birds. *Archaefructus* (Fig. 14.21) is preserved almost completely, and seems to be a

Figure 14.20 The fly orchid *Ophrys insectifera* has a flower that resembles a female digger wasp. Undiscriminating and optimistic male digger wasps may be fooled into trying to mate with the flower, and in the process pollinating one flower after another. Photograph by Ian Capper and placed into Wikimedia.

Figure 14.21 *Archaefructus*, a very early angiosperm from the Early Cretaceous of China. Courtesy Professor David A. Dilcher of the University of Indiana.

water-dwelling weed. There are no petals, but the plant has closed carpels with seeds inside, a classic feature of angiosperms. Cladistic analyses of *Archaefructus* place it as the most primitive as well as the earliest angiosperm. *Archaeanthus* (Fig. 14.18) is reconstructed with a large reproductive axis containing many pollen-bearing and seed-bearing organs.

Angiosperms were diverse by the Middle Cretaceous, especially in disturbed environments such as riverbanks. But how does the rise of angiosperms fit into the larger picture of Mesozoic ecology?

At the end of the Jurassic, we see a reduction of the sauropod dinosaurs that probably had been high browsers, and the rise of low-browsing ornithischians. More seedlings would now have been cropped off before reaching maturity, and any plant that could reproduce and grow quickly would have been favored.

Conifers reproduce slowly. It takes two years from fertilization until the seed is released from the cone, and wind dispersal typically does not take the seed very far. The whole reproductive system of conifers depends on wind and works best in a group situation such as a forest.

On the other hand, most angiosperms are adapted for pollination by animals, especially insects; for rapid germination and growth; and for rapid release of seeds (within the year). An angiosperm is much more likely to succeed

as a weed, rapidly colonizing any open space, and is more likely to be widely distributed because of its dispersal method. The earliest angiosperms were small, weedy shrubs, exactly the kind of plant that could survive heavy dinosaur browsing. A conifer forest, once broken up by dinosaur browsing or natural accident, would most likely have been recolonized by shrubs and weeds that could invade and grow rapidly (look at the results of clear-cutting in a conifer forest today). The weeds themselves would have reproduced quickly, so would have been more resistant to browsing than were young conifer seedlings.

Even without dinosaur browsing, angiosperms would have found habitats where they would have been very successful. In Middle Cretaceous rocks, for example, angiosperm leaves dominate sediments laid down in river levees and channels. Shifting and changing riverbank areas favor weeds because large trees are felled by storms and frequent floods. Most Middle Cretaceous pollen, on the other hand, comes from sediments laid down in lakes and near-shore marine environments. This is the windblown pollen from stable forests on the shores and on lowland plains away from violent floods, and it is dominantly conifer pollen.

Angiosperms did not take over the entire Cretaceous world, however. They were very slow to colonize high latitudes. (I suspect this reflects their greater dependence on insect pollinators, which drop off in both number and diversity in higher latitudes.)

Furthermore, Late Cretaceous fossil floras preserved in place under a volcanic ash fall in Wyoming show that even if angiosperms dominate a local flora in diversity of species, they may make up only a small percentage of the biomass. In the Big Cedar Ridge flora, angiosperms made up 61% of the species, but covered only 12% of the ground. We have to be careful in distinguishing between the diversity, the abundance, and the ecological importance of angiosperms. They cannot really be said to have dominated the ecology of any Cretaceous area.

Nevertheless, one can argue, as Bruce Tiffney and others have done, that the angiosperm radiation provided the basis for the radiations of the 5-ton ornithischians of the later Cretaceous. They seem to have lived in much larger herds than Jurassic dinosaurs, up to several thousand in the case of *Maiasaura*, and later Cretaceous dinosaurs were much more diverse as well as more abundant than their predecessors.

Ants and Termites

The success of angiosperms benefited pollinators and seed dispersers, and vice versa, and the later evolution of angiosperms was related to the ecology of large animal browsers. But today, some of the most effective tropical herbivores are leaf-cutting ants, and most terrestrial vegetation litter is broken down by termites. One-third of the animal biomass in Amazonia is made up of ants and termites. In the savannas of West Africa there are more like

2000 ants per square meter! There may be 20 million individuals in a single colony of driver ants, but the world record is held by a supercolony of ants in northern Japan, which has 300 million individuals, including a million queens, in 45,000 interconnected nests spread over 2.7 square kilometers (one square mile).

The higher social insects (bees, ants, termites, and wasps) began a major evolutionary radiation in the Late Cretaceous, as angiosperms became dominant in terrestrial ecosystems. The earliest known bee, found in Cretaceous amber from New Jersey, is a female worker bee adapted for pollen gathering. Bee society already had a sophisticated structure.

Angiosperm Chemistry

As we have seen, many angiosperms attract animals to themselves for pollination and seed dispersal. The plant usually pays a price in the production cost of substances such as nectar and in the cost of seeds eaten. Browsing animals and plant- and sap-eating insects often eat more plant material than they return in the form of services to the plant, and attracting such creatures results in a net loss of energy.

Angiosperms have therefore evolved an amazing variety of structures and chemicals that act to repel herbivores. These can be as simple and as effective as spines and stings, they can be contact irritants as in poison ivy and poison oak, or they can be severe or subtle internal poisons. Cyanide is produced by a grass on the African savanna when it is grazed too savagely. Many of our official and unofficial pharmacological agents were originally designed not for human therapy but as plant defenses. More than 2000 species of plants are insecticidal to one degree or another. Caffeine, strychnine, nicotine, cocaine, morphine, mescaline, atropine, quinine, ephedrine, digitalis, codeine, and curare are all powerful plant-derived chemicals, and it is not a coincidence that many of them are important insecticides or act strongly on the nervous, reproductive, or circulatory systems of mammals (some are even contraceptive and would act directly to decrease browsing pressure). Every day 150 million pyrethrum flowers are harvested, to fill a demand for 25,000 tons of "natural" insecticide per year. A million tons of nicotine per year were once used for insect control, until it was found that the substance was extremely toxic to mammals (self-destructive humans still smoke it!). Other plant chemicals are powerful but can be used to flavor foods in low doses. All our kitchen flavorings and spices are in this category. Garlic keeps away insects as well as vampires and friends.

For paleobiologists, the problem of angiosperm chemistry is its failure to be preserved in the fossil record. Clearly, the increasing success of angiosperms in the Late Cretaceous and Early Cenozoic occurred in the face of intense herbivory by the radiating mammals and insects of that time. The chemical defenses of angiosperms probably evolved very early in their history.

Further Reading

Marine Reptiles

Buchholtz, E. A. 2001. Swimming styles in Jurassic ichthyosaurs. *Journal of Vertebrate Paleontology* 21: 61–73.

Erickson, G. M. and C. A. Brochu 1999. How the "terror crocodile" grew so big. *Nature* 398: 205–206.

Lindgren, J. et al. 2010. Convergent evolution in aquatic tetrapods: insights from an exceptional fossil mosasaur. *PLoS ONE* 5(8): e11998. Available at http://www.plosone.org/article/info%3Adoi%2F10.1371%2Fjournal.pone.0011998

Lindgren, J. et al. 2011. Landlubbers to leviathans: evolution of swimming in mosasaurine mosasaurs. *Paleobiology* 37: 445–469.

Lindgren J. et al. 2011. Three-dimensionally preserved integument reveals hydrodynamic adaptations in the extinct marine lizard Ectenosaurus (Reptilia, Mosasauridae). *PLoS ONE* 6(11): e27343. Available at http://www.plosone.org/article/info%3Adoi%2F10.1371%2Fjournal.pone.0027343

Lingham-Soliar, T. 1992. A new mode of locomotion in mosasaurs: subaqueous flight in *Plioplatecarpus marshi*. *Journal of Vertebrate Paleontology* 12: 405–421.

Motani, R. et al. 1999. Large eyeballs in diving ichthyosaurs. *Nature* 402: 747. Available at http://mygeologypage.ucdavis.edu/motani/pdf/Motanietal1999.pdf

Motani, R. 2000. Rulers of the Jurassic seas. *Scientific American* 283 (6): 52–59. [Ichthyosaurs.]

Motani R. 2009. The evolution of marine reptiles. *Evolution: Education and Outreach* 2: 224–235. Available at http://escholarship.org/uc/item/6qf0t40w

Rieppel, O. 2002. Feeding mechanics in Triassic stem-group sauropterygians: the anatomy of a successful invasion of Mesozoic seas. *Zoological Journal of the Linnean Society* 135: 33–63. More convincing detail than I have space to discuss.

Robinson, J. A. 1975. The locomotion of plesiosaurs. *Neues Jahrbuch für Geologie und Paläontologie Abhandlungen* 149: 286–332.

Robinson, J. A. 1976. Intracorporal force transmission in plesiosaurs. *Neues Jahrbuch für Geologie und Paläontologie Abhandlungen* 153: 86–128.

Sander P. M. et al. 2011. Short-snouted toothless ichthyosaur from China suggests Late Triassic diversification of suction feeding ichthyosaurs. *PLoS ONE* 6(5): e19480. Available at http://www.plosone.org/article/info%3Adoi%2F10.1371%2Fjournal.pone.0019480

Switek, B. 2011. *Polycotylus*—the good mother plesiosaur? Blog at http://www.wired.com/wiredscience/2011/08/polycotylus-the-good-mother-plesiosaur/

Flowering Plants

Bernhardt, P. 1999. *The Rose's Kiss: A Natural History of Flowers*. Washington, D.C.: Island Press.

Bond, W. J. 1989. The tortoise and the hare: ecology of angiosperm dominance and gymnosperm persistence. *Biological Journal of the Linnean Society* 36: 227–249.

Buchmann, S. L. and G. P. Nabhan 1996. *The Forgotten Pollinators*. Washington, D.C.: Island Press. Read this book! You will look at the world around you in a different way.

Endress, P. K. 2010. The evolution of floral biology in basal angiosperms. *Philosophical Transactions of the Royal Society*

B 2010 365, 411–421. Available at http://www.ncbi.nlm.nih.gov/pmc/articles/PMC2838258/

Paxton, R. J. and Tengö, J. 2001. Doubly duped males: the sweet and sour of the orchid's bouquet. *Trends in Evolution and Ecology* 16: 167–169. Outrageous swindling of its pollinators by an orchid.

Regal, P. J. 1977. Ecology and evolution of flowering plant dominance. *Science* 196: 622–629.

Ren, D. et al. 2009. A probable pollination mode before angiosperms: Eurasian, long-proboscid scorpionflies. *Science* 326: 840–847. Available at http://www.ncbi.nlm.nih.gov/pmc/articles/PMC2944650/

Sun, G. et al. 2002. Archaefructaceae, a new basal angiosperm family. *Science* 296: 899–904, and comment, p. 821. Available at http://www.flmnh.ufl.edu/paleobotany/dldpdfs/2002sunetalarchaefructus.pdf

Tang, W. 1987. Insect pollination in the cycad *Zamia pumila* (Zamiaceae). *American Journal of Botany* 74: 90–99.

Thien, L. B. et al. 2009. Pollination biology of basal angiosperms (ANITA grade). *American Journal of Botany* 96: 166–182. Available at http://www.amjbot.org/content/96/1/166.full

Questions for Thought, Study, and Discussion

1. Choose one of the large marine reptiles that lived in Mesozoic seas, and describe the similarities and differences between that group and living whales and dolphins, in terms of anatomy, locomotion, and ecology. If you see some general principles at work, describe them.
2. The rise of the flowering plants coincided with a rise in many insect lineages. But plants and insects had both lived on Earth for many tens of millions of years. Why was special about the parallel rise in flowering plants and insects?
3. Describe the science behind this limerick:

We're proud of humanity's powers,
But these potions and medicine of ours,
Coffee, garlic, and spices
Evolved as devices
So that insects would stop bugging flowers.

The Origin of Mammals

The Derived Features of Mammals

The origin of mammals had practically no significance for Mesozoic ecology. Mammals were small, rare members of Mesozoic land communities. Yet they evolved into us and the great array of mammals that dominate the large- and small-bodied vertebrate faunas of the world today.

Living reptiles and living mammals are very different, with no surviving intermediates, and this requires us to make some mental adjustments as we try to understand how their ancestral counterparts, the diapsids and synapsids, evolved in such divergent ways in the Triassic.

Living mammals suckle their young, and they are warm-blooded: endothermic and homeothermic. They have hair,

History of Life, Fifth Edition. Richard Cowen.
© 2013 by Richard Cowen. Published 2013 by Blackwell Publishing Ltd.

not scales. They have only one bone along their lower jaw, instead of the reptilian four bones, and the jaw hinges between this lower jaw, the dentary, and the squamosal, replacing the joint of earlier synapsids (and diapsids), which had been between the articular and quadrate (Fig. 15.1).

Mammalian teeth are not replaced continuously during life. Typically, milk teeth are replaced only once, and other teeth, such as the big molars or wisdom teeth, are formed only once. Mammalian teeth meet very accurately and work very efficiently, at the cost of severe problems if teeth are damaged, lost, or worn out.

The three bones that were once in the lower jaw evolved into the middle ear of mammals, giving mammals particularly acute hearing at high frequency (squeaks and insect buzzing). In addition, the mammal brain is enlarged and specialized. The forebrain has huge lobes that wrap around older parts of the brain and contain a completely new structure, the **neocortex**, found only in mammals. The parts of the brain that are greatly increased in volume provide improved sensitivity to hearing, smell, and touch, and they are divided into the left and right lobes that psychologists talk about so much.

It is impossible to imagine all these differences arising overnight, but we can see some of them evolving gradually within the therapsids that were the ancestors of mammals. The fossil record of the transition is richest in jaws and teeth. The **dentary** bone in the therapsid jaw, originally the small section at the front, came to dominate the jawbone while other bones were reduced to little nubbins near the hinge (Fig. 15.1). The teeth became even more differentiated, and, in particular, the teeth behind the canines became larger and more complex in their shape and structure. This may suggest that tooth replacement during life became slower, but that is difficult to judge from the fossil record. Later therapsids evolved the **secondary palate**, the division between the mouth and the nasal passages that allows mammals, including us, to breathe and chew at the same time.

In terms of soft parts and thermoregulation, we have no direct evidence and must make indirect deductions. Temperature control of some sort probably evolved among therapsids long before their bony characters became mammalian.

Therapsids were abundant and diverse at the beginning of the Triassic (Chapter 10). But by the end of the Triassic the few surviving lineages of therapsids were rather small. These are the **cynodonts**, the last major therapsid group to appear. At least six separate lineages of cynodonts evolved some mammalian characters, and one of them, a small-bodied carnivorous cynodont group, evolved into the first mammals late in the Triassic. I give only a general account of the evolution of mammalian characters in cynodonts, to show that the changes were gradual ones that produced more efficient cynodonts.

Evolving Mammalian Characters

There's a paradox about the evolutionary transition from therapsid to mammal: it is very well known and complete. Everyone agrees that the therapsids are a clade, that cynodonts are a clade within therapsids, and that mammals are a clade within cynodonts (Fig. 15.2). So paleontologists argue over which particular cynodont was actually the first mammal (and which animals should be included in the clade Mammalia).

Most paleontologists use a *crown-group* definition of Mammalia: for them, the first member of Mammalia would be the latest common ancestor of all living mammals. This

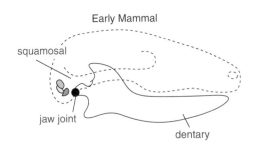
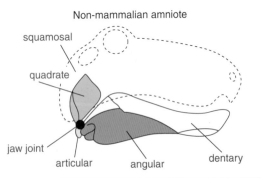

Figure 15.1 Jaw joints in a) an early mammal, and b) an earlier synapsid. The ancestral condition in amniotes is to have several bones making up the lower jaw. In mammals the lower jaw is entirely formed by the dentary, and the quadrate and articular are miniaturised into the incus and malleus and included in the middle ear. Diagram by Philcha and placed into Wikimedia.

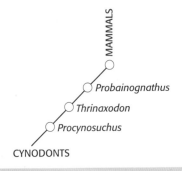

Figure 15.2 Cynodonts mentioned in the text, showing their relationship to the origin of mammals.

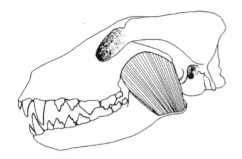

Figure 15.3 The masseter muscle is set into the angle of the lower jaw in mammals.

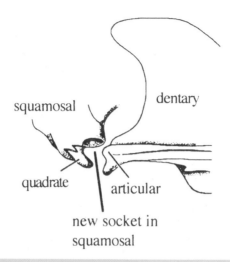

Figure 15.4 The structure of the back of the jaw in the advanced cynodont *Probainognathus*. Only a small transition would be needed to change the hinge from the articular and quadrate, as normal in early synapsids, to the dentary and squamosal, as in mammals. After Romer with permission from the Museum of Comparative Zoology, Harvard University.

definition would exclude from Mammalia a lot of Triassic and Jurassic creatures that had many "mammalian" characters, such as single lower jaw bone, jaw joint between dentary and squamosal, expanded brain, and so on. I suspect that these "nonmammals" also had fur, and suckled their young, and ecologically would have looked and behaved like living mammals. One has to grit one's teeth and place these early creatures in Mammaliaformes, a larger clade that includes the crown-group Mammalia.

Jaws

The secondary palate, which allows chewing and breathing at the same time, evolved in other therapsids as well as cynodonts. But cynodonts evolved a key innovation involving the rearrangement of the jaw: the **masseter**, a large muscle that runs from the skull under the cheekbone to the outer side of the lower jaw (Fig. 15.3). In living mammals it is the most powerful muscle that closes the jaw. (Put your fingers on the angle of your own jaw, clench your teeth and relax again, and you will feel one end of the masseter at work.) The evolution of the masseter had several important effects.

First, jaw movements were easier to control and could become more precise and complex. There was much more accurate lateral and back-and-forward movement of the lower jaw in chewing. Second, biting became more powerful. Third, the force of the bite was transmitted more directly through the teeth rather than indirectly by leverage around the jaw hinge. The lower jaw was slung in a cradle of muscles, and stresses acting on the jaw joint during chewing were much reduced.

In reptiles and in earlier therapsids, the lower jaw is made of several bones, but as chewing efficiency improved, the dentary bone, the most forward bone in the lower jaw, became the largest. The other bones became smaller and were crowded back towards the jaw joint. Stresses on the jaw joint itself were reduced as the masseter evolved, and the bones behind the dentary on each side became specialized for transmitting vibrations to the stapes rather than strengthening the back of the jawbone. Eventually the dentary became the only bone in the lower jaw, and the others became part of the ear.

As this happened, the jaw joint was gradually remodeled. In reptiles and early therapsids, the jaw is hinged between the articular and quadrate bones, but in living mammals, the jaw hinges between the dentary on the lower jaw and the squamosal bone of the upper jaw. Many people have worried about the apparent jump of the jaw joint from one pair of bones to another, since evolution is a gradual process. However, all the relevant bones in cynodont skulls were small and close together, allowing a major structural shift without major displacement of the jaw hinge (Fig. 15.4).

Probainognathus, from the Middle Triassic of South America, is very close to the cynodont ancestor of mammals (Fig. 15.2, Fig. 15.4). Later changes still needed to complete the transition included smaller size; completing the change in jaw structure to hinge only on the squamosal and dentary; completing the middle ear from the "excess" bones on each side of the lower jaw; enlarging the brain; forming definite premolars and molars in the jaw and reducing tooth replacement to only two sets of teeth; better sculpture of the molars, with the mammalian jaw movements that go with it; and changing the backbone to make it more flexible in curling up during mammalian springing and hopping. None of these changes would have been difficult or unlikely.

Teeth and Tooth Replacement

Cynodonts had teeth as well differentiated as those of many later mammals. They had complex, multi-cusped teeth behind the canines, which implies more complex food processing than in other therapsids. The jaw changes gave greater biting forces near the hinge and smaller errors in occlusion. The teeth themselves, meeting their counterparts accurately, came to be exquisitely sculptured to perform their functions precisely. Different cynodonts, presumably with different diets, evolved shearing, crushing, or shredding actions. Shearing is well seen in later carnivorous cynodonts, and there may have been limited self-sharpening of the teeth. Among herbivores, the teeth were organized for crushing; even here, slightly worn (self-wearing) opposing surfaces made a better crushing surface than new tooth surfaces. Look at a newly exposed permanent tooth of a child to see how irregular an edge it has when it first erupts.

Reptiles replace their teeth often during life, and although the process has some systematic pattern to it, any adult reptile has a mixture of larger, older teeth and smaller, newer teeth along its jaw. This means that top and bottom teeth cannot be relied upon to meet precisely against one another, so that tooth functions are comparatively crude. In advanced cynodonts, however, the jaw was slung in a rearranged set of muscles so that jaw control could be more precise; the teeth also show precise occlusion between top and bottom jaws. Tooth replacement must have been more controlled and less frequent among cynodonts than in other therapsids, and the fossil record confirms that. Cynodont teeth were replaced precisely, to maintain good occlusion of different, specialized teeth along a growing jaw. Thus the molarlike teeth of young animals were replaced by canines, while new molarlike teeth were added to the back of the jaw.

Hearing

Early cynodonts had a hearing system that transmitted ground-borne vibrations through the fore limbs and shoulder girdle to the brain, by way of the bones of the lower jaw and a massive stapes. As therapsid feeding came to emphasize chewing and slicing, it was important for teeth to be arranged all the way along the jaw (actually along the dentary bone), far back toward the hinge. The three bones on each side of the jaw behind the dentary became smaller, and so did the stapes, especially as therapsid body size became smaller. The hearing system evolved to detect and transmit airborne sound, and included former jaw bones evolved into parts of the middle ear (Fig. 15.5, Fig. 15.6).

Clearly, airborne sound became increasingly important to late cynodonts and early mammals. Perhaps they hunted insects at least partly by sound. The middle ear bones were linked to the jaw in very early mammals, but later they came to be suspended from the skull. As the hearing pathway was separated from the jaw, the mammal no longer had to listen to its own chewing so much, so would have had much better hearing. It took some time, into the Jurassic, to reorganize the other bones into the "mammalian" middle ear. Only advanced mammals evolved the complex spiral inner ear.

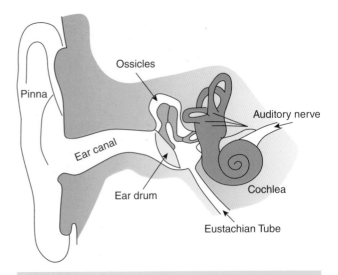

Figure 15.5 Anatomy of the ear in a living mammal (human). Incoming sound is transmitted via the eardrum and then through three small bones, two of which once formed the jaw joint in the ancestral synapsids. Diagram by Iain and placed into Wikimedia.

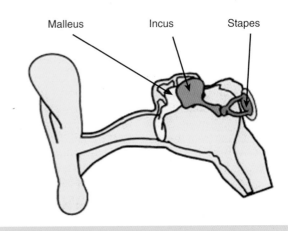

Figure 15.6 The bones of the mammal inner ear that transmit sound inward from the eardrum. United States National Cancer Institute.

Brains

The huge increase in brain size and complexity between advanced cynodonts and mammals occurred at the same time as the changes in the jaw and ear structure. Tim Rowe suggested that these changes were connected. Essentially, he said, a regulatory growth clock was reset, allowing the brain to keep growing longer than the structures around it. As the skull and jaw adjusted to accommodate a bigger brain, other changes could occur. In the living opossum, the ear bones reach adult size after three weeks, while the brain grows for twelve weeks.

Rowe's suggestion does not explain the changes in jaw and ears, but it sets up an evolutionary situation in which

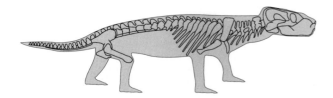

Figure 15.7 Body and skeletal outline of the early cynodont *Thrinaxodon*. Note also the expanded ribs. Based on skeletal reconstruction by Farish Jenkins.

the changes could happen. It provides an ecological and/or behavioral context in which a relatively large, more complex brain evolved, and it encourages us to ask why such a brain would have been important to a mammal.

Locomotion

Cynodonts still had *wheelbarrow locomotion* (Chapter 10): the hind limbs provided propulsion while the fore limbs gave only passive support. Cynodont hind limbs evolved to become semi-erect, whereas the fore limbs remained sprawling (Fig. 15.7). The change in the hind limbs brought the feet closer together, and the ankle changed enough to give more direct propulsion along the line of travel. Some improvement in the shoulder joints allowed better locomotion, but it was only a better wheelbarrow style. The spine shows adaptations toward greater stiffness, so that power was transmitted more efficiently from the hind limbs. Late cynodonts also evolved more flexible neck vertebrae, so that the head could swivel freely on the stiffened body. Even with these changes among cynodonts, truly erect limbs were not evolved by the first mammals but came much later.

Thermoregulation and Metabolic Level

Because their jaws and teeth show such an emphasis on efficient food processing, cynodonts probably had higher metabolic rates than pelycosaurs. This does not mean that cynodonts reached the metabolic levels of modern mammals, especially as their limbs (especially the fore limbs) were semi-erect at best. The spine of therapsids still flexed laterally rather than up and down (but note the strangely widened ribs on *Thrinaxodon* in Figure 15.7, which perhaps were retrofitted devices that cut down on lateral flexing).

Several lines of evidence suggest that therapsids, and cynodonts in particular, were evolving toward endothermy. Mammals have a diaphragm as an important part of the breathing system. This sheet of muscle forces the lungs to expand, helping respiration. A diaphragm can work only when there are no ribs around the abdomen, so its evolution can be detected in fossil vertebrates. It seems to have evolved within early cynodonts, which lost their abdominal ribs (*Thrinaxodon*, Fig. 15.7).

Primitive mammals today have comparatively low metabolic levels, and they thermoregulate at temperatures far below those of most other mammals. Therapsids were mostly medium-sized, with stocky bodies. Perhaps they operated at a body temperature of 28°–30°C (82°–86°F), a little less than primitive mammals today. In other words, they could have been moderately warm-blooded, with at least primitive thermoregulation.

Whatever therapsid body temperature was, they did not evolve great performance. They improved their breathing enough to maintain a fairly high basal metabolic rate (diaphragm, perhaps the ribs of *Thrinaxodon*), but they were not erect athletes the way that dinosaurs were, and they could not support sustained high speed because of Carrier's Constraint. Therapsids could have evolved limited endothermy without solving Carrier's Constraint. Therapsid physiology probably differed dramatically from that of living mammals, from that of living reptiles, and from that of dinosaurs.

As therapsids evolved into mammals, they became smaller. A therapsid with endothermy would have found this difficult, because small bodies lose heat faster than large ones, even with hair/fur. A possible solution is suggested by the thermal ecology of the little Australian marsupial *Pseudantechinus*, which today forages for insects at night in the Australian desert. This is not a problem in the summer, but desert temperatures at night in the winter are usually below freezing.

Pseudantechinus is so small that it cannot maintain its body temperature in freezing air. So in winter, it forages until it is cold, then goes into shelter and allows its body temperature to drop into torpor, 10°C or more below "normal". It wakes up, basks to regain body heat and digest its food, and ventures out at dusk to forage while it is still warm. This strategy may also have been used by the first mammals, until they achieved full homeothermy later in the Mesozoic.

Other Mammalian Characters

Thrinaxodon had pits on the bones in its snout, which probably contained the roots of whiskers (Fig. 15.8). That implies that it had hair, since whiskers are modified hairs. *Procynosuchus*, another early cynodont (Fig. 15.2), had lower incisors arranged in a horizontal comb. A similar arrangement occurs in living lemurs, who use the incisors to groom the fur of other members of the troop. These two observations suggest (weakly) that most or all cynodonts had hair.

Mammalian Reproduction

There are major biological differences between living reptiles and living mammals in other characters as well as the skeleton. Reptiles have larger eggs with a large energy store, and their young hatch as independent juveniles capable of

Figure 15.8 A CT scan of the skull of the early cynodont *Thrinaxodon*. The skull is about 7 cm long. The snout is pitted, which probably suggests it had whiskers (which are modified hair). This implies that many cynodonts probably had hair. Courtesy Professor Timothy Rowe and the Digimorph Project at the University of Texas at Austin.

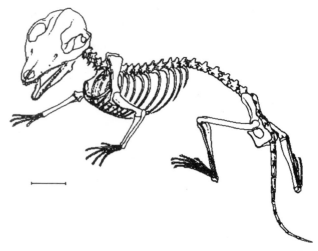

Figure 15.9 Reconstruction of an early mammaliaform. The scale bar is 2 cm. After Jenkins and Parrington.

living without parental care. Mammals have smaller eggs, and their young depend on parental care. Other major differences are physiological: most living mammals have high body temperatures and hair to insulate them, while reptiles lack hair and are cold-blooded.

Small, warm-blooded animals have a high ratio of body surface to volume, and this is especially true for young (tiny) individuals. If therapsids were warm-blooded, how did they deal with this problem for their offspring? Tiny, warm-blooded animals need very large quantities of food compared with their body size.

We can find clues from other small living warm-blooded vertebrates, the birds. Many nestling birds are helpless and cold-blooded. They depend on their parents for food and for warmth, and they have very low metabolic rates. But because they do not have to find their own food to keep warm, nestlings can devote all their food intake to rapid growth. Helpless nestlings have very large digestive tracts for their size. Warm blood, temperature control, and the ability to make coordinated movement come later and gradually. This strategy avoids the energy problem of warm blood at small size, and nestlings are essentially cold-blooded until they have grown to considerable size. Furthermore, most birds cut down environmental temperature fluctuations in their nestlings by caring for them in nests designed to maintain a uniform temperature. But the system demands intensive care by one or both parents.

Most likely, some similar strategy was followed by late therapsids and mammals, but in burrows rather than nests. The little therapsid *Diictodon* was digging burrows by the end of the Permian (Fig. 10.19). As cynodonts evolved toward very small size in the Late Triassic, the need for parental care would have become more and more acute. As the pelvis became smaller, eggs would necessarily have become smaller and smaller, with less and less yolk, and the young would have hatched earlier and been more helpless. The parents were now freed from the anatomical problem of laying large eggs, but were now committed to providing

a steady supply of food to the young after hatching, like birds and unlike most reptiles. On the other hand, smaller eggs and the rapid growth rates of helpless hatchlings gave an opportunity for very rapid reproduction in closely spaced litters (or clutches).

Suckling

Living monotremes still have the kind of reproduction that we infer for advanced cynodonts and early mammals. The platypus lays and hatches tiny eggs in a nest inside a burrow. Monotremes also nourish their hatchlings by suckling, rather than collecting food for them. This behavior has advantages: the parent does not have to leave the hatchling to search for suitable food for it, because any normal adult food can be converted into milk. The hatchling digests milk easily, and its parent is never far away, providing protection and warmth.

Charles Darwin suggested how suckling might have evolved in mammals, even before Western science discovered monotremes. His theory survives with only minor modifications. Let's assume that mammalian ancestors were already caring for eggs by incubating them. A special gland may have secreted moisture to keep the eggs humid during incubation. Hatchlings that licked the incubation gland benefited by gaining water to help deal with the food brought back by the parents, and perhaps the secretions had the added advantage of being antibacterial.

The adaptation was selective as long as the fluid helped hatchlings to survive and grow. Gradually, as the secretions came to contain mineral salts and trace elements and then nutritious organic compounds (**milk**) as well as water, the mother's excursions for food could be reduced and the hatchlings benefited even more by her increased attendance. Rapid evolution of full lactation from specialized nipples followed, with efficient suckling by the hatchlings.

The mammalian system is interesting because only the female parent is specialized to have milk glands, so that the male may take little or no role in caring for the young. Male mammals have nipples, of course, and there is no obvious biochemical reason why baby mammals should suckle only from the female, so the reason is probably genetic. The development of milk glands in mammals is controlled by a set of the Hox genes that are universal among metazoans, typically laying out nerves, vertebrae, segments, limbs and other body systems. Almost certainly, the lactation system is switched on, under genetic control, as the female goes through pregnancy and delivery. The switching system is complex, and has components from three of the four separate Hox gene clusters that mammals carry (Duboule 1999). Even in females, occasional mutations may upset this complex system. Since males do not go through pregnancy, they would never receive the signals to switch on lactation genes.

The development of suckling can be dated indirectly. Cynodonts had a secondary palate, so could chew while still breathing, but even the tiniest baby cynodont had teeth and so probably did not suckle. Perhaps the parent brought food to the nest or into the burrow. But the first mammals had very limited tooth replacement, possibly related to their small size and short lifespans, and they probably suckled in some fashion.

The evolutionary transition from licking to suckling was not as simple in baby mammals as it might seem: suckling demands full and flexible cheeks. Cheeks must have evolved, along with many other "mammalian" characters, among Triassic cynodonts. (Some sort of cheek would have been needed to cover the newly evolving masseter muscle [Fig. 15.3].)

Live Birth

Suppose mammals had reached the point of being reproductively like monotremes: they laid eggs but suckled their hatchlings. What would cause or encourage the evolution of live birth or **viviparity**?

There is nothing unusual about live birth. It has evolved independently many times in fishes, amphibians, reptiles, and mammals: in fact, in every living vertebrate group except archosaurs (birds and crocodiles). It has evolved independently in at least 90 different groups of lizards and snakes, and some insects have evolved it too. But how, why and when did it evolve in the mammal lineage?

Laying eggs is a difficult proposition below a critical body size. The egg must be laid through a pelvic opening, and a shelled egg with a reasonable amount of yolk must have a certain minimum size to be viable. Constraints on the pelvis that would forbid laying a large shelled egg may not apply to a fetus, which is structurally and physiologically more flexible than an egg. A fetus does not need a yolk or shell during its development, and it can be squeezed

through a birth canal more safely than can an eggshell. Inside a thermoregulating mother, a fetus develops at a more uniform temperature than in a nest. The growing fetus has an unlimited supply of water and oxygen, and an easy way of getting rid of CO_2 and other wastes, all of which are problems for an embryo inside an eggshell. There is far less chance of predation or infection. Finally, if suckling has already been evolved, the young never need be separated from the care and protection of the mother, even if they are helpless at birth.

Egg-laying monotremes survive today, proving that viviparity is not essential for mammals in spite of the list we have just compiled. But all other living mammals have live birth. The necessary evolutionary steps would include the gradual improvement of ways to transport material between mother and fetus, the beginnings of the placenta. The first viviparity would have been on the marsupial pattern, with or without a pouch to contain the young, but it need not have been as specialized a process as it is in living marsupials. We do not know when mammals evolved live birth, but indirect evidence suggests that it took place in the Cretaceous.

Early Mammaliaformes

Mammaliaforms were tiny, and their fossils are rare and difficult to collect except by washing and sieving enormous volumes of soft sediment. But after years of effort we now have fragments (mostly teeth) from many localities in many continents, beginning close to the Triassic-Jurassic boundary. The problem with teeth is that there are specific advantages to having particular types of teeth, and it is becoming clear that even complex tooth patterns have evolved more than once over time, much to the confusion of mammalian classification. We simply stay calm and do the best we can . . .

Among Late Triassic mammaliaforms the **morganucodonts** (Fig. 15.9), named after *Morganucodon*, could be ancestral to most later groups. Morganucodonts are fairly well known from two nearly complete skeletons found in South Africa. They were small animals, perhaps only 10 cm (4 inches) to the base of the tail, and weighing only about 25 g, about an ounce, much like modern shrews. They had small but nasty teeth and were obviously little carnivores, probably eating insects, worms, and grubs. They had relatively longer snouts and much larger brains than cynodonts. The skeletons show that they were agile climbers and jumpers. The neck was very flexible, as in living mammals, and the spine could have flexed up and down in addition to the lateral bending of therapsids.

The jaw joint was still like that of late cynodonts. But the teeth were fully differentiated, and the molars had double roots. As in most mammals today, the front teeth were replaced once, and there was only one set of molars. The molar teeth had three cusps in a line, so the name **triconodont** is used for this structure. Triconodont molars worked by shearing vertical faces up and down past one another

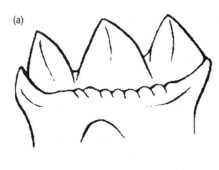

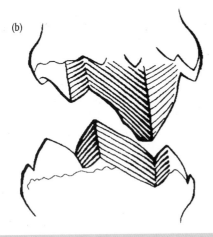

Figure 15.10 a) triconodont teeth have three cusps in a row. After Simpson. b) they had an action rather like that of pinking shears; after Jenkins.

Figure 15.11 The skull of the Jurassic mammaliaform *Hadrocodium*, so far the closest fossil to the ancestry of Mammalia. Diagram by Philcha, and placed into Wikimedia.

(Fig. 15.10), giving a zigzag cut exactly like that of pinking shears in dressmaking. This is efficient, especially for thin or soft material, but requires precise up-and-down movement. Triconodont teeth evolved more than once, confusing the picture of early mammal evolution.

A tiny fossil from the Early Jurassic of China, *Hadrocodium* (Fig. 15.11), has some advanced features of the skull, despite its early age. It is the nearest mammaliaform yet to the direct ancestry of Mammalia, the living mammals. All Jurassic mammals were small and probably nocturnal. They were carnivorous, insectivorous, or perhaps omnivo-

rous: only a few had teeth that could chew up fibrous vegetation.

Therians and Non-Therians

The three living clades of mammals are the **monotremes, marsupials,** and **placentals**. The easy distinction between them today is reproductive: monotremes lay eggs and suckle their young, while marsupials and placentals have live birth. Marsupials and placentals are classed together as therian mammals, the **Theria**. But how do we deal with their extinct ancestors, which leave us no direct clues about their reproduction?

Another shared character separates therians from monotremes: therians have **tribosphenic** molar teeth. These are complex in shape and can perform a large variety of functions as upper and lower teeth interact. They evolved from simpler teeth by adding new surfaces that shear past one another as the jaw moves sideways in a chewing motion. Tribosphenic molars are particularly well suited for puncturing and shearing, and especially for grinding, superbly fitting mammals for a diet of insects and high-protein seeds and nuts. Living monotremes do not have tribosphenic molars.

But there is a problem with this character. The tribosphenic type of molar is so efficient that teeth looking very much like them evolved independently in four different groups (Davis 2011). For example, basal monotremes evolved tribosphenic-type molars in South America (then part of Gondwana), as early as the Jurassic, though surviving monotremes today have lost them.

Other extinct clades of northern creatures were part of a Laurasian radiation, and they did not have the tribosphenic molar: they were mammals but not therians. Two of these clades were triconodonts and multituberculates. We can draw a provisional evolutionary diagram that shows these new pieces of evidence (Fig. 15.12).

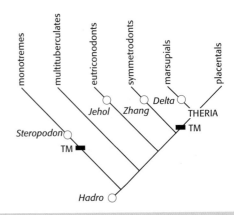

Figure 15.12 Relationships among early mammals. Jehol = *Jeholodens*; Sino = *Sinodelphys*; J = *Juramaia*. Eo = *Eomaia*; TM = tribosphenic molar.

Triconodonts were successful into Early Cretaceous times. Several well-preserved specimens come from the same remarkable rocks in China that also yielded feathered theropod dinosaurs (Chapter 12), many early birds (Chapter 13), the earliest therians, and the earliest flowering plant (Chapter 14). *Repenomamus giganticus* evolved to be the size of a raccoon, the largest Mesozoic mammal so far discovered. One specimen was found to have a baby dinosaur inside it, presumably its last meal (Fig. 15.13). *Jeholodens* was a smaller triconodont: strangely, it had sprawling hind limbs but erect fore limbs (Fig. 15.14).

Multituberculates are the most advanced nontherian mammals, and were successful in the Late Jurassic, Cretaceous, and early Cenozoic. They often make up more than half the mammals in Late Cretaceous faunas. They survived the great extinction at the end of the Cretaceous and reached their greatest diversity in the Paleocene, before being replaced by more modern mammals, especially the true rodents. Multituberculates evolved superficially rodentlike teeth, so are sometimes called the "rodents of the Mesozoic," but their ecology may not be so simple toreconstruct.

Figure 15.13 *Repenomamus*, a raccoon-sized triconodont from the Early Cretaceous of China, shown capturing a baby dinosaur. Art by Nobu Tamura, and placed into Wikimedia.

Figure 15.14 *Jeholodens*, a triconodont from the Early Cretaceous of China. It had sprawling hind limbs but erect fore limbs, a reversal from the usual structure. Art by Nobu Tamura, and placed into Wikimedia.

The incisor teeth of multituberculates were usually specialized for grasping and puncturing, rather than gnawing, but there were six in the upper jaw and only two in the lower. The very large, sharp-edged premolars were designed for holding and cutting, while the molars were grinding teeth. The system looks well suited for cropping and chewing vegetation with a back-and-forward jaw action (Fig. 15.15).

Although their radiation corresponds with the general rise of the flowering plants, many multituberculates were probably omnivores, like rats rather than guinea pigs. Specific forms can be interpreted more precisely. Some incisors were ever-growing and self-sharpening, well designed for gnawing. (Gnawing teeth may not have evolved for chewing nuts and seeds, but to open up wood to get at insects.) Other multituberculates had long, thin, saberlike incisors, like some modern insectivores that use them to impale insects. Still others probably used the shearing premolars and the crushing molars to eat fruits or seeds.

The range in tooth style and body size (mouse- to rabbit-sized) indicates a fairly wide ecological range among multituberculates. Some later forms from the Early Cenozoic were clearly tree dwellers. *Ptilodus* had a prehensile tail and squirrel-like hind feet that could rotate backwards for climbing downward (Fig. 15.16).

Kryptobataar, from the Late Cretaceous of Mongolia (Fig. 15.17), is an important multituberculate because we can say confidently that it had live birth. It had a narrow, rigid pelvis that was incapable of widening during birth. Thus the birth canal would have been at most only 3 to 4 mm wide. The animal could not have laid any reasonable-sized egg, but it could have borne a very small fetus (newborn marsupials weigh about 1 gram).

Therian Mammals

Formally, therians include **metatherians** and **eutherians**. Metatherians consist of the common ancestor of living

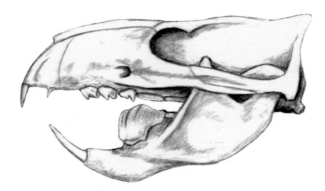

Figure 15.15 The skull of *Ptilodus*, a multituberculate from the Paleocene of North America. Art by Nobu Tamura, and placed into Wikimedia.

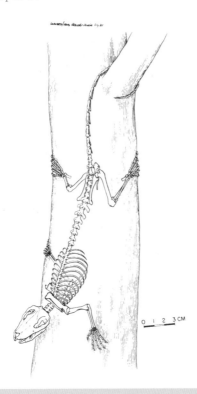

Figure 15.16 Reconstruction of *Ptilodus* as a tree-climbing multituberculate. Courtesy of Professor David Krause of Stony Brook University.

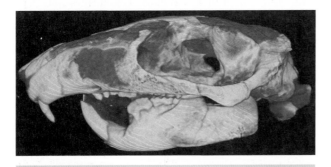

Figure 15.17 CT scan of the skull of the multituberculate mammal *Kryptobataar*, from the Cretaceous of Mongolia. Skull about 3 mm long. Courtesy Timothy Rowe and the Digimorph Project at the University of Texas at Austin.

marsupials and all its descendants; eutherians consist of the common ancestor of living placentals and all its descendants.

Early mammals probably reproduced by delivering small, helpless young once they had evolved beyond the monotreme stage of egg-laying. Mammals diverged into separate marsupial and placental clades in the Jurassic, but we do not know when their styles of reproduction diverged. Each style of reproduction, in its own way, solves some of the problems of the mammalian way of life. Marsupial and placental styles of reproduction are now quite distinct, but they probably both evolved from a state that we would now identify as simple but largely marsupial. (It's usually impossible to infer the reproductive style of any given Cretaceous mammal, especially when the pelvic regions are not well preserved.)

Even today more than 90% of all mammals weigh less than 5 kg (11 pounds) as adults. All small mammals give birth to tiny helpless young, probably because they do not have enough body volume to pack into their babies all the requirements for fully independent mammal life. Tiny mammal babies are cold-blooded at first and absolutely dependent on parental care. Mammals with large bodies can accommodate and give birth to larger, more competent offspring. This factor may have been the key to the success of the large placental mammals as opposed to large marsupials, but it doesn't apply to small placentals and marsupials.

Living marsupials bear fetuses surrounded by a membrane like the eggshell membrane of a bird or a monotreme. Its most important component is the **trophoblast**, a cell layer that allows very close contact between fetal and maternal tissue yet prevents the passage of substances that would cause the mother to reject the foreign body growing within her. Only a limited amount of nutrition can be passed to the growing embryo from the mother, and after a certain gestation time it is better for the fetus to be born so that it can take nutrition more efficiently by suckling. Living marsupials, therefore, have short gestation periods followed by long suckling periods, often with the young in a pouch.

Sometime early in the Cretaceous, a line of small mammals evolved a new derived character, the true **placenta**. This is a specialized structure built in the uterus jointly by the fetus and the mother. The placenta has an enormous surface area (fifty times the skin area of a newborn human), and it is used to supply the fetus with nutrition, oxygen, and hormones, and to pass waste products from the fetus to the mother for disposal. Essentially, the placenta is a large, discriminatory, two-way pump. The trophoblast of placental mammals is much more effective than it is in marsupials, allowing the placenta to support a growing fetus much longer. As a result, placental mammals can evolve a long gestation period, so they can have shorter lactation periods before the young reach a stage where they are independent of the mother.

Marsupials never evolved a placenta or a trophoblast as efficient as that of placental mammals, so they cannot supply the fetus with all its needs past a certain stage of development. Their trophoblast separates mother from fetus but allows only a limited range of materials to pass between them. As a result, marsupial newborns are fetuses that must be agile enough to reach the nipple.

None of this means that marsupials are inferior to placentals. A marsupial mother who experiences a natural crisis can easily abandon her young while they are fetuses, because she already carries them as an external litter. She may be ready to breed again quickly. A few placentals can absorb their fetuses, but most placental mothers must carry their internal young to term for a comparatively long gestation period, even during a flood, drought, or harsh winter, often at the risk of their own lives. Marsupial females can delay fetal development after implantation, whereas placental females rarely can.

The marsupial reproductive system stresses flexibility in the face of an unpredictable environment, so it may sometimes be superior to the placental system. Native marsupials and introduced placentals of the same body weight in the same environments in Australia (wallabies versus rabbits, for example) take on average about the same time to rear their young successfully. We still don't know the relative energy cost of the two methods. Placental and marsupial styles of reproduction, each in their own way, reduce the hazards of rearing young at small body size, but one is not always more efficient than the other.

Note that the flexibility of marsupials in abandoning their young is comparable with that of birds, who may abandon a nest in a crisis, even if there are eggs or young in it. Herons and storks will abandon a single chick if there is enough time left in the year for them to start another clutch of eggs that gives them a greater chance of rearing several chicks.

In contrast, a principle called the Concorde Fallacy seems to operate in human affairs. If a great deal has been invested in a project, then a great deal more will be invested in order to see it through to the bitter end, even after it is clear that the project will never repay its cost. The supersonic Concorde airliner was one case, but there are many others, such as the Vietnam War, the Space Shuttle, and nuclear power plants.

Animals operating under natural selection cannot afford to waste anything and must be ruthless in cutting their losses as soon as they detect eventual failure. Lions and cheetahs should (and do) abandon the chase as soon as they see they cannot catch their prey, and prospective parents should abandon their young if they cannot be reared successfully. In these terms, the allegedly superior placental reproductive system is more likely to result in wasteful expense than either the egg-laying of birds or the marsupial system. It is simply a bigger gamble than the others. In the long run, the three methods must be about equal in their results, because different animals practice them all successfully.

Other major differences between marsupials and placentals today are in thermoregulation and metabolic rate. Size for size, placentals thermoregulate at slightly higher temperatures and have slightly higher metabolic rates. They are "faster livers," as one writer has put it. This need not affect reproduction, because female marsupials increase their metabolic rate during pregnancy and lactation, up to placental levels. The brain grows faster in fetal placental mammals than in marsupials, and there is a small but significant difference in adult brain size, weight for weight, between the two groups. In turn, the metabolically active brain uses more oxygen in placentals, partly accounting for their higher energy budget. In spite of the metabolic differences, however, there is no systematic difference in at least one vitally important aspect: locomotion. Marsupials can run at about the same maximum speeds as equivalent placentals, and they have about the same stamina.

The lineages leading to marsupials and placentals diverged in the Jurassic. *Juramaia*, from the Jurassic of China at about 160 Ma, is the earliest eutherian, with important features (mostly in the teeth) only found in later placentals. In the Early Cretaceous of China at about 125 Ma, two mammals from the same formation are basal placental and marsupial mammals. *Eomaia* lies along the placental line, distinct from the marsupial lineage. It has kneecaps, for example! It was tiny, probably weighing less than 25 grams, less than an ounce (Fig. 15.18). *Sinodelphys* is about the same size as *Eomaia*, but it is unmistakably a metatherian, on the marsupial lineage. Notice that if *Juramaia* is definitely on the placental line, that there must be marsupial "ghost ancestors" for *Sinodelphys* stretching back at least 35 m.y. It underlines how rare early mammals and mammaliaforms are, compared with contemporary archosaurs.

These three earliest eutherians may have been the first mammals to explore tree-dwelling as a preferred habitat: we will need more evidence to see whether that is more than just a good idea. At first marsupials and placentals would not have been greatly different ecologically. Early placental mammals would still have had tiny, helpless young. The evolution of precocious young such as colts, calves, and fawns, which are large and can run soon after birth, had to wait until placental mammals reached large size (and probably came to live on the ground); not until then did placental mammals become more successful than marsupials in their distribution and diversity. Placental mammals may well have little or no advantage over marsupials when both are small, but large precocious

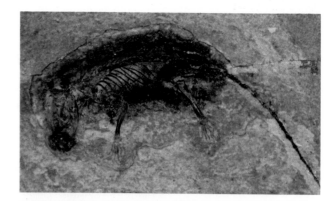

Figure 15.18 The little mammal *Eomaia* from the Cretaceous of China is an early fossil in the eutherian line: that is, the lineage leading to placental mammals. This specimen is preserved with its hair intact. Photograph by Laikayu and placed into Wikimedia.

young are not an option for marsupials, while they are for placentals.

The Inferiority of Mammals

If cynodonts were moderately warm-blooded, their evolution to smaller size would almost automatically have produced adaptations such as insulation, parental care, and so on. But why did they evolve to smaller size? Given our ideas about dinosaur biology and physiology (Chapter 12), it was probably because of competition from archosaurs, which certainly had solved Carrier's Constraint. Ecologically squeezed between the first dinosaurs (fast-moving predators with sustained running) and the small, lizardlike reptiles of the Triassic (running on cheap solar energy with a low resting metabolic rate), late cynodonts may have escaped extinction only by evolving into a habitat suitable for small, warm-blooded animals and no-one else: the night. In doing so, they underwent the radical changes in body structure, physiology, and reproduction that resulted in the evolution of mammals.

By the end of the Triassic, archosaurs had replaced and probably outcompeted the therapsids, driving them underground, deep into forests, or into nocturnal habits all over the world. And as the last few therapsids became extinct or were confined to tiny body size, the dinosaurs evolved into one of the most spectacular vertebrate groups of all time.

Burrowing in the dark, the mammals lived in a habitat that required much greater sensitivity to hearing, smell, and touch. This requirement may have selected for a relatively large, complex brain and sophisticated intelligence. So why didn't they take over the Cretaceous world? It may have depended on the competition.

With the spread of flowering plants in the Early Cretaceous, herbivorous dinosaurs, insects, and mammals all increased in diversity. The increase in food in the form of insects, seeds, nuts, and fruits provided a great ecological opportunity for small mammals. Mammals did increase in diversity through the Cretaceous, but not in a spectacular way, and probably in environments that do not yield many fossils: the forest canopy.

Forest ecosystems had flourished since Carboniferous times (Chapter 9). Mesozoic mammals, small-bodied and insectivorous, were clear candidates to invade them, but it seems as if eutherians were the first to do it. By the end of the Cretaceous, it is easy to envisage a diverse set of mammals occupying many small-bodied ways of life in the forest, particularly at night. The ancestors of primates and bats most likely evolved their special characters in the forest canopy. Small mammals are very important in the canopy even today: the equatorial forest has many species of birds active by day and mammals at night, each with a small-bodied way of life, eating insects, seeds, nuts, and fruits. But mammals simply could not compete mechanically with dinosaurs: their locomotion and probably their metabolism simply did not allow them to survive on open ground by day.

Immediately after the end of the Cretaceous and the disappearance of the dinosaurs, mammals began a tremendous radiation into all body sizes and many different ways of life. The inverse relationship between the success of Mesozoic archosaurs, especially dinosaurs, and Mesozoic therapsids and mammals is probably not a coincidence. It reflects some real inability of the mammalian lineage to compete successfully in open terrestrial environments at the time. The extinction at the Cretaceous-Tertiary boundary that finally seems to have "released" the evolutionary potential of mammals must be seen in the context of the rest of the world's life in the Mesozoic, and we shall look at that in the next chapter.

Further Reading

Davis, B. M. 2011. Evolution of the tribosphenic molar pattern in early mammals, with comments on the "dual-origin" hypothesis. *Journal of Mammalian Evolution* 18: 227–244.

Duboule, D. 1999. No milk today (my Hox have gone away). *PNAS* 96: 322–323. Short comment on an accompanying detailed paper. Available at http://www.pnas.org/content/96/2/322.full

Geiser, F. et al. 2002. Was basking important in the evolution of mammalian endothermy? *Naturwissenschaften* 89: 412–414. Yes. At least it's important for the tiny living Australian marsupial *Pseudantechinus*. Available at http://www.une.edu.au/esnrm/pdf/fritz%20geiser/Pseudantechinus%20NaWi02.pdf

Hu, Y. et al. 2005. Large Mesozoic mammals fed on young dinosaurs. *Nature* 433: 149–152. [*Repenomamus*.] Available at http://www.bi.ku.dk/dna/course/papers/J2.hu.pdf

Ji, Q. et al. 1999. A Chinese triconodont mammal and mosaic evolution of the mammalian skeleton. *Nature* 398: 326–330, and comment, pp. 283–284; also comment in *Science* 283: 1989–1990. [*Jeholodens*.]

Ji, Q. et al. 2002. The earliest known eutherian mammal. *Nature* 416: 816–822, and comment, pp. 788–789. [*Eomaia*, but it is no longer the earliest!] Available at http://www.carnegiemnh.org/assets/science/vp/JiEtAl(Eomaia-Nature).pdf

Kielan-Jaworowska, Z. et al. 1987. The origin of egg-laying mammals. *Nature* 326: 871–873.

Luo, Z-X. et al. 2001. A new mammaliaform from the Early Jurassic and evolution of mammalian characteristics. *Science* 292: 1535–1540, and comment, pp. 1496–1497. [*Hadrocodium*.] Available at http://www.oeb.harvard.edu/faculty/crompton/pdfs/luocromptonsun2001.pdf

Luo, Z-X. et al. 2002. In quest for a phylogeny of Mesozoic mammals. *Acta Palaeontologica Polonica* 47: 1–78. [Long detailed paper which explains the problems in interpreting early mammal evolution.]

Luo, Z-X. et al. 2003. An Early Cretaceous tribosphenic mammal and metatherian evolution. *Science* 302: 1934–1940, and comment, pp. 1899–1900. [*Sinodelphys*.] Available at http://www.carnegiemnh.org/assets/science/vp/Luo-et-al(2003).pdf

Luo, Z-X. et al. 2011. A Jurassic eutherian mammal and divergence of marsupials and placentals. *Nature* 476: 442–445. [*Juramaia*.] Available at http://www.dtabacaru.com/secret.pdf

Meng, J. and A. R. Wyss 1995. Monotreme affinities and low-frequency hearing suggested by multituberculate ear. *Nature* 377: 141–144, and comment, pp. 104–105. These may be ancestral similarities.

Rowe, T. 1996. Coevolution of the mammalian middle ear and neocortex. *Science* 273: 651–654. Good idea: may not be right. See Wang et al. 2001. Available at https://hpc.hamilton.edu/~lablab/Rowe_1996.pdf

Wang, Y. et al. 2001. An ossified Meckel's cartilage in two Cretaceous mammals and origin of the mammalian middle ear. *Science* 294: 257–361.

Questions for Thought, Study, and Discussion

1. As you know now, the first mammals were small (tiny), and not very fast. They must have evolved into a world that already had efficient small animals on land: lizards, for example, and other small reptiles that are now extinct. Discuss how mammals might have been able to survive and flourish in the face of this competition.

2. We humans have only two sets of teeth: "baby teeth" and "permanant teeth". Reptiles grow new teeth all their lives. Old mammals may starve to death as their teeth wear out, especially if they grind food in their molars. So why have no mammals ever evolved to replace their "permanent teeth" with another new set?

3. Describe the science behind this limerick:

   Early mammals all suckled their brood
   They breathed in and out as they chewed
   Their molar tooth facets
   Were masticatory assets
   But their locomotion was crude.

SIXTEEN
The End of the Dinosaurs

In This Chapter

The extinction of all dinosaurs (except birds), all plesiosaurs, all ichthyosaurs, all mosasaurs, and all pterosaurs at the end of the Cretaceous was a huge ecological disaster. (Many lineages survived, but the tops of the food chain were destroyed.) Thirty years ago, an astonishing new paper suggested that the extinction was caused by the impact of a 6-km wide asteroid on to Earth. There was great resistance to the idea at first, but evidence has built up now to the point where the impact and its size are not in doubt. The impact crater (in Mexico) has been found, along with dozens of sites where debris was scattered over the Earth. This chapter describes the effects of the impact, and discusses a huge volcanic eruption that took place at the same time, this one centered on India. I then try to sort through the mass of data to an explanation of the extinction.

The Extinction at the End of the Cretaceous

Almost all the large vertebrates on Earth, on land, at sea, and in the air—all dinosaurs, plesiosaurians, mosasaurs, and pterosaurs—suddenly became extinct about 65 Ma, at the end of the Cretaceous Period. At the same time, most plankton and many tropical invertebrates, especially reef-dwellers, became extinct, and many land plants were severely affected. This extinction event was recognized in 1860 by John Phillips (Fig. 6.5), and he used it to mark a major boundary in Earth's history, the Cretaceous–Tertiary or K–T boundary, and the end of the Mesozoic Era. The K–T extinctions were worldwide, affecting all the major continents and oceans. There are still arguments about just how short the extinction event was. It was certainly sudden in geological terms and may have been catastrophic by anyone's standards. And it's not just the number of genera and families that became extinct. As we saw in Chapter 6, the K–T extinction was a huge ecological disaster, second only to the Permo-Triassic extinction.

Despite the scale of the extinctions, however, we must not be trapped into thinking that the K–T boundary marked a disaster for all living things. Most major groups of organisms survived. Insects, mammals, birds, and flowering plants on land, and fishes, corals, and molluscs in the ocean went on to diversify tremendously soon after the end

History of Life, Fifth Edition. Richard Cowen.
© 2013 by Richard Cowen. Published 2013 by Blackwell Publishing Ltd.

of the Cretaceous. The K–T casualties included most of the large creatures of the time, but also some of the smallest, in particular the plankton that generate most of the primary production in the oceans.

There have been many bad theories to explain dinosaur extinctions. More bad science is described in this chapter than in all the rest of the book. For example, even in the 1980s a new book on dinosaur extinctions suggested that they spent too much time in the sun, got cataracts, and because they couldn't see very well, fell over cliffs to their doom. But no matter how convincing or how silly they are, any theory that tries to explain only the extinction of the dinosaurs ignores the fact that extinctions took place in land, sea, and aerial faunas, and were truly worldwide. The K–T extinctions were a global event, so we look for globally effective agents to explain them: geographic change, oceanographic change, climatic change, or an extraterrestrial event (Chapter 6). The most recent work on the K–T extinction has centered on two hypotheses that suggest a violent end to the Cretaceous: a large asteroid impact and a giant volcanic eruption. We have an enormous amount of data to assess these two hypotheses.

An Asteroid Impact?

An asteroid hit Earth precisely at the end of the Cretaceous. The evidence for the impact was first published by Walter Alvarez and his colleagues (Alvarez et al. 1980), who found that rocks laid down precisely at the K–T boundary contain extraordinary amounts of the metal **iridium**. It doesn't matter whether the boundary rocks were laid down on land or under the sea. In the Pacific Ocean and the Caribbean the iridium-bearing clay forms a layer in ocean sediments (Fig. 16.1); it is found in continental shelf deposits in

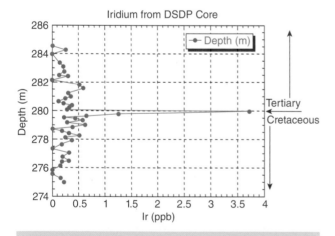

Figure 16.1   The iridium layer marking the Cretaceous-Tertiary boundary is often identified in drill holes in the ocean floor. This is a typical iridium anomaly or "spike" from a hole drilled during the Deep Sea Drilling Program. From a NASA document http:/rst.gsfc.nasa.gov/Sect18/Sect18_4.html

Europe; and in much of North America it occurs in coal-bearing rocks laid down on floodplains and deltas. The dating is precise, and the iridium layer has now been identified in several hundred places worldwide. This has allowed geologists to agree that the K–T boundary should be defined by the impact layer. Cretaceous rocks lie under it. The layer itself belongs to the Tertiary or Cenozoic (Fig. 2.6). In marine sediments, the iridium occurs just above the last Cretaceous microfossils, and the sediments above it contain Paleocene microfossils from the earliest Cenozoic.

The iridium is present only in the boundary rocks and therefore was deposited in a single large **spike**: a very short event. Iridium occurs in normal seafloor sediments in microscopic quantities, but the K–T iridium spike is very large. Iridium is rare on Earth. Chemical processes in a sediment can concentrate iridium to some extent, but the K–T iridium spike is so large that it must have arisen in some unusual way. Iridium is much rarer than gold on Earth, yet in the K–T boundary clay iridium is usually twice as abundant as gold, sometimes more than that. The same high ratio is found in meteorites. The Alvarez group therefore suggested that iridium was scattered worldwide from a cloud of debris that formed as an **asteroid** struck somewhere on Earth.

An asteroid big enough to scatter the estimated amount of iridium in the worldwide spike at the K–T boundary may have been about 10 km (6 miles) across. Computer models suggest that if such an asteroid collided with Earth, it would pass through the atmosphere and ocean almost as if they were not there and blast a crater in the crust about 100 km across. The iridium and the smallest pieces of debris would be spread worldwide by the impact blast, as the asteroid and a massive amount of crust vaporized into a fireball.

If indeed the spike was formed by a large impact, what other evidence should we hope to find in the rock record? Well-known meteorite impact structures often have fragments of **shocked quartz** (Fig. 16.2) and **spherules** (tiny glass spheres) (Fig. 16.3) associated with them. Shocked quartz is formed when quartz crystals undergo a sudden pulse of great pressure, yet do not melt. The shock causes peculiar and unmistakable microstructures. The glass spherules are formed as the target rock is melted in the impact, blasted into the air as a spray of droplets, and almost immediately frozen. Over geological time, the glass spherules may decay to clay.

All over North America, the K–T boundary clay contains glass spherules, and just above the clay is a thinner layer that contains iridium along with fragments of shocked quartz. It is only a few millimeters thick, but in total there must be more than a cubic kilometer of shocked quartz in North America alone.

The K–T impact crater has been found. It is a roughly egg-shaped geological structure called **Chicxulub**, deeply buried under the sediments of the Yucatán peninsula of Mexico (Fig. 16.4). The structure is about 180 km across, one of the largest impact structures so far identified with

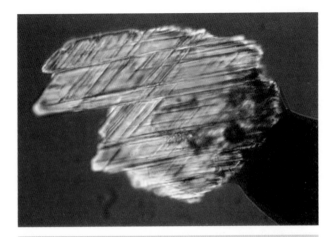

Figure 16.2 Shocked quartz: a crystal that has been caught up in a meteorite impact has characteristic shock marks in its crystal structure. From a NASA document http://rst.gsfc.nasa.gov/Sect18/Sect18_4.html

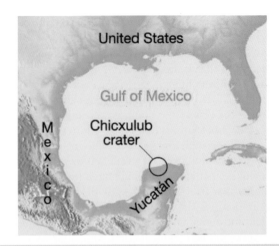

Figure 16.4 The location of the Chicxulub crater in the Yucatan peninsula of Mexico. It has very few surface indicators of its presence or its origin. It was discovered during drilling for oil, but its recognition as the K–T impact site was a stroke of insight. Every piece of evidence from the structure points to a massive impact precisely at the K–T boundary, about 65.5 Ma. From a NASA document http://rst.gsfc.nasa.gov/Sect18/Sect18_4.html

Figure 16.3 Tiny glass spherules picked out of the K–T boundary layer and glued on to a specimen card. Scale in millimeters. From a NASA document http://rst.gsfc.nasa.gov/Sect18/Sect18_4.html.

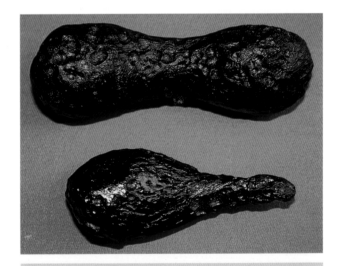

Figure 16.5 Two tektites, showing two shapes molded while the tektites were still molten and spinning through the atmosphere. These are not from the K–T boundary. Photograph by Brocken Inaglory and placed into Wikimedia.

confidence on Earth. A borehole drilled into the Chicxulub structure hit 380 meters (more than 1000 feet) of igneous rock with a strange chemistry that could have been generated by melting together a mixture of the sedimentary rocks in the region. The igneous rock contains high levels of iridium, and its age is 65.5 Ma, exactly coinciding with the K–T boundary.

On top of the igneous rock lies a mass of broken rock, probably the largest surviving debris particles that fell back on to the crater without melting, and on top of that are normal sediments that formed slowly to fill the crater in the shallow tropical seas that covered the impact area.

Well-known impact craters often have tektites associated with them as well as shocked quartz and tiny glass spher-

ules. **Tektites** are larger glass beads with unusual shapes and surface textures. They are formed when rocks are instantaneously melted and splashed out of impact sites in the form of big gobbets of molten glass, then cooled while spinning through the air (Fig. 16.5).

In Haiti, which was about 800 km from Chicxulub at the end of the Cretaceous, the K–T boundary is marked by a normal but thick (30 cm) clay boundary layer that consists mainly of glass spherules (Fig. 16.3). The clay is overlain by a layer of **turbidite**, submarine landslide material that contains large rock fragments. Some of the fragments look like shattered ocean crust, but there are also spherical pieces of yellow and black glass up to 8 mm across that are unmistakably tektites. The tektites were formed at about 1300°C from two different kinds of rock; and they are dated precisely at 65 Ma. The black tektites formed from continental volcanic rocks and the yellow ones from evaporite sediments with a high content of sulfate and carbonate. The rocks around Chicxulub are formed dominantly of exactly this mixture of rocks, and the igneous rocks under Chicxulub have a chemistry of a once-molten mixture of the two. Above the turbidite comes a thin red clay layer only about 5–10 mm thick that contains iridium and shocked quartz.

One can explain much of this evidence as follows: an asteroid struck at Chicxulub, hitting a pile of thick sediments in a shallow sea (Fig. 16.6). The impact melted much of the local crust and blasted molten material outward from as deep as 14 km under the surface. Small spherules of molten glass were blasted into the air at a shallow angle, and fell out over a giant area that extended northeast as far as Haiti, several hundred kilometers away, and to the northwest as far as Colorado. Next followed the finer material that had been blasted higher into the atmosphere or out into space and fell more slowly on top of the coarser fragments.

The egg-shape of the Chicxulub crater shows that the asteroid hit at a shallow angle, about 20°–30°, splattering more debris to the northwest than in other directions. This accounts in particular for the tremendous damage to the

Figure 16.6 The Chicxulub asteroid impact as envisaged by Donald E. Davis, 1998. Image from a NASA press release, http://www.jpl.nasa.gov/releases/98/yucatan.html, in the public domain.

North American continent, and the skewed distribution of shocked quartz far out into the Pacific.

Other sites in the western Caribbean suggest that normally quiet, deep-water sediments were drastically disturbed right at the end of the Cretaceous, and the disturbed sediments have the iridium-bearing layer right on top of them. At many sites from northern Mexico and Texas, and at two sites on the floor of the Gulf of Mexico, there are signs of a great disturbance in the ocean at the K–T boundary. In some places, the disturbed seafloor sediments contain fossils of fresh leaves and wood from land plants, along with tektites dated at 65 Ma (Fig. 16.4). Around the Caribbean and at sites up the Eastern Atlantic coast of the United States, existing Cretaceous sediments were torn up and settled out again in a messy pile that also contains glass spherules of different chemistries, shocked quartz fragments, and an iridium spike. All this implies that a series of **tsunami** or tidal waves affected the ocean margin of the time, washing fresh land plants well out to sea and tearing up seafloor sediments that had lain undisturbed for millions of years. The resulting bizarre mixture of rocks has been called "the Cretaceous-Tertiary cocktail." The tsunami were generated by the shock of the impact, which has been estimated as the equivalent of a magnitude 11 earthquake, with 1000 times more energy than any "normal" earthquake recorded on Earth (about 100 million megatons, if you like measuring in terms of hydrogen bomb blasts!).

We now have found hundreds of localities containing K–T boundary sediments, so we can put together the history of the impact, sometimes down to the minute. Larger fragments of solid rock and molten lava were blasted outward from the crater at lower angles, but not very far, and were deposited first and locally as mixed masses of rock fragments within 500 km of the crater: 100 meters of rock rubble close to the crater rim, for example. Up to about 1000 km, round the Gulf of Mexico and in the Caribbean, we find meters of rock rubble with glass spherules, and because these were coastal areas, they are mixed with tsunami debris. Up to 5000 km away, we find the glass spherule layer that was blasted out at low angles (about 15 minutes travel time to Colorado, for example), and on top of them, smaller fragments, including shocked quartz crystals and iridium, that had been lofted higher and fell more slowly (about 30 minutes to Colorado). Then over the rest of the Earth we find the bulk of the mass in the fireball that had been vaporized to form molten debris high above the atmosphere. It was deposited last, slowly drifting downward as frozen droplets and dust particles to form a thin layer, now perhaps only 2 mm thick, that is usually made of clay rich in iridium.

Some scenarios have come to be extensively quoted. Thus the dust that was blasted out into space and then fell back to Earth has been envisaged as forming millions of meteor-like trails in the atmosphere, which together would have heated the Earth's surfaces as if it was in a microwave oven. The heat was first envisaged as starting enormous forest fires worldwide, producing smoke, soot, and carbon dioxide in prodigious quantities. As the heat shock was absorbed into the ocean, it was thought that it would produce "hypercanes," gigantic hurricanes far larger than Earth's normal climate can generate. Sulfur-bearing aerosols were suggested as causing acid rain with the strength of battery acid. Overall, global ecological cycles would have been devastated, with primary production of plankton in the ocean and plants on land cutting off all food supplies to consumers on land and in the sea. Climatic damage would have heated the Earth, or cooled it, or both, further crippling biological systems.

All these scenarios now look too extreme. There was no global forest fire, for example, and there is no evidence of hypercanes. The environmental insults that followed the impact were severe enough to cause global extinctions without us exaggerating their magnitude.

A Giant Volcanic Eruption?

Exactly at the K–T boundary, a new plume (Chapter 6) was burning its way through the crust close to the plate boundary between India and Africa. Enormous quantities of basalt flooded out over what is now the Deccan Plateau of western India to form huge lava beds called the **Deccan Traps**. Some of these flows are the longest and largest ever identified on Earth (Self et al. 2008). A huge extension of that lava flow on the other side of the plate boundary now lies underwater in the Indian Ocean (Fig. 16.7). The Deccan Traps cover 500,000 sq km now (about 200,000 square miles), but they may have covered four times as much before erosion removed them from some areas. They have a surviving volume of 1 million km3 (240,000 cubic miles) and are over 2 km thick in places. The entire volcanic

volume that erupted, including the underwater lavas, was much larger than this.

The date of the Deccan eruptions cannot be separated from the K–T boundary. The peak eruptions may have lasted only about one million years, but that short time straddled the boundary. The rate of eruption was at least 30 times the rate of Hawaiian eruptions today, even assuming it was continuous over as much as a million years; if the eruption was shorter or spasmodic, eruption rates would have been much higher. The Deccan Traps probably erupted as lava flows and fountains like those of Kilauea, rather than in giant explosive eruptions like that of Krakatau. The Deccan plume is still active; its hot spot now lies under the volcanic island of Réunion in the Indian Ocean.

I discussed the Siberian Traps eruptions in Chapter 6, and the Deccan Traps eruptions of the K–T are comparable in size. We do not think the Deccan Traps were erupted through anything like the same kinds of sediments that caused so much damage at the P–Tr boundary, because the crust under India was so much older, but any eruptions of that scale must have had global effects.

Thus the K–T boundary coincided with two very dramatic events. The Deccan Traps lie across the K–T boundary and were formed in what was obviously a major event in Earth history. The asteroid impact was exactly at the K–T boundary. The asteroid impact, or the gigantic eruptions, or both, would have had major global effects on atmosphere and weather (Fig. 16.8).

How do we assess the catastrophic scenarios that have been suggested for the K–T boundary? Naturally, we compare them with the evidence from the geological record, and we look at the survivors as well as the victims.

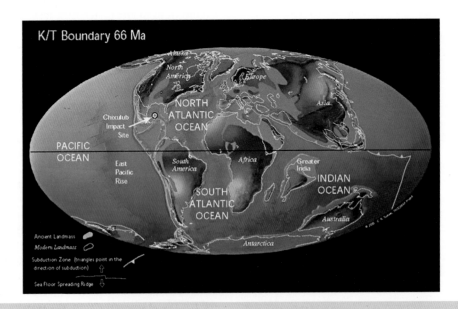

Figure 16.7   World paleogeography at the K–T boundary. Gondwana and Laurasia are split into pieces, with Australia just leaving Antarctica. The Chicxulub impact is marked, and the great eruptions took place on and between India and the East coast of Africa. Paleogeographic map by C.R. Scotese © 2012, PALEOMAP Project (www.scotese.com).

ASKHAM BRYAN
COLLEGE
LEARNING RESOURCES

Figure 16.8 Dinosaurs doomed by the Deccan Traps eruptions. This dramatic image was produced by Zina Deretzky for the National Science Foundation (in the public domain).

Birds, tortoises, and mammals live on land and breathe air: they survived the K–T boundary event.

The most persuasive (reasonable) scenarios of the K–T extinction are quickly summarized. Regionally, there is little doubt that the North American continent would have been absolutely devastated by dust, hot toxic debris, and tsunami. Globally, even a short-lived catastrophe among land plants and surface plankton at sea would drastically affect normal food chains. Pterosaurs, dinosaurs, and large marine reptiles would have been vulnerable to food shortage, and their extinction after a catastrophe seems plausible. Lizards and primitive mammals, which survived, are small and often burrow and hibernate; for a certain amount of time, they would have found plenty of nuts, seeds, insect larvae, and invertebrates buried or lying around in the dark. In the oceans, invertebrates living in shallow water along the shore would have suffered greatly from cold or frost, or perhaps from CO_2-induced heating. But deeper-water forms are insulated from heat or cold shock and have low metabolic rates; they therefore would be able to survive even months of starvation. High-latitude faunas in particular were already adapted to winter darkness, though perhaps not to extreme cold. Thus, tropical reef communities could have been decimated, but deep-water and high-latitude communities could have survived much better. All these patterns are observed at the K–T boundary.

Whatever the actual events at the time of the impact, the killing agent or agents were transient, operating for only a short time geologically. Only if there was already a major and ongoing environmental insult from the gigantic Deccan Traps eruptions could one envisage a sustained period of major crisis for life on Earth. Those are ecological scenarios that would cause the recovery to be lengthened over a period of a million years or more.

Paleobiological Evidence from the K–T Boundary

The paleontological evidence from the K–T boundary is ambiguous. Many phenomena are well explained by an impact or a volcanic hypothesis, but others are not. The fossils do provide us with real evidence about the K–T extinction events, instead of inferences from analogy or from computer models.

The best-studied terrestrial sections across the K–T boundary are in North America. Immediately this is a problem, because we know that the effects of the asteroid impact were greater here than in most parts of the world. Perhaps this has given us a more catastrophic view of the boundary event that we would gather from, say, comparable careful research in New Zealand. Even so, it is obvious that life, even in North America, was not wiped out: many plants and animals survived the K–T event.

Land Plants

North American land plants were devastated from Alberta to New Mexico at the K–T boundary. The sediments below the boundary are dominated by angiosperm pollen, but the boundary itself has little or no angiosperm pollen and instead is dominated by fern spores in a **spore spike** analogous to the iridium spike. Normal pollen counts occur immediately after the boundary layer. The spore spike therefore coincides precisely with the iridium spike in time and is equally intense and short-lived. A fern spore spike also occurs in New Zealand, suggesting that the crisis was widespread.

The spore spike could be explained by a short but severe crisis for land plants, generated by an impact or an eruption, in which all adult leaves died off (some mix of lack of light, or prolonged frost, or acid rain). Perhaps ferns were the first plants to recolonize the debris, and higher plants returned later. This happened after the eruption of Krakatau in 1883. Ferns quickly grew on the devastated island surfaces, presumably from windblown spores, but they in turn were replaced within a few decades by flowering plants as a full flora was reestablished.

Evidence from leaves confirms the data from spores and pollen. Land plants recovered from the crisis, but many Late Cretaceous plant species were killed off.

Angiosperms were in the middle of a great expansion in the Late Cretaceous (Chapter 14), and the expansion continued into the Paleocene and Eocene. Climate (and plant diversity) were fluctuating before the K–T boundary, but not to the extent that the plant crisis at the boundary can be blamed on climate. Yet there were important and abrupt changes in North American floras at the K–T boundary. In the Late Cretaceous, for example, an evergreen woodland grew from Montana to New Mexico in a seasonally dry, subtropical climate. At the boundary the dominantly evergreen Late Cretaceous woodland changed to a largely deciduous Early Cenozoic swamp woodland growing in a wetter climate. The fern spike marks a period of swampy

ASKHAM BRYAN COLLEGE LEARNING RESOURCES

mire at the boundary itself. Deciduous trees survived the K–T boundary events much better than evergreens did; in particular, species that had been more northerly spread southward. More significantly, there were very few changes in the high northern (polar) floras of Alaska and Siberia.

These changes are probably best explained by a catastrophe that wiped out most vegetation locally, with recolonization from survivors that remained safe during the crisis as seeds and spores in the soil, or even as roots and rhizomes. Other survivors came from larger refuges such as the high Arctic.

Freshwater Communities

Some ecological anomalies at the K–T boundary are not easily explained by a catastrophic scenario. Freshwater communities were less affected than terrestrial ones. For example, turtles and other aquatic reptiles survived in North Dakota while dinosaurs were totally wiped out. Freshwater communities are fueled largely by stream detritus, which includes the nutrients running off from land vegetation. It has been suggested that animals in food chains that begin with detritus rather than with primary productivity would survive a catastrophe better than others.

Unsolved Puzzles

ESD. Many living reptiles have environmental sex determination (ESD). The sex of an individual with ESD is not determined genetically, but by the environmental temperatures experienced by the embryo during a critical stage in development. Often, but not universally, the sex that is larger as an adult develops in warmer temperatures. This pattern probably evolved because, other things being equal, warmer temperatures promote faster growth and therefore larger final size (at least for ectotherms). Female turtles are larger than males because they carry huge numbers of large eggs, so baby turtles tend to hatch out as females if the eggs develop in warm places and as males in cooler places. (This makes turtle farming difficult.) Crocodiles and lizards are just the reverse. Males are larger than females because there is strong competition between males, so eggs laid in warmer places tend to hatch out as males. ESD is not found in warm-blooded, egg-laying vertebrates (birds and monotreme mammals), and it didn't occur in dinosaurs if they too were warm-blooded.

ESD is found in such a wide variety of ectothermic reptiles today that it probably occurred also in their ancestors. If so, a very sudden change in global temperature should have caused a catastrophe among ectothermic reptiles at the K–T boundary. But it did not. Crocodilians and turtles were hardly affected at all by the K–T boundary events, and lizards were affected only mildly.

High-Latitude Dinosaurs. Late Cretaceous dinosaurs lived in very high latitudes north and south, in Alaska and in South Australia and Antarctica. These dinosaurs (and their ecosystem) would have been well adapted to strong seasonal variation, including periods of darkness and very cool temperatures. An impact scenario would not easily account for the extinction of such animals at both poles, no matter what time of year the asteroid hit.

Birds. The survival of birds is the strangest of all the K–T boundary events, if we are to accept the catastrophic scenarios. Smaller dinosaurs overlapped with larger birds in size and in ecological roles as terrestrial bipeds. How did birds survive while dinosaurs did not? Birds seek food in the open, by sight; they are small and warm-blooded, with high metabolic rates and small energy stores. Even a sudden storm or a slightly severe winter can cause high mortality among bird populations. Yet an impact scenario, according to its enthusiasts, includes "a nightmare of environmental disasters, including storms, tsunamis, cold and darkness, greenhouse warming, acid rains and global fires." There must be some explanation for the survival of birds, turtles, and crocodiles through any catastrophe of this scale, or else the catastrophe models are wrong.

Where Are We?

The K–T impact was sudden, and coincided precisely with the asteroid impact. The Deccan Traps eruption was massive and geologically short-lived, though it did last perhaps a million years. The clear implication is that the asteroid impact triggered the mass extinction. Yet the eruptions were not environmentally benign. The unusual ecological severity of the K–T extinction, and its global scope, may have happened because an asteroid impact and a gigantic eruption occurred when global ecosystems were particularly vulnerable to a disturbance of oceanic stability. However, that is a difficult argument to test, so most scientists naturally stress impact as the major cause of the K–T extinction.

Steve D'Hondt has suggested that climatic change connects the impact and the extinction: the impact upset normal climate, with long-term effects that lasted much longer than the immediate and direct consequences of the impact, because of the effects of the eruptions.

There are interesting patterns among the survivors. Hardly any major groups of organisms became entirely extinct. Even the dinosaurs survived in one sense (as birds). Planktonic diatoms survived well, possibly because they have resting stages as part of their life cycle. They recovered as quickly as the land plants emerged from spores, seeds, roots, and rhizomes. The sudden interruption of the food chains on land and in the sea may well have been quite short, even if full recovery of the climate and full marine ecosystems took much longer. On one hand, climatic modellers and paleobotanists have concluded that land plants recovered to full production in perhaps ten years; yet D'Hondt and his colleagues suspect that normal surface productivity took a few thousand years to re-establish in the oceans. However, it took about three million years for the full marine ecosystem to recover, probably because so many marine predators (crustaceans, molluscs, fishes, and marine reptiles) disappeared, and had to be replaced by evolution among surviving relatives.

We still do not have a full explanation for the demise of the victims of the K–T extinction, while so many other groups survived. We will probably gain a better perspective on the K–T boundary as we gather more information about the P–Tr extinctions. Perhaps mass extinctions also require a tectonic or geographic setting that makes the global ecosystem vulnerable.

Further Reading

Alvarez, L. W. et al. 1980. Extraterrestrial cause for the Cretaceous-Tertiary extinction. *Science* 208: 1095–1108. The paper that started it all. Available at http://www2.coloradocollege.edu/dept/GY/ISES/docs/alvarez_et_al_1980.pdf

Alvarez, W. 1997. *T. rex and the Crater of Doom*. Princeton: Princeton University Press. [The best of many books on the extinction.]

Bourgeois, J. et al. 1988. A tsunami deposit at the Cretaceous–Tertiary boundary in Texas. *Science* 241: 567–570.

Bralower, T. J. et al. 1998. The Cretaceous-Tertiary boundary cocktail: Chicxulub impact triggers margin collapse and extensive sediment gravity flows. *Geology* 26: 331–334. Available starting from http://www3.geosc.psu.edu/people/faculty/personalpages/tbralower/index.html

D'Hondt, S. 2005. Consequences of the Cretaceous/Paleogene mass extinction for marine ecosystems. *Annual Review of Ecology, Evolution, and Systematics* 36: 295–317.

Goldin, T. J. and H. J. Melosh. 2009. Self-shielding of thermal radiation by Chicxulub impact ejecta: firestorm or fizzle? *Geology* 37: 1135–1138.

Hildebrand, A. R. et al. 1991. Chicxulub Crater: a possible Cretaceous/Tertiary boundary impact crater on the Yucatán Peninsula, Mexico. *Geology* 19: 867–871. The recognition of Chicxulub as the K-T asteroid crater.

Hofmann, C. et al. 2000. 40Ar/39Ar dating of mineral separates and whole rocks from the Western Ghats lava pile: further constraints on duration and age of the Deccan traps. *Earth and Planetary Science Letters* 180: 13–27. Confirms that most of the Deccan Traps were erupted within 1 m.y., at a time that cannot be distinguished from the K-T boundary. Available at http://seismo.berkeley.edu/~manga/LIPS/hofman00.pdf

Johnson, K. R. and B. Ellis. 2002. A tropical rainforest in Colorado 1.4 million years after the Cretaceous-Tertiary boundary. *Science* 296: 2379–2383. Available at ftp://ftp.soest.hawaii.edu/engels/Stanley/Textbook_update/Science_296/Johnson-02b.pdf

Jones, T. P. and B. Lim. 2002. Extraterrestrial impacts and wildfires. *Palaeogeography, Palaeoclimatology, Palaeoecology* 164, 57–66. [There was no global wildfire.]

Kring, D. A. 2007. The Chicxulub impact event and its environmental consequences at the Cretaceous–Tertiary boundary. *Palaeogeography, Palaeoclimatology, Palaeoecology* 255: 4–21. Available at http://www.ela-iet.com/EMD/Kring2007ChicxulubK-TReview.pdf

Schulte, P. et al. 2010. The Chicxulub asteroid impact and mass extinction at the Cretaceous-Paleogene boundary. *Science* 327: 1214–1217. Available at https://lirias.kuleuven.be/bitstream/123456789/264213/2/Schulte_et_al_2010(KTB-Chicxulub_SOM_Science).pdf

Self, S. et al. 2006. Volatile fluxes during flood basalt eruptions and potential effects on the global environment: a Deccan perspective. *Earth and Planetary Science Letters* 248: 518–532. Available at http://seismo.berkeley.edu/~manga/LIPS/self06.pdf

Self, S. et al. 2008. Correlation of the Deccan and Rajahmundry Trap lavas: Are these the longest and largest lava flows on Earth? *Journal of Volcanology and Geothermal Research* 172: 3–19.

Sheehan, P. M. and T. A. Hansen. 1986. Detritus feeding as a buffer to extinction at the end of the Cretaceous. *Geology* 14: 868–870.

Spicer, R.A. and A. B. Herman. 2010. The Late Cretaceous environment of the Arctic: A quantitative reassessment based on plant fossils. *Palaeogeography, Palaeoclimatology, Palaeoecology* 295: 423–442. Available at http://oro.open.ac.uk/20876/1/Spicer_and_Herman_2010.pdf

Vajda, V. and S. McLoughlin. 2004. Fungal proliferation at the Cretaceous-Tertiary boundary. *Science* 303: 1489. Available at ftp://ftp.soest.hawaii.edu/engels/Stanley/Textbook_update/Science_303/Vajda-04.pdf

Question for Thought, Study, and Discussion

One disputed calculation suggests that the asteroid that impacted the Earth at 65 m.y. ago was formed in a major collision in the "asteroid belt" sometime in the Jurassic. Tens of millions of years later it arrived in Earth's orbit and collided with it. This is about as random as geological processes get. So the question is, what difference would it have made in the great sweep of life on Earth if it has arrived say 50 million years before, or 50 million years after it did. This is a question asking for thoughtful speculation.

SEVENTEEN
Cenozoic Mammals

In This Chapter

One of the immediate results of the KT extinction was the radiation of the mammals, which included their evolution to fill some of the ecological niches vacated by the dinosaurs. I use this example to offer some general principles found in all evolutionary radiations. It's also a good opportunity to compare the results of examining the mammalian radiation by looking at the genetics of living mammals, and examining it by looking at the fossils left by early mammals. (It is sometimes difficult to reconcile the two!) Mammals radiated on continents that had been separated by continental drift, and sometimes much the same adaptations arise in parallel evolution in different lineages on different continents. I quickly review the history of the early Cenozoic.

As climate changed, so did plant life. One enlightening episode is the way that different mammals on different continents reacted to the appearance of large grasslands called savannas, with specialized grazing groups such as horses playing an important role in helping our understanding of the "Savanna Story."

Finally, as an example of major evolution among mammals, I outline the history of whales, evolving from small land-going browsers to the giant blue whale that is the largest mammal of all time. We now have enough fossils to document the transitions along the way, and to understand them in terms of the biology of the animals.

Evolution Among Cenozoic Mammals

The end of the Cretaceous Period was marked by so many changes in life on the land, in the sea, and in the air that it also marks the end of the Mesozoic Era and the beginning of the Cenozoic. Survivors of the Cretaceous extinctions radiated into a very impressive and varied set of organisms,

beginning in the Paleocene epoch, the first 10 m.y. of the Cenozoic. In the marine fossil record, the Cenozoic is dominated by molluscs, especially by bivalves and gastropods, the clams and snails of beach shell collections.

On land, the Cenozoic is marked by the dominance of flowering plants, insects, and birds, and in particular by the radiation of the mammals from insignificant little insecti-

History of Life, Fifth Edition. Richard Cowen.
© 2013 by Richard Cowen. Published 2013 by Blackwell Publishing Ltd.

vores into dominant large animals in almost all terrestrial ecosystems. Cenozoic mammals have a very good fossil record. There are thousands of well-preserved skeletons, and we understand their evolutionary history very well. I shall not try to give anything close to an overall survey of mammalian evolution. Instead, I shall use the mammal record to illustrate the ways in which evolution has acted on animals, because the same effects can be seen (more dimly) throughout the rest of the fossil record.

Evolution is the result of environmental factors acting on organisms through natural selection. But it is easier to understand evolutionary processes if we can isolate some of the different aspects involved. In this chapter and the next, I shall describe how successive groups of mammals evolved to replace dinosaurs and marine reptiles, and discuss some of the major evolutionary events of the Cenozoic. I shall look at four major aspects of evolution among Cenozoic mammals, and in each case try to identify the various opportunities that allowed or encouraged evolutionary change to occur:

- The ecological setting of evolution.
- Improving or changing well-defined adaptations.
- Geographical influences on evolution.
- Climatic influences on evolution.

Much of the turnover in the fossil record consists of the ecological replacement of one group of animals by another. An older group may disappear, for various reasons, offering an ecological opportunity for a new set of species that evolves and replaces the older set. Sometimes a new group outcompetes an older group, driving it to extinction. The replacement group often evolves much the same adaptations as its predecessor, providing wonderful examples of **parallel evolution**: certain body patterns are apparently well suited for a particular way of life, so they evolve again and again in different continents at different times. Understanding these processes helps us to sort through the complexity of catalogs of fossils.

Then we look at a smaller-scale phenomenon, **evolution by improvement**. Given that a particular body plan is well suited for executing a particular way of life, we often see changing morphology through time within a single group of organisms. These evolutionary changes can often be interpreted as a series of increasingly good adaptations to the characteristic way of life, or as a set of alternative adaptations within the general way of life. **Coevolution**, as in arms races between predator and prey, in the relationship between plant and herbivore, or between plant and pollinator, can lead to increasingly efficient adaptation. In a successful, long-lived group that survives, one can trace the various adaptations that eventually led to the derived characters of the survivors.

Obviously, one must first have a good idea of the evolutionary relationships within the group (a reliable phylogram, in other words). In almost all cases, the evolutionary and adaptive pattern of a group is not a straight line but a winding pathway through time. But the attempt to trace a lineage through the complexity of evolution can be instructive, showing how the adaptations correspond to environmental opportunities.

The Radiation of Cenozoic Mammals

The surviving major groups of terrestrial creatures after the Cretaceous extinction were mammals and birds (which are highly derived dinosaurs, of course). Crocodilians were amphibious rather than terrestrial. Most Mesozoic mammals had been small insectivores, probably nocturnal, many of them tree-dwellers or burrowers, and usually with limbs adapted for agile scurrying rather than fast running. Flying birds must be small, but there is not the same constraint on terrestrial birds. There was probably intense competition between ground-dwelling birds and mammals in a kind of ecological race for large-bodied ways of life during the Paleocene, with crocodilians playing an important secondary role in some areas. The mammals evolved explosively, their diversity rising from eight to 70 families.

Molecular Studies

The fossil record suggests that there was "explosive radiation" among mammals in the early Cenozoic. This receives a ready explanation that I used in Chapter 15: the dinosaurs had been dominant in terrestrial ecosystems, worldwide, for over 100 million years, and had effectively suppressed any ecological radiation of Mesozoic mammals. With the disappearance of the (nonbird) dinosaurs, new ecological roles suddenly became available for mammals (and birds), and dramatic adaptive radiation was a predictable response to that ecological opportunity. We see very few mammals in Cretaceous rocks, and they are all small.

But was that explosive radiation an evolutionary (genetic) explosion, or was it an ecological explosion? Perhaps the different groups of mammals had already diverged genetically, at small body size, long before the end of the Cretaceous, but radiated ecologically after the K–T extinction. (This question is exactly the same as the one we asked in Chapter 4 about the radiation of metazoans in the late Precambrian relative to the Cambrian explosion.)

How would we detect and describe a Cretaceous radiation of major mammal lineages? We could look more carefully at Cretaceous mammals, to try to find advanced characters among them. But the record is so poor that this approach has been very difficult. In any case, if the ancestors of, say, horses were mouse-sized, they would not look like, eat like, run like, or behave like horses, so they would also lack most of the skeletal characters that we use to recognize horses. Genetics is not ecology. An alternative approach is to look at **molecular evidence**.

Molecular geneticists proposed that under certain circumstances, evolutionary changes in DNA and proteins might be selectively neutral, unaffected by natural selection. Such molecular changes should happen at a random rate that is fairly constant through time. In theory, reliable **molecular clocks** of evolutionary change, based on proteins, or specific genes, or DNA sequences from the nucleus or from mitochondria, could allow us to calculate the times of divergence of living animal groups without ever having to look at, or look for, their fossil ancestors.

The concept that there should be reliable molecular clocks often disagreed with the facts of the fossil record, though it allowed geneticists to publish many papers quickly that essentially added nothing to our understanding but a lot to our confusion. Most geneticists now accept that molecular clocks do not run reliably, and have found ways to analyze their data that throws away that assumption. Their results have become much more compatible with the real fossil record, and finally genetics has come into its own as a wonderful complement to fossil morphology as we try to work out the history of groups of organisms. We shall see a particularly good example of that in the primate fossil record (Chapter 19).

So let us return to the question of the timing of the mammalian radiation. The Paleocene radiation of mammals that we see in the fossil record apparently occurred so fast that we cannot identify the very great number of branching events that resulted in the great morphological diversity of the major groups of mammals alive today. Meanwhile, molecular geneticists are painting a vivid picture of early branching events among Cretaceous placental and marsupial mammals. Are these two viewpoints compatible? The answer is yes.

Molecular Results

Stripped of "clock" assumptions, many molecular results are reasonable (that is, they agree with fossil evidence!), which gives confidence in the methods. Thus, marsupials and monotremes always fall outside the groups that form the placental mammals. However, some results within placentals came as a surprise. Some of these are very exciting, because they give insights into the mammalian radiation that had not been discovered by standard morphological comparison.

We saw in Chapter 15 that classical methods in paleontology defined a southern (Gondwana) origin of monotremes from stem mammals (mammaliaforms), and a northern (Laurasian) origin of therians (marsupials and placentals). But molecular methods have established another set of landmarks in mammal history. In the early Cretaceous, marsupials and placentals arrived in Gondwana as it was breaking up, and founded lineages, one in Africa and two in South America, that evolved in those regions separately from mammalian evolution elsewhere.

Marsupials have been established in South America for a long time, and they almost certainly reached Australia across Antarctica before the South Pole froze up.

More important, a large group of African placentals, the **Afrotheria**, forms a clade separate from other placentals. The clade today includes elephants, sea-cows, hyraxes, aardvarks, elephant shrews, and golden moles: a tremendous array of mammals with different body plans, sizes, and ecologies (Fig. 17.1). The reality of the clade is unanimously supported by all molecular evidence, yet we probably never would have discovered it by analyzing fossil skeletons. The reality of Afrotheria implies that Africa became isolated in Cretaceous times, carrying a cargo of early placentals that evolved to fill all these ecological roles, separated from evolution on other continents, in an astounding case of an adaptive radiation. I shall return to this story in Chapter 20.

Second, a South American clade of placental mammals that has long been recognized also turns out to have very deep roots. The **Xenarthra**, which includes living sloths and armadillos (and many extinct mammals) also shows prominently in molecular analyses (Fig. 17.2).

The rest of the placentals are left as a northern clade, which, on molecular evidence, splits into two groups, one the ancestors of ungulates, carnivores, and bats, the other the ancestors of primates and rodents.

When did these branching events take place? Using assumptions that do not include a "clock", the molecular results imply that the Afrotheria became separate at perhaps 105 Ma (Middle Cretaceous) and the Xenarthra perhaps 95 Ma. These estimates coincide roughly with the major break-up of Gondwana to form the southern continents, and since the molecular evidence and the geological evidence are completely independent, this again adds credibility to the analyses.

Molecular evidence suggests that the northern placentals split into their major clades during the later Cretaceous, so that perhaps twenty or so separate lineages survived the K-T extinction to radiate in the Paleocene. Many paleontologists are happy with the *pattern* of these molecular results, but are dubious about the inferences on the *timing* of the branches.

The discussion may have a simple resolution. Molecular results measure gene changes. They do not, and cannot, measure ecological changes. So most likely there were genetic branches in the Late Cretaceous, but they produced sets of Cretaceous mammals that were ecologically restricted, so would have been anatomically restricted too. There is no rule that Cretaceous mammals should look like, or have ecological roles like, their eventual descendants of today, or even of Cenozoic times. Simply stated, the major groups of placental mammals diverged from one another in the latter half of the Cretaceous. Their radiation into many families and even more genera was dominantly a Paleocene and Eocene *ecological* radiation. Nevertheless, there are Cretaceous mammals still to be discovered, interpreted, and placed into the framework; and since world geography was changing in the Cretaceous as continents moved about, there will be many opportunities to test ideas about evolutionary divergence against geological evidence.

By the end of the Cretaceous there were mammals with varied sets of genes but muted variation in morphology. The principle is clear: now the scientists involved should tone down the rhetoric and try to work out what actually happened!

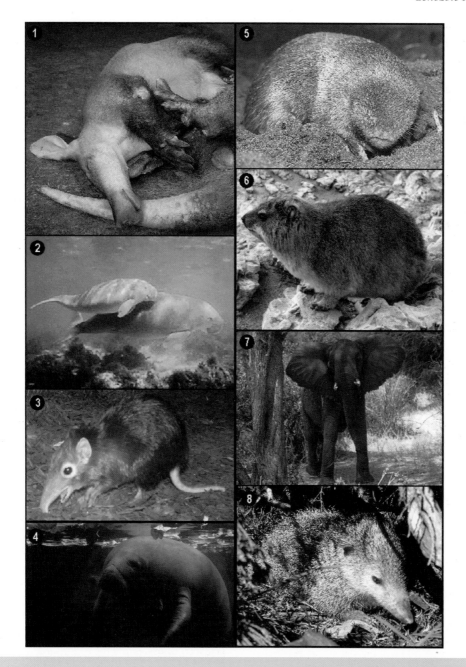

Figure 17.1 The Afrotheria, a clade of mammals that are grouped together with certainty by the DNA that they share. The skeletal and ecological differences between them are so great that we would never have identified the clade using morphological evidence. Collage of Wikimedia images assembled by Esculapio and placed into Wikimedia.

Now let us return to the fossil record, which, remember, contains many clades that played major roles in evolution and ecology, but cannot be assessed by molecular analyses because they are extinct.

The Paleocene

By Paleocene times, mammals included recognizable ancestors of a great many living groups, including marsupials, shrews, rabbits, modern carnivores, elephants, primates, whales, and hedgehogs. The ancestors of the peculiar South American fauna were already isolated on that continent.

Among all this diversity, the dominant group of Paleocene mammals was a set of generalized, rapidly evolving early "ungulates," most of them herbivores of various sizes, such as *Phenacodus* (Fig. 17.3).But arctocyonids had low, long skulls with canines and primitive molars, and were probably raccoonlike omnivores. *Chriacus* had much the same size and body plan as the tree-climbing coati, but *Arctocyon* itself was the size of a bear and probably had much the same omnivorous ecology. Mesonychids were probably otter-like carnivores or scavengers, but some of them were running predators on land. For those interested in the largest of anything, the mesonychid

Figure 17.2 The Xenarthra, a South American clade of mammals, again showing a wide variety of ecologies after a long time of radiation on the continent. Collage of Wikimedia images assembled by Xvazquez and placed into Wikimedia.

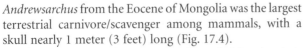

Figure 17.3 *Phenacodus*, a Paleocene mammal that probably behaved much like living sheep. Art by Heinrich Harder, in the public domain.

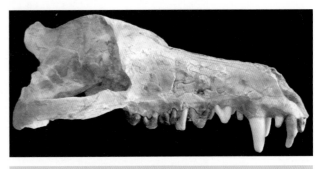

Figure 17.4 *Andrewsarchus*, the largest terrestrial mammalian scavenger/carnivore that has ever lived, a mesonychid from the Paleocene of Mongolia. The skull is about 1 meter long. Photograph by Ryan Somma, and placed into Wikimedia.

Andrewsarchus from the Eocene of Mongolia was the largest terrestrial carnivore/scavenger among mammals, with a skull nearly 1 meter (3 feet) long (Fig. 17.4).

Paleocene mammals are generally primitive in their structure, but after a drastic turnover at the end of the epoch, many new groups appeared in the Eocene that survive to the present.

The Eocene

The turnover at the end of the Paleocene is partly related to a chance event: climatic change briefly allowed free migration of mammals across the northern continents of Eurasia and North America (more detail in Chapter 18). Roughly the same fossil faunas are found across the Northern Hemisphere in North America and Eurasia. In contrast, South America, Africa and Arabia, India, and Australasia were island continents to the south of this great northern land area (Fig. 17.5).

Many modern groups of mammals appeared very early in the Eocene, including rodents, advanced primates, and modern artiodactyls and perissodactyls. There are some disputes about molecular evidence linked with this radiation. For example, molecular evidence suggests that whales form a clade with artiodactyls (antelope, cattle, deer, pigs, and so on), while carnivores are linked with perissodactyls

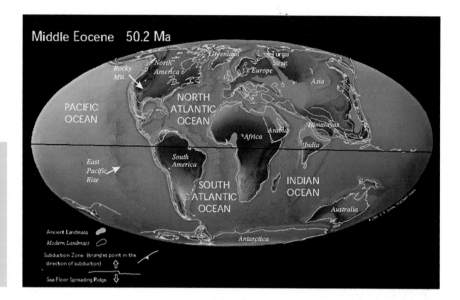

Figure 17.5 Paleogeography of the Eocene. The arrangement of the continents was an important controlling influence on the distribution of mammals (see text). Paleogeographic map by C.R. Scotese © 2012, PALEOMAP Project (www.scotese.com)

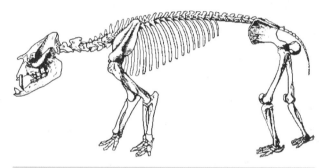

Figure 17.6 Various 5-ton vegetarian mammals evolved in the early Eocene on different continents. This is *Astrapotherium*, which lived in South America. After Riggs.

(horses, rhinos, tapirs, and so on). In contrast, zoologists and paleontologists have always linked artiodactyls and perissodactyls within a large group of herbivores, the ungulates. (If the molecular evidence holds, then "ungulates" are perhaps an ecological group, but not a clade.)

Some of the problem may lie in the fact that living mammals are merely the survivors of a massive radiation that included extinct mammals that have left us fossil skeletons but which cannot be sampled for molecular studies. These results need a lot of further analysis and debate.

By the end of the Early Eocene, digging, running, climbing, leaping, and flying mammals were well established at all available body sizes. Above all, Eocene faunas record the evolution of many different groups of mammals into herbivores of all sizes. Many of these early herbivores were small or medium-sized, including the earliest known horses, but soon there were large-bodied herbivores that ranged up to 5 tons. In North America the large herbivores were uintatheres, followed by titanotheres; in South America they were astrapotheres (Fig. 17.6); and in the Old World, especially in Africa, they were arsinoitheres.

Perissodactyls and artiodactyls appeared abruptly at the base of the Eocene in North America. Small at first, both groups evolved long, slim, stiff legs and other adaptations for fast running.

Proboscideans (elephants and related groups) and sea cows, which belong to the Afrotheria, evolved along the African shores of the tropical ocean that spread east-west between Africa and Eurasia (Fig. 17.5). Many other herbivores evolved in isolation in South America. Whales were evolving from land mammals, possibly along the southern coasts of Eurasia.

Mammals did not evolve quickly into large carnivores. Some early carnivorous mammals, the mesonychids, arctocyonids, and creodonts, were probably the equivalents in size and ecology of hyenas, coyotes, and dogs. Larger mammals seem to have been omnivores or herbivores.

On land, the carnivorous mammals were outclassed in body size in the northern continents by large, flightless birds with massive heads and impressive beaks, the diatrymas (Fig. 13.23). Similar carnivorous birds called phorusrhacids (Fig. 13.24) evolved independently in South America, and both groups were likely the dominant predators in their respective ecosystems for some time. At the same time, some crocodiles became important predators on land: for example, the pristichampsid crocodiles of Europe and North America evolved the high skulls, serrated teeth, and rounded tails of terrestrial carnivorous reptiles. One Eocene crocodile evolved hooves!

The End of the Eocene

Toward the end of the Eocene, many families on land and in the sea became extinct and were replaced by others. Naturally, the mammals that became extinct have come to be called archaic and the survivors modern, but this does not necessarily imply that there were major functional differences between them. The event has been called La

Grande Coupure, "the great cut-off," and it has been well documented in Europe and Asia. Even so, the extinction was much less abrupt than the K–T event. It was gradual rather than catastrophic and was accompanied by changes in climate and ocean currents, so agents here on Earth were probably responsible.

The Oligocene

As Antarctica became isolated and began to refrigerate, the Earth's climate began to cool on a global scale. It seems that the cooling took place in sharp steps, occasionally reversing for a while, so that there may have been a series of climatic events, each of which set up stresses on the ecosystems of the various continents. For example, a rapid cooling in southern climates in the mid-Oligocene seems to have had global effects, and there were some abrupt extinctions among North American mammals. Later events were even more severe, however.

The Later Cenozoic

In the Miocene the refrigeration of the Antarctic deepened, and its ice cap grew to a huge size, affecting the climate of the world. Vegetation patterns changed, creating more open country, and a major innovation in plant evolution produced many species of grasses that colonized the open plains. The mammals in turn responded, and a grassland ecosystem evolved on many continents, continuing with changes to the present. The Savanna Story receives separate treatment later in this chapter.

Climatic and geographical changes allowed exchanges of mammals between continents, often in pulses as opportunities occurred. A favorite example is *Hipparion*, a horse that migrated out of North America, where horses had originally evolved and spent most of their evolutionary history. It trotted across the plains of Eurasia about 11 Ma, leaving its fossils as markers of a spectacular event in mammalian history.

By the end of the Miocene, the mammalian fauna of the world was essentially modern. Two further events demand special attention: the great series of ice ages that have affected the Earth over the last few million years (Chapter 21), and the rise to dominance of animals that greatly changed the faunas and floras of Earth—humans (Chapters 20 and 21).

Ecological Replacement: The Guild Concept

Ancient mammal communities may have included some strange-looking animals, but nevertheless certain ways of life are always present in a fully evolved tropical ecosystem. Plant life is abundant and varied, and provides food for browsers and grazers, usually medium to large in size. Small animals feed on high-calorie fruits, seeds, and nuts. Pollen and nectar feeding is more likely to support really tiny animals. Carnivores range from very small consumers of insects and other invertebrates to medium-sized predators on herbivorous mammals; scavengers can be any size up to medium. There may be a few rather more specialized creatures, such as anteaters, arboreal or flying fruit-eaters, or fishing mammals.

Easily categorized ways of life that have evolved again and again among different groups of organisms are called **guilds**, and their recognition helps to make sense of some of the complexity of evolution on several continents over more than 60 m.y.

For example, the **woodpecker guild** includes many creatures that eat insects living under tree bark. Woodpeckers do this on most continents. They have specially adapted heads and beaks for drilling holes through bark (Fig. 17.7), and very long tongues for probing after insects. But there are no woodpeckers on Madagascar, where the little lemur *Daubentonia*, the aye-aye, occupies the same guild. It has ever-growing incisor teeth, like a rodent, and instead of using beak and tongue like a woodpecker, it gnaws with its teeth and probes for insects with an extremely long finger (Fig. 17.8).

On New Guinea, where there are no primates and no woodpeckers, the marsupial *Dactylopsila* has evolved specialized teeth and a very long finger for the same reasons (Fig. 17.9). Because these three species all belong to the same guild, understanding the adaptations of any one of them helps us to interpret other members. In the Galápagos Islands, the woodpecker finch *Camarhynchus* does not have a long beak but uses a tool, usually a cactus spine, to probe into crevices (Fig. 17.10). In Australia, some cockatoos fill the woodpecker niche, but they rely on the brute strength of their beaks to rip off bark, and they have not evolved the sophisticated probing devices of the others. Another small mammal evolved woodpecker devices fifty million years

Figure 17.7 A Nubian woodpecker from Kenya. Photograph by Brad Schram and placed into Wikimedia.

Figure 17.9 The marsupial *Dactylopsila* of New Guinea. Its long fingers make it a member of the woodpecker guild there. Painting by Joseph Wolf in 1858, in the public domain.

Figure 17.8 The lemur *Daubentonia*, the aye-aye of Madagascar. Its long fingers make it a member of the woodpecker guild there. Photograph of a mounted specimen by Matthias Kabel, and placed into Wikimedia.

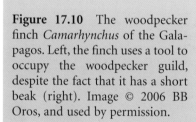

Figure 17.10 The woodpecker finch *Camarhynchus* of the Galapagos. Left, the finch uses a tool to occupy the woodpecker guild, despite the fact that it has a short beak (right). Image © 2006 BB Oros, and used by permission.

Woodpecker Finch
Camarhynchus pallidus
Gemelos, Santa Cruz, Galapagos, Ecuador
22 June 2006

ago. *Heterohyus*, from the Eocene of Germany, had powerful triangular incisor teeth, and the second and third fingers of each hand were very long, with sharp claws on the ends.

Some guilds are unexpected (to me). For example, there is a recognizable guild of small mammals that live among rocks. From marmots and rock hyraxes to chinchillas, pikas, and rock wallabies, small rock-dwelling mammals on several different continents look alike, behave alike, and even sound alike.

Of course, there is no guarantee that a guild will be occupied by only one major group. In the tropics today, small arboreal animals that feed at night are almost all mammals, but the daytime feeders are almost all birds.

Most medium sized predators and scavengers are mammals, but raptors are very effective at smaller body weights.

Cenozoic Mammals in Dinosaur Guilds

All Mesozoic mammals were small. Mammals with small bodies can play only a limited number of ecological roles, mainly insectivores and omnivores. But when dinosaurs disappeared at the end of the Cretaceous, some of the Paleocene mammals quickly evolved to take over many of their ecological roles, particularly as omnivores and vegetarians. Others continued to occupy the same small-bodied guilds that Mesozoic mammals had occupied for 100 million years. Even today, 90% of all mammal species weigh less than 5 kg (11 pounds).

Dinosaurs dominated many guilds in the Cretaceous, including that of large browsers. Most of them, such as the ceratopsians, hadrosaurs, and iguanodonts, weighed about 5 to 7 tons as adults. The K–T extinction wiped out all these creatures, and it was not until the late Paleocene that the guild was occupied again, by large mammals.

Although some birds are large herbivores (ostriches are omnivorous, but much of their food is browse), mammals are the dominant browsers and grazers today. Even at the very beginning of the Paleocene, the mammals were dominated not by insectivores but by the largely herbivorous early ungulates. Very late in the Cretaceous, some mammals had evolved molars even more complex than tribosphenic molars. The new teeth permitted or even required complex jaw motions, but they allowed much more shearing and grinding than before. The capacity for grinding more and tougher food allowed mammals to turn to low-calorie vegetarian diets.

There seems to be something special about the 5- to 7-ton range for large land herbivores. This limit applied to all dinosaurs except for sauropods, and it has apparently applied to almost all mammals since, including living elephants and rhinos. Presumably there is some metabolic reason for this limit, associated with the fact that browse and forage is low in calories. The 5- to 7-ton size was approached by different mammalian groups in the different continents of the Paleocene and Eocene (Fig. 17.5). The best record is in North America, where uintatheres and titanotheres followed the dinosaurs.

Uintatheres were most successful in the northern continents. They had massive skeletons and gradually increased in size through the Paleocene and Eocene. *Uintatherium* itself (Fig. 17.11) was as large as a rhino. They had large canine teeth modified into cutting sabers, but they were not carnivores. The large flattened molar teeth were used for grinding vegetation. (The sabers were probably for fighting between adults: compare some of the large-bodied synapsids, Fig. 10.14 and Fig. 10.15).

These large herbivores were replaced in the large herbivore guild in North America, and later in Asia, by perissodactyls called **titanotheres** (or brontotheres). These were small in the Early Eocene, but by the Middle Eocene they were large, and at the end of the Eocene they were very large indeed (Fig. 17.12, Fig. 17.13). They evolved massive blunt horns as they evolved to larger body size. The horns have been interpreted as ramming devices, but most of them have a shape and a position on the head that would have been much better designed for pushing and wrestling (Fig. 17.13). Titanotheres became extinct at the end of the Eocene, and their guild was filled by modern-looking rhinos in Eurasia and North America. Later, rhinos were joined in the guild by elephants, which had evolved in Africa but did not leave that continent until the Miocene.

The Savanna Story: Modern Savannas

Research by Samuel McNaughton and his colleagues on the savanna grazing ecosystem of East Africa revealed patterns that may also be true for other ecosystems in space and time.

Herbivores tend to graze off the tops of any plants they can reach, because the top of the plant contains the most tender, juicy parts, and is less well protected by any mineral or chemical compounds the plant produces. Grazing thus promotes the survival and evolution of plants that tend to grow sideways rather than upward. If grazing is continuous, such plants are selected because they lose less of their foliage. They are not shaded out by competitors that grow upward, because the grazing animals remove those competitors. Low plants tend to occupy a smaller area than high plants, so there is space for more plants in a grazed environment. This may often translate into more species as well as more individuals of one species.

Figure 17.11 *Uintatherium*, which gives its name to the uintatheres, a large clade of large Paleocene and Eocene North American herbivores. Reconstruction by Bob Guiliani. © Dover Publications Inc., New York. Used by permission.

Figure 17.12 The large Eocene titanothere *Brontops*. Reconstruction by Bob Guiliani. © Dover Publications Inc., New York. Used by permission.

(a)

(b)

Figure 17.13 Titanotheres evolved to very large body size between the Early Eocene and the Early Oligocene. a) an early small titanothere, *Eotitanops* (i), and b) a gigantic late one, *Brontotherium* (ii) in the act of displaying. The lower diagram shows *Brontotherium* walking naturally with the head held low. The huge double horns look to me like wrestling structures rather than ramming devices. (From Osborn.)

McNaughton fenced off savanna areas to protect them from grazing. It turned out that grazed areas actually had much more available vegetation per cubic centimeter than fenced areas. Plants that are not grazed grow tall and airy, not low and bunched. This happens on lawns too, where mowing is artificial grazing. In areas that are grazed, therefore, food resources are densely packed. A grazing animal can get more food per bite than in ungrazed areas, and it feeds more efficiently.

For example, a cow needs a certain level of nutrition per mouthful in order to survive, considering the energy that is required to move, bite, chew, and digest that mouthful. If the cow lived on the Serengeti Plains of East Africa, it could not survive if it had to crop grassland that had grown more than about 40 cm (16 inches) high, but the same environment, already grazed down to 10 cm (4 inches) high, provides a very rich food supply.

Grazed plants react in more sophisticated ways than by simply altering their growth habit. After some time, they

coevolve with the grazers to produce different reproductive patterns and structures. For example, plants that can regrow from the base rather than the growing tip will be favored, as will plants that reproduce by runners.

All this has important consequences. It implies that a grazing ecosystem is balanced evolutionarily so that the herbivores are controlling the type and density of their food resources, but at the same time the response of the plants forces certain behavioral patterns and perhaps social structures on the herbivores. The ecosystem will tilt out of balance unless the grazing pressure is maintained at a minimum level to keep the low-growing plants at an advantage over possible competitors.

Grazing animals probably can't do this if they are solitary. Solitary grazers have two problems. They have to spend energy to defend a territory, and in open country they are liable to predation from running carnivores. Living in herds is an efficient solution to this problem, because it removes the need to spend energy on defense of a territory, it increases the chance of early warning of the approach of a predator, it allows group defense, and it provides a better guarantee of the heavy and continuous grazing that maintains a healthy ecosystem.

Furthermore, with a seasonal and local variation in food supply, it is easy to envisage the evolution of a set of grazing species, each specializing in a different part of the available food. In the Serengeti, for example, three different grazers eat grass and herbs. Zebras eat the upper parts of the blades of grass and the herbs, wildebeest follow up and eat the middle parts, and the Thomson's gazelle eats the lower portions. The teeth and digestive systems of each animal are specialized for its particular diet. Thus, a succession of animals grazes the plain at different times, each species modifying the plants in a way that (by chance) permits its successor to graze more efficiently. A great diversity of grazers is encouraged: today there are ten separate tribes of bovid antelopes on the savannas of Africa.

Because these principles are so general, they have probably operated at least since grasslands spread widely in the Miocene. There were low-plant ecosystems even before the evolution of the first grasses at the end of the Oligocene. Early horses seem to have grazed in open country in the Paleocene, for example. If dinosaurs were warm-blooded, they probably faced similar problems related to feeding requirements per mouthful. Even if dinosaurs were cold-blooded, with lower metabolic requirements, they would still have faced similar problems.

Similar principles probably apply to browsers too. Obviously, the rules will be rather different, because the defense of many plants against browsing is to grow tall quickly. And finally, herbivores, whether they are grazers or browsers, are a food resource for predators and scavengers. The animals of the African savannas are in a delicate and interwoven ecological network.

McNaughton's work explicitly defined principles that many workers had guessed at previously. It is a breakthrough not only for understanding modern savanna ecosystems but in interpreting past ones too.

The Savanna Story: Savannas in the Fossil Record

A major climatic change in the Miocene was apparently triggered by the refrigeration of the Antarctic and the growth of its huge ice cap. The cooler climate encouraged the spread of open woodland in subtropical latitudes, at the expense of thicker forests and woods. In California, for example, this occurred around 12 Ma, when the climate changed from wet summers to dry summers. There had been open woodland on Earth ever since the Permian, but the plants that grew in the open had been ferns and shrubs. The new feature of Miocene open country was the spread of grasses, with their high productivity.

Savanna ecosystems produce a great deal of edible vegetation, even though grasses have high fiber and low protein. Grasses are adapted to withstand severe grazing; they recover quickly after being cropped because they grow throughout the blade instead of mainly at its growing tip. They have evolved tiny silica fragments, or **phytoliths**, that make them tough to chew and cause significant tooth wear in grazing animals.

The spread of grasses was perhaps encouraged at first by intense grazing pressure, but the whole savanna ecosystem quickly stabilized, no doubt through mechanisms like those suggested by McNaughton. There was a rapid and spectacular evolutionary response, especially from the mammals, which evolved many different grazing forms. This event and its continuation into plains ecosystems today is called the **Savanna Story** by David Webb, who has done the most to document it. The change in vegetation was worldwide, and although the North American evidence is the most complete, similar trends can be traced on all the continents with subtropical land areas. On each continent, the savanna fauna evolved from animals that lived there before the major climatic change.

The animals that were particularly successful in savanna ecosystems were grazers or browsers on these open plains with only scattered woodland patches. Deer and antelope evolved to great diversity. Often their teeth evolved to become very long for their height, or **hypsodont**, with greatly increased enamel surfaces. Elephants, rodents, horses, camels, and rhinos, for example, all evolved jaws and teeth with adaptations for better grinding. Presumably, hypsodont teeth wore longer and permitted a grazer to chew tough fibers and resist the abrasion of phytoliths.

The larger savanna animals also showed changes in size and locomotion consistent with life in open country where there was nowhere to hide. They became taller and longer-legged, well adapted for running fast. Some of these Miocene plains animals were gigantic, and they included the largest land mammal that has ever lived, the giant hornless Eurasian rhino *Paraceratherium* (Fig. 17.14), as well as the tallest camel.

In the Miocene of North America, the grazers were at first native ruminants such as camelids and horses. Eurasian deer arrived and radiated during the Miocene. The Late Miocene savanna fauna of North America was very rich, peaking at about 50 genera of ungulates and large

Figure 17.14　The gigantic Miocene rhinoceros *Paraceratherium*. It probably weighed over 10 tonnes, far beyond the "usual" size range for large herbivores. Reconstruction by Bob Giuliani. © Dover Publications Inc., New York. Reproduced by permission.

carnivores, dominated by hypsodont horses, camelids, and pronghorns.

Starting about 9 Ma, this North American savanna ecosystem suffered a series of shocks, including a great extinction that began about 6 Ma. New genera evolved and new immigrants arrived, but they did not come close to replacing the losses.

The extinction patterns are interesting. All nonruminant artiodactyls disappeared except for one peccary, *Platygonus*, which evolved shearing teeth and shifted toward a coarser, more fibrous diet. Grazing horses flourished, but browsing horses disappeared. Only hypsodont camelids survived, while short-toothed forms became extinct. The casualties included a giant camel *Aepycamelus*, 3.5 meters (12 feet) high (Fig. 17.15), which was probably a giraffelike browser. Among the ruminants, the major survivors were the pronghorns, which are hypsodont.

The common ecological pattern is adaptation to coarse fodder and more open country, and it presumably reflects a change from savanna to steppe grassland. Perhaps rain shadow effects were produced by major uplift in western North America; but overall, global climates became colder in the Late Cenozoic.

Evolution by Improvement

The fossil record of mammals is so good that we can trace related groups of mammals through long time periods, and often across large areas and across geographic and climatic barriers. In many cases, we can see considerable evolutionary change in the groups, and because we understand the biology of living mammals rather well, we can interpret the changes confidently. Often the changes can be linked with

Figure 17.15 I love this painting of the giant Miocene camel *Aepycamelus* by Heinrich Harder (1920). Image in the public domain.

Figure 17.16 *Basilosaurus* was recognzed as a huge fossil whale as early as the 1840s. Known from Eocene rocks, it ranged up to 18 meters (60 feet) in length. Reconstruction by Pavel Riha and placed into Wikimedia.

specific biological functions and can be seen as allowing the animals to perform those functions in a more effective way.

Many people do not like the concept of improvement, or of evolutionary progress, which is another way of saying the same thing. However, there are parameters that we can measure that show clearly that many clades of organisms do get better through time at doing what they do. The easiest way to show this is to use simple mechanics (biomechanics, if applied to animals or plants). So, for example, dinosaurs moved in a mechanically better way than their ancestors, and so do living mammals. Horses in particular have evolved clever mechanical couplings between bone, muscle, and tendon that give living horses, including racehorses, much better running performance than their predecessors. The examples are too numerous to list here, and one could make the same arguments about physiology, biochemistry, reproduction, and so on. I simply want to make the point that it is legitimate to write about "progress" as applied to evolution. I will deal here with whales as an example. (I treat horses on the Web site for this book).

Whales

Whales, dolphins, and porpoises are well-known marine mammals today. They are beautifully adapted to carnivorous ways of life in the sea, with streamlined bodies, fore limbs turned into flippers, hind limbs lost, and tail flukes. Whales give birth at sea, and many of them have sophisticated sonar, and physiological adaptations for deep diving. Fossil whales are not common, but *Basilosaurus*, from the Eocene of the American South, was recognized as a huge

fossil whale over 150 years ago. (It is now the State fossil of both Mississippi and Alabama.) But even in the 1960s, textbooks on vertebrate paleontology did not try to identify the ancestors of whales.

Over the past 20 years, new fossil discoveries, and genetic analyses, have worked together to give new insight into whale ancestors, and on the adaptive pathways that evolved whales from land-going ancestors.

Astoundingly, whale ancestors were artiodactyls, whose modern members are herbivores or (rarely) omnivores: cattle, deer, antelope, hippos, camels, giraffes, and pigs. The evidence from genetics first alerted us: cladograms based on genetic similarity placed whales as the close relatives of artiodactyls, and of hippos in particular. This caused consternation among paleontologists. However, the ancestors of whales and hippos would have diverged during the Eocene at the latest, so we should not expect that the ancestor would look anything like their very different distant descendants. (Remember that the closest relative of birds among living animals is a crocodile; and neither of these animals looks like the early archosauromorph that was their common ancestor.)

At the same time, Eocene whales were turning up from Egypt, India, and Pakistan, and I will describe some of them, roughly in morphological order, since they show transitions between land- and water-based ecologies.

Indohyus is from the early and middle Eocene of India, and is the size of a very small deer (Fig. 17.17). It is well enough preserved to have ankle bones that show it was an

Figure 17.17 Reconstruction of *Indohyus*, from the Eocene of India. Art by Nobu Tamura, and placed into Wikimedia.

Figure 17.18 Reconstruction of *Pakicetus*, from the Eocene of Pakistan. Illustration by Carl Buell, and taken from http://www.neomed.edu/DEPTS/ANATThewissen/publ.html Used by permission.

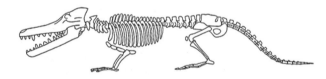

Figure 17.19 Reconstruction of *Ambulocetus* from the Eocene of Pakistan. The body was alligator-sized, about 12 feet long. Taken from http://www.neomed.edu/DEPTS/ANAT/Thewissen/whale_origins/index.html, and used by permission.

artiodactyl. It also had ear bones that were thicker than normal (as in whales), and its limb bones were dense for their size, as they are in hippos. The dense bones make walking in shallow water more stable. Most likely, then, *Indohyus* was grazing on water plants, like the living water chevrotain, a tiny deer from central Africa.

Pakicetus (Fig. 17.18), from the Eocene of Pakistan, looked superficially like a wolf, but its skeleton still had artiodactyl characters. Its teeth were more like a predator's, and it probably waded in shallow water, eating any available prey.

Ambulocetus (Fig. 17.19), the "walking whale," is also from the Eocene of Pakistan. It was much larger, up to 12

Figure 17.20 Reconstruction of *Kutchicetus*, from the Eocene of India, a whale that looked like a giant otter. Illustration by Carl Buell and taken from http://www.neocom.edu/Depts/Anat/Thewissen/whale_origins/whales/Remi.html. Used by permission.

feet long, and looked superficially like an alligator, with short powerful limbs and a long flat head with fish-eating teeth. It clearly could swim powerfully, though slowly, probably with an otter-like vertical bending of the spine. But it could also walk perfectly well on land. *Ambulocetus* marks a major step toward life in water: it clearly fed in the water, even if it came to land for sleeping and breeding.

Kutchicetus (Fig. 17.20) was a whale from the Eocene of India that probably behaved like a giant otter, even better adapted to feeding and swimming in water than previous whales. It had a long beak-like set of jaws, with the skull mounted on a relatively stiff neck. The tail was relatively long and powerful, and may have been the major power producer in swimming.

The next whales we see in the fossil record are distributed in a great tropical belt from southern Asia westward to North America, and that means that whales had become long-distance ocean-going swimmers. The protocetids include more than a dozen genera. They are varied in size and surely represent a radiation of early whales away from their presumed origins in the Indian subcontinent. There are two particularly compelling protocetids, partly because they are very well preserved, and partly because of the persuasive reconstructions by the artist John Klausmeyer. *Rodhocetus* (Fig. 17.21), from the Eocene of Pakistan, still has strong limbs, though its skull is very whale-like, with a blow-hole halfway up the snout. If one had to search for a living comparison in ecological terms, one might choose sea-lions, which migrate at sea extensively, but still go on land for mating and giving birth.

Maiacetus, from the Eocene of Pakistan, is known from an almost complete skeleton, allowing a reconstruction in swimming position (Fig. 17.22). The sea-lion analogy seems reasonable for this animal, too. But the most compelling aspect of this specimen is that it was a female with a late-stage fetus inside the body cavity (Fig. 17.23).

It is clear that the head of the fetus faces backward, as it does in large mammals that give birth on land. The head is delivered first so that the fetus can breathe air even while the rest of the body is still inside the mother. However, in all living whales, the tail is delivered first, and the head last. In this way, the whale fetus can still get some oxygenated blood through the umbilical cord, and does not drown as it is pushed out into the water. Nurse whales in the pod help the newborn to the surface as soon as the head emerges, so that the newborn can take its first breath. The evidence from this single specimen of *Maiacetus* tells us that these whales gave birth on land (Fig. 17.24), as we could deduce indirectly from the fact that they had strong limbs.

Figure 17.21 The Eocene protocetid whale *Rodhocetus*. Reconstruction by John Klausmeyer, and used courtesy of the University of Michigan Museum of Natural History.

But *Basilosaurus* was different. As we have seen, it was a huge whale, and it had lost any functional hind limbs. Its fore limbs were reduced to hydrofoils, and all the propulsion came from the tail (Fig. 17.16). There is no way that it could have emerged on to land: it spent its entire life at sea. *Basilosaurus* and a related smaller genus *Dorudon* had fearsome fish-eating teeth set in a long jaw (Fig. 17.25).

Huge specimens of *Basilosaurus* have been found exposed on the surface in the desert basin of the Fayum, in Egypt. Cleaning the spine revealed tiny nubbins of bone that were the evolutionary remnants of the pelvis and hind limbs, obviously of no practical use to the animal. We see this also in some snakes that still have tiny remnants of the hind limbs they once had.

Seen from the heaving deck of an Eocene boat, these creatures would have looked very whale-like, even though they did not yet have the particular derived features of living whales. Modern whales appear in the Oligocene. **Odontocetes** are the toothed whales, and they perfected the echolocation that is such an important factor in their ability to locate prey in dark or turbid water. The baleen whales or **mysticetes** evolved from odontocetes as they changed their feeding habits to use a filtering system based on modified bone, and they live on small prey such as the tiny crustaceans called krill. As we have come to expect in evolution, the first whales that are recognizably mysticetes have not yet evolved baleen! They still had teeth but did not have echolocation (Fitzgerald 2006, 2012). It seems that they used a widely opening mouth to suck in dozens of fish at a time, and it could be that the filter-feeding of later baleen whales evolved from that sort of prey capture. We need more fossils from that time period (Oligocene).

Overall, the evolution of whales from small herbivores on land to carnivorous predators at sea is the most spectacular, and now one of the best understood, transitions in the mammalian fossil record. But as I write this in 2012, research on the very earliest whales has come to a halt, because the very areas in western Pakistan that have provided most of the animals described in this chapter are remote areas under the unstable control of tribal leaders and the Taliban.

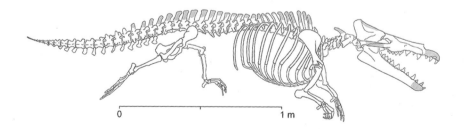

0   1 m

Figure 17.22 The swimming position of the protocetid *Maiacetus*. Part of Figure 1 in Gingerich et al. (2009): http://www.plosone.org/article/info%3Adoi%2F10.1371%2Fjournal.pone.0004366. Drawn by Bonnie Miljour. Published in PLoS and thereby placed in Wikimedia.

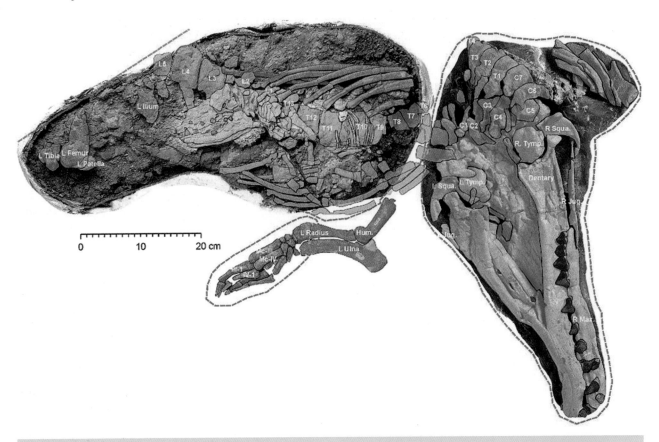

Figure 17.23 The single fossil of *Maiacetus*, showing the fetus (in blue) inside the body cavity. Figure 2 in Gingerich et al. (2009): http://www.plosone.org/article/info%3Adoi%2F10.1371%2Fjournal.pone.0004366. Published in PLoS and thereby placed in Wikimedia.

Figure 17.24 A female *Maiacetus* and her newborn pup. Art by John Klausmeyer. Used courtesy of the University of Michigan Museum of Natural History.

Figure 17.25 *Basilosaurus* and a smaller relative *Dorudon*. Photograph by Dr. Philip Gingerich, used courtesy of the University of Michigan Museum of Natural History.

Further Reading

Archibald, J. D. et al. 2001. Late Cretaceous relatives of rabbits, rodents, and other extant eutherian mammals. *Nature* 414: 62–65. The branching of northern placentals was under way in the Cretaceous, but not as intensively as molecular evidence suggests. Available at http://www.bio.sdsu.edu/faculty/archibald/ArchibaldEtAl01Nat414p62.pdf

Bianucci, G. and P. D. Gingerich. 2011. *Aegyptocetus tarfa*, n. gen. et sp. (Mammalia, Cetacea), from the middle Eocene of Egypt: clinorhynchy, olfaction, and hearing in a protocetid whale. *Journal of Vertebrate Paleontology* 31: 1173–1188.

Fitzgerald, E. M. G. 2006. A bizarre new toothed mysticete (Cetacea) from Australia and the early evolution of baleen whales. *Proceedings of the Royal Society B* 273: 2955–2963. [*Janjucetus*] Available at http://www.ncbi.nlm.nih.gov/pmc/articles/PMC1639514/

Fitzgerald, E. M. G. 2012. Archaeocete-like jaws in a baleen whale. *Biology Letters* 8: 94–96.

Gingerich, P. D. et al. 2001. Origin of whales from early artiodactyls: hands and feet of Eocene Protocetidae from Pakistan. *Science* 293: 2239–2242, and comment, pp. 2216–2217. [*Rodhocetus*] Available at http://www-personal.umich.edu/~gingeric/PDFfiles/PDG381_Artiocetus.pdf

Gingerich, P. D. et al. 2009. New protocetid whale from the middle Eocene of Pakistan: birth on land, precocial development, and sexual dimorphism. *PLoS ONE* 4: e4366.

[*Maiacetus*]. Available at http://www.plosone.org/article/info%3Adoi%2F10.1371%2Fjournal.pone.0004366

Murphy, W. J. et al. 2001. Molecular phylogenetics and the origins of placental mammals. *Nature* 409: 614–618.

O'Leary, M. A. and M. D. Uhen. 1999. The time of origin of whales and the role of behavioral changes in the terrestrial-aquatic transition. *Paleobiology* 25: 534–556.

Springer, M. S. et al. 2003. Placental mammal diversification and the Cretaceous–Tertiary boundary. *PNAS* 100: 1056–1061. [Fine synthesis of molecular data up to that time.] Available at http://www.pnas.org/content/100/3/1056.full

Thewissen, J. G. M. et al. 2001. Skeletons of terrestrial cetaceans and the relationship of whales to artiodactyls. *Nature* 413: 277–281, and comment, pp. 259–260. Available at http://www.faculty.virginia.edu/bio202/202-2002/Lectures%2020202/thesissen%20et%20al%202001.pdf

Thewissen, J. G. M. et al. 2007. Whales originated from aquatic artiodactyls in the Eocene epoch of India. *Nature* 450: 1190–1194. [*Indohyus*.] Available at http://repository.ias.ac.in/4642/1/316.pdf

Thewissen, J. G. M. et al. 2009. From land to water: the origin of whales, dolphins, and porpoises. *Evolution: Education and Outreach*. Available at http://www.springerlink.com/content/whn1654v74t64301/fulltext.html

Van Valkenburgh, B. 1999. Major patterns in the history of carnivorous mammals. *Annual Reviews of Earth & Planetary Sciences* 27: 463–493.

Questions for Thought, Study, and Discussion

1. Briefly describe the strengths and weaknesses of reconstructing the evolution of mammals based only on molecular and genetic data. Then describe the strengths and weaknesses of reconstructing the evolution of mammals based only on the skeletons of living an fossil mammals. What happens when the results give different answers?

2. The concept of a guild is rather simple in principle. And it's an easy question (so answer it) to give some examples of guilds that once did not live on Earth, but are present today. But it's more difficult (so try it) to pick out guilds that once lived on Earth but no longer do. And it's even more difficult to ask why those guilds disappeared (so answer it for the examples you chose).

3. Whales are not the only mammal group that lives successfully on the sea. For example, seals, sealions, sea otters, manatees and even an extinct group of sloths all evolved to take on a successful life in the sea. Find out more about these other animal groups and then point out the features that make whales different: better adapted to marine life, if you like. And finally, do you think that whales are different because they "got there first"?

EIGHTEEN

Geography and Evolution

In This Chapter

Here I look at ways that changing geography influenced evolution during the Cenozoic. At the end of the Paleocene, an extraordinary warm climate opened up the Arctic to mammal migration, resulting in a mixing of mammals that had evolved independently in Eurasia and in North America. This unique "tipping point" had permanent effects on the later history of Cenozoic mammals. Australia, meanwhile, was an isolated continent with a limited fauna on it. The marsupials that had reached Australia radiated into an amazing array of creatures that nevertheless occupied many of the ecological niches in Australia that placental mammals occupied elsewhere, and Australia's reptiles are equally impressive. New Zealand became isolated with no mammals at all. Bats flew there, but birds became the dominant land animals, including the giant moas. South America had a radiation of its own mammals, the Xenarthra, but the ecosystem for a long time had placentals at fairly small sizes, and marsupials as the top predators (including *Thylacosmilus*, a marsupial sabertooth). With geography changing all the time, South America drifted close enough to North America that animals were exchanged, and in that process many of the South American animals became extinct. I take a brief look at Africa from the same point of view, and then turn to two more island examples to conclude the chapter.

Holarctica and the PETM

Holarctica is a useful term for the Northern continents of Eurasia and North America (with Greenland). Today they form an almost continuous land mass across the high Arctic, with two breaks: a small one across the Bering Strait between Alaska and Siberia, and a larger one between Greenland and Norway. But in the Eocene, the gaps were smaller, and the only thing that prevented free flow of animals across the Arctic was the fact that it was a cold region (Fig. 18.1).

The Eocene was a warm period in Earth history, and there were no ice-caps either in the northern continents or on Antarctica. However, these areas were cold tundra. They would have had accumulations of peat, with a permanently frozen soil (**permafrost**) under the surface, just as we find over huge areas of Holarctica today (Fig. 18.2). But in the Eocene, the permafrost would have extended over Greenland and Antarctica as well.

Permafrost contains water ice, of course, but in addition it contains methane hydrate or **clathrate**, a strange substance that contains rather a lot of methane frozen into

History of Life, Fifth Edition. Richard Cowen.
© 2013 by Richard Cowen. Published 2013 by Blackwell Publishing Ltd.

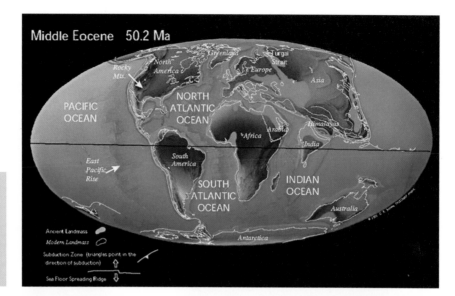

Figure 18.1 Eocene geography across the Southern Hemisphere. Paleogeographic map by C. R. Scotese © 2012 PALEOMAP Project (www.scotese.com)

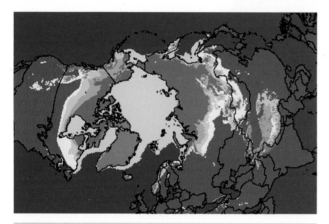

Figure 18.2 Permafrost in the Northern Hemisphere today. Areas of occasional or permanent sea ice are light blue. The Greenland ice cap is violet. Areas of various thicknesses of permafrost are in shades of brown. Image from the U.S. NSDIC, in the public domain.

Figure 18.3 Methane clathrate, the ice that burns. The clathrate crystal structure is shown in the inset. Image from the U.S. Naval Research Office, in the public domain.

water ice. Clathrates occur in marine sediments as well as permafrost, and they may become valuable sources of natural gas if technical problems of extracting them can be solved. They are bizarre, being made of an ice substance that burns (Fig. 18.3).

At the end of the Paleocene, a very unusual set of circumstances combined to give the Earth a sudden, short-lived heat shock. For millions of years, permafrost (and methane clathrate) had been accumulating at both ends of the Earth. Global temperatures rose through the Paleocene, until finally the permafrost began to thaw. That released methane, which is a powerful greenhouse gas. That raised the temperature more, which released more methane, until much of the permafrost thawed, and the world became about 6° warmer, probably in only a few thousand years.

This is a geological instant, of course, but it did not cause any major change in sea-level, because there were no ice-caps to melt. Instead, the environment of Holarctica became much warmer, and more hospitable to plants and animals that had lived in temperate regions to the south.

Almost immediately, plants and animals were able to move across Holarctica. Fossil forests and large herbivores have been found as far north as Ellesmere Island, in the Canadian Arctic. In the North American fossil record, there is a major influx of mammals from Asia, and the change in faunas marks the beginning of the Eocene.

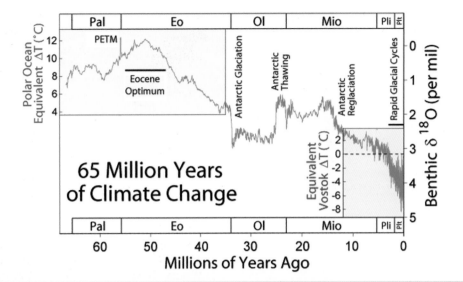

Figure 18.4 Climate change over the past 65 million years. For the PETM, note the rising temperatures during the Paleocene (Pal) that triggered the PETM at the beginning of the Eocene. An event as large, sudden, and short-lived as the PETM never happened again. Diagram by Robert A. Rohde of the Global Warming Art Project, (http://www.globalwarmingart.com/) and placed into Wikimedia.

Once the methane dissipated, the climate cooled again, and permafrost built up. But the normal small variations in Earth's climate were enough to trigger small permafrost melts before a large methane reservoir had built up. So there was never again such a dramatic warming as the **PETM**: the **Paleocene-Eocene Thermal Maximum** (Fig. 18.4). (This account follows the latest syntheses of the PETM, by DeConto et al. 2012 and Abels et al. 2012.)

But the biogeographic changes had been made, and the fossil record of Holarctica shows a dramatic "tipping point" at the PETM. The PETM is a striking example of a single unlikely event that had permanent effects on the fossil record. One can think of others: the K-T asteroid impact, the combination of factors that caused the P-Tr extinction; and the Great Oxidation Event. Like human history, the history of life on Earth has been subject to unexpected and unforeseeable events!

Australia

Australia is linked in people's minds with exotic creatures such as kangaroos and jillaroos, but they are only a part of the story of evolution on this isolated continent. Australian plants, insects, amphibians, reptiles, birds, and mammals are all unusual. Australia and New Zealand were part of Gondwana in Cretaceous times, joined to Antarctica in high latitudes (Fig. 16.7). The climate was mild, however, and pterosaurs, dinosaurs, and marine reptiles have been found there. In the early Cenozoic the two land masses broke away from Antarctica and began to drift northward and diverge. Both Australia and New Zealand became isolated geographically and ecologically from other land masses, and evolution among their faunas and floras led to interesting parallels with other continents.

Among amphibians, Australia has (or had) at least two species of frogs that brood young in their stomachs. Instead of the colubrid snakes and vipers that are abundant elsewhere, Australia had a radiation of elapid snakes (cobras and their relatives) into 75 species, all of them virulently poisonous. The largest Australian predators are the saltwater crocodiles (the world's largest surviving reptiles), which lurk along northern rivers and shorelines, and giant monitor lizards the size of crocodiles, related to the Komodo dragon of Indonesia. Monitors are ambush predators, the largest living Australian monitor being 2 meters (over 6 feet) long. Smaller monitors dig for prey like the badgers of larger continents. In contrast, most Australian mammals are herbivores.

Extinct Australian reptiles included giant horned tortoises that weighed up to 200 kg (450 pounds), a monitor 7 meters (23 feet) long that weighed perhaps a ton, and competed with large terrestrial crocodiles of about the same size and weight. The giant snake *Wonambi* was 6 meters (19 feet) long and must have weighed 100 kg (220 pounds). Extinct Australian birds included *Dromornis*, the heaviest bird that has ever evolved (Chapter 13).

Australia is the only continent with living **monotremes**. They have been in Australia since the Early Cretaceous (Chapter 15), and a single early Cenozoic monotreme tooth from Argentina shows that they once ranged more widely over Gondwana. The surviving monotremes are egg-laying mammals, including the duckbilled platypus and the spiny echidna of Australia and New Guinea. Many aspects of monotreme biology are bizarre: for example, the platypus swims in muddy water with its eyes, ears, and

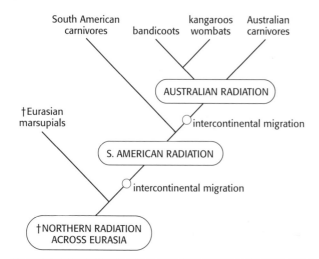

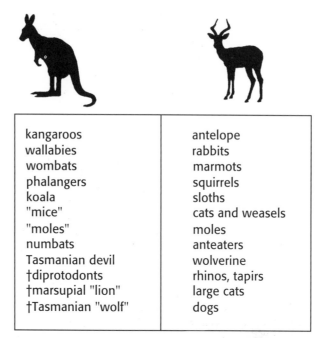

Figure 18.5 The biogeographic evolution of marsupials. An early (Cretaceous) radiation in Asia and North America was followed by a Late Cretaceous and Cenozoic radiation in South America, and a Cenozoic dispersal through Antarctica to Australia, where a spectacular radiation occurred.

kangaroos	antelope
wallabies	rabbits
wombats	marmots
phalangers	squirrels
koala	sloths
"mice"	cats and weasels
"moles"	moles
numbats	anteaters
Tasmanian devil	wolverine
†diprotodonts	rhinos, tapirs
†marsupial "lion"	large cats
†Tasmanian "wolf"	dogs

Figure 18.6 A gallery of Australian marsupials, each of which has a placental ecological counterpart on other continents.

nostrils tightly shut, searching for its crustacean prey with electrical sensors in its beak. Since monotremes have evolved to include the specialized platypus and ant-eating echidnas, it's likely that their fossil record will eventually show us many other surprises.

Marsupials had originally evolved in the northern continents, but as we shall see later in this chapter, they reached South America and radiated there in the Cenozoic. Marsupial fossils have now been discovered in Eocene rocks in Antarctica and Australia, so it is likely they reached Australia from South America across Antarctica when the region was much warmer than it is now (Fig. 18.1, Fig. 18.5), well before the refrigeration of Antarctica in Oligocene times (Fig. 18.4).

By the Late Cenozoic, marsupials had evolved to fill most of the ecological roles in Australia that are performed by placental mammals on other continents (Fig. 18.6). Wallabies and kangaroos are grazers comparable with antelope and deer, wombats are large burrowing "rodents" rather like marmots, and koalas are slow-moving browsers like sloths. The cuscus is like a lemur, and the numbat is a marsupial anteater. There are marsupial cats, marsupial moles, and marsupial mice, and at least six gliding marsupials can be compared with flying squirrels. The honey possum *Tarsipes* is the only nonflying mammal that lives entirely on nectar and pollen, which it gathers with a furry tongue (Fig. 18.7). The small marsupial *Dactylopsila* of New Guinea has evolved specialized teeth and a very long finger to become a marsupial woodpecker (Fig. 17.9). The Tasmanian wolf and the Tasmanian devil are marsupial carnivores comparable in size and ecology to wolf and wolverine. They once ranged over the main continent of

Figure 18.7 *Tarsipes*, the honey possum, which feeds on pollen and nectar with a long hairy tongue. Painting by John Gould, 1863, in the public domain.

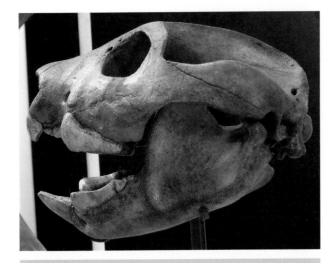

Figure 18.8   *Thylacoleo*, an extinct Australian marsupial carnivore the size of a leopard. Photograph by Ghedoghedo and placed into Wikimedia.

Figure 18.9   A diprotodont, one of several large extinct quadrupedal marsupials in Australia. Diprotodonts weighed up to 3 tons. (Reconstruction by Bob Giulani. © Dover Publications Inc., New York. Reproduced by permission.)

Australia. The Tasmanian devil is now confined to Tasmania, and the Tasmanian wolf is probably extinct.

The fossil record of extinct Australian marsupials is even more impressive. Entire families of marsupials are now extinct. Many were very large, including giant kangaroos and giant wombats that each weighed 200 kg or so (450 pounds). *Thylacoleo* was a Pleistocene carnivore whose name means the marsupial lion. It was the size of a leopard, and had efficient stabbing and cutting teeth. It was better adapted for cutting off chunks of flesh than any living carnivore is, and its bite is reconstructed as one of the most powerful for its size ever evolved by a mammal (Wroe et al. 2005) (Fig. 18.8). It also had very powerful retractable claws, which is not usually a marsupial character.

Diprotodonts were quadrupedal Pleistocene marsupials about the size of tapirs and rhinoceroses (Fig. 18.9). They were the largest marsupials ever: the largest diprotodont was the size of a small elephant, almost 3 meters (10 feet) long and 2 meters (over 6 feet) high at the shoulder, weighing probably close to 3 tonnes. Discoveries of enormous numbers of Miocene bats and marsupials at Riversleigh, in Queensland, will eventually allow a better description of the radiation of these Australian mammals.

People often talk of marsupials as primitive and inferior to placentals, and it's true that today they are outclassed in diversity and range by placentals. But marsupials do not always have inferior adaptations (Chapter 15). For example, a kangaroo is rather clumsy as it hops slowly around on the ground, using its tail as an extra limb in what is really a five-footed movement. It does use more energy than a placental at this speed. But at high speed a kangaroo is not only very fast (up to 60 kph, or 40 mph), but its incredibly long leaps are much more efficient than the full stride of a four-footed runner of the same weight.

Figure 18.10   *Dromornis*, perhaps the heaviest bird that has ever evolved, and certainly the heaviest goose, from the Miocene of Australia. Image by Nobu Tamura and placed in Wikimedia.

Dromornithids (mihirung in aboriginal legend) are giant extinct Australian birds that evolved large body size and flightlessness there (Fig. 18.10, Fig. 18.11). Their nearest relatives are basal geese. *Dromornis* was probably as large as *Aepyornis*, the elephant bird of Madagascar, and rivals it for the heaviest bird of all time (Chapter 13). The living Australasian emu and cassowary are large ground-running ratites, related to the ratites on other remnants of Gondwana (Fig. 13.24).

The isolated position of Australia has meant that only very mobile birds and placental mammals (bats and

Figure 18.11   A dromornithid, or mihirung, a giant extinct bird from the Pleistocene of Australia, next to a living emu for scale.

Figure 18.12   South America drifted away from Africa, first west and then west-north-west during the Cenozoic, and for most of that time it was an island continent accessible only to lucky or mobile immigrants.

humans) have reached it. Humans brought with them a host of other invaders, such as rats, cats, dogs, sheep, cattle, rabbits, cactus, fish, and cane toads, with serious results for the Australian ecosystem. Captain Cook's first reaction to a kangaroo was to set his dog on it! More recently, other bizarre introductions have helped to restore a little of the damage—for example, the organism that causes the rabbit disease myxomatosis, and the dung beetles that keep Australian grasslands from being buried in cattle dung. The biogeographic story of Australia is still in an active phase.

New Zealand

New Zealand was part of Gondwana until the Cretaceous, and it had a normal fauna at that time. But it had no land mammals until humans arrived, and the rest of its prehistoric fauna suggests that migration into the region was difficult. The native fauna includes only four amphibian species, primitive frogs that hatch as miniature adults from the egg with no tadpole stage. There are only a few native reptile species: 11 geckos that all have live birth, 18 skinks, of which 17 have live birth, and the tuatara, an ancient and primitive reptile (Chapter 11). New Zealand has no snakes and no normal lizards. The only native mammals are two species of bats.

The dominant prehistoric creatures of New Zealand were birds. The kiwis survive as nocturnal insectivores, but the major vegetarians were moas, very large ratites. The largest moa (females were much larger than males) was 3.5 meters (11 feet) in height (Fig. 13.24).

Moas coevolved with New Zealand plants so that 10% of the native woody plants have a peculiar branching pattern called *divarication*—they branch at a high angle to form a densely growing plant with interlaced branches that are difficult to pull out or break. There are few leaves on the outside, and the largest, most succulent leaves are on the inside. But nine species of divaricating plants that grow more than 3 meters (10 feet) tall look more like normal trees once they reach that height, and other divaricating

species grow more normally on small offshore islands. The only reasonable explanation of divarication is that it evolved as a defense against browsing moas, the largest of which was about 3 meters tall.

Other vegetarian guilds that were filled by small mammals on other land masses were partly occupied by moas and other birds and partly by huge flightless insects—enormous weevils and wetas (giant grasshoppers). It's not easy to identify the major prehistoric predators, but they were present. The largest surviving New Zealand birds (the kiwi, for example) are well camouflaged, although there is no obvious surviving predator on them. But extinct New Zealand raptors include a bird that was the largest goshawk that ever evolved (3 kg or 7 pounds in weight) and a huge extinct eagle that weighed about 13 kg (30 pounds).

South America

South America is in many ways more interesting than Australia for mammalian evolution because we know its history in more detail. South America split away from Africa in the Late Cretaceous (around 80 Ma) to become an island continent (Fig. 18.12).

In Cretaceous times the South American mammals and dinosaurs included unique forms belonging to basal Jurassic groups that had become extinct everywhere else but continued to evolve in South America. Examples include the giant dinosaur *Megaraptor*, a large sphenodont, and early mammals. Triconodonts, symmetrodonts, and multituberculates (Chapter 15) have all been collected from Cretaceous rocks in South America, yet therian mammals (marsupials and placentals) are not found.

Around the end of the Cretaceous, marsupials and placental herbivores arrived in South America, presumably from North America, and South America probably provided the gateway to a route across Antarctica for marsupials to reach Australia.

The climatic changes at the end of the Eocene seem to coincide with the arrival of a further few immigrants into South America: rodents and monkeys, tortoises, and colubrid snakes. The same climatic changes led to the spread of Oligocene grasslands over much of South America, and the early expansion of the South American placentals into a guild of open-country grazers.

Apart from these brief periods of immigration, Cenozoic evolution in South America took place in isolation for over 60 m.y. The strange South American mammals in particular are well known, and they divided up available ecological roles in the usual way. Charles Darwin noticed peculiar fossil mammals in Argentina during his voyage on the *Beagle*, and later expeditions to Argentina have found hundreds of beautifully preserved Cenozoic fossils.

From Early Cenozoic times, the South American marsupials took on the roles of small insectivores (and still do). There is a living aquatic marsupial with webbed feet and a watertight pouch. *Argyrolagus* was a rabbit-sized marsupial that looked like a giant kangaroo rat. It hopped and had ever-growing molars for grazing coarse vegetation. The arrival of the placental rodents did not affect these small marsupials. One of the most successful marsupials in the world, even in the face of intense competition from placentals, is the small omnivorous opossum, *Didelphis*.

The placental grazers of South America had evolved by the Miocene into a bewildering variety of forms ranging from rhino-sized to rabbit-sized. *Thoatherium* and *Diadiaphorus* had an uncanny resemblance to horses, with long faces, horselike front teeth, grinding molars, straight backs, and slender legs ending in one or three toes (Fig. 18.13). Some of their relatives looked like camels. Large vegetarians such as *Toxodon* had large grinding ever-growing molars (Fig. 18.14).

Armadillos, sloths, and anteaters are also characteristic South American mammals. Armadillos and their relatives evolved heavy body armor for protection and became highly successful opportunistic insectivores and scavengers. The Pleistocene armadillo *Glyptodon* was very large, probably a vegetarian, 1.5 meters (5 feet) long. It had a thick armored skullcap as well as body armor, and some glyptodont species had a spiked knob at the end of the tail (Fig. 18.15). Glyptodonts were certainly too big to burrow like the smaller armadillos, and they had to be heavily armored and armed to survive out on the surface. Naturally, their skeleton was very strong to support all the weight of the armor.

Sloths now live in trees, eating leaves and moving with painful slowness. But remains of huge ground sloths have been found in South America, including one that must have been almost as large as an elephant. Anteaters evolved from the same group of ancestors but are now specialized to an amazing extent for eating termites, beginning by tearing apart their nests with tremendously powerful clawed forearms.

The most impressive South American creatures were the larger carnivores. None of them were placental mammals,

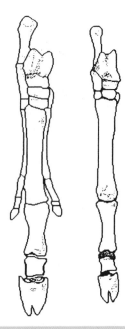

Figure 18.13   The hind feet of two Cenozoic mammals from South America, *Diadiaphorus* (left), and *Thoatherium* (right), the thoat. These are strikingly similar to the hind feet of horses, but they evolved in parallel to horses and are not related to them. After Scott.

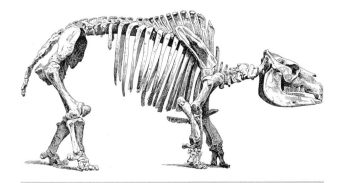

Figure 18.14   *Toxodon*, a large vegetarian mammal from the Cenozoic of South America. Image in the public domain.

and most were marsupials. This is not surprising, considering how savage the surviving little marsupial insectivores are, but it is unusual compared with other continents. Borhyaenids were basically like wolves, but were generally larger. *Proborhyaena* was as big as a bear and probably had a similar way of life. *Borhyaena* itself was a wolf-sized Miocene marsupial with canine teeth adapted for stabbing and molars that had evolved into meat-slicing teeth (Fig. 18.16). It was a successful medium-sized carnivore, but it was the last of the large borhyaenid carnivores. They were replaced by invading placentals from the north and by giant predatory birds.

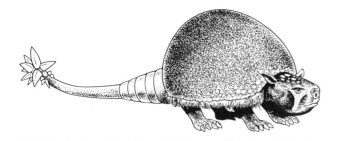

Figure 18.15 A glyptodont, a giant, heavily armored extinct relative of living armadillos. This one was close to 3 meters (9 feet) long. Reconstruction by Bob Giuliani. © Dover Publications Inc., New York. Reproduced by permission.

Figure 18.17 The skull, jaws, and teeth of the South American marsupial sabertooth *Thylacosmilus*. Art by Dmitry Bogdanov and placed into Wikimedia.

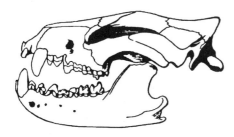

Figure 18.16 *Borhyaena*, a wolf-sized marsupial carnivore from the Cenozoic of South America. After Sinclair.

Thylacosmilids looked like large cats. *Thylacosmilus* was a marsupial sabertooth, but its savage stabbing canines were better designed than those of the placental sabertooth cats of North America. In *Thylacosmilus* the sabers were longer, slimmer, more securely anchored in huge, recessed tooth cavities extending far up the face; thus, they were better protected from damage than those of true cats (Fig. 18.17). The sabers were ever-growing and self-sharpening, and they were backed by more powerful neck and head muscles. Presumably they were adapted to killing large (placental) herbivores by stabbing and slashing deep into the soft tissues of throat or belly. The cheek teeth were not as powerful as those of placental cats, however.

These amazing marsupials had unusual competitors for mastery of the carnivorous guild, the phorusrhacids: flightless, ostrich-sized birds equipped with very powerful tearing beaks as well as foot talons (Fig. 13.22). It seems that the phorusrhacids eventually gained the upper hand over the carnivorous marsupials.

South America had its own group of crocodiles, the sebecids. They apparently evolved in Gondwana in the Cretaceous, survived the K–T extinction, and radiated in the Early Cenozoic in South America to become powerful terrestrial predators. Unlike aquatic crocodiles, they had high, deep skulls and snouts. Other crocodilians in South America also evolved into unusual morphologies; for example, a duckbilled caiman is known from the Miocene of Colombia.

The South American ecosystem gained new immigrants in Oligocene times, around 25 Ma, with the arrival of rodents and primates, probably from Africa by way of islands in the widening Atlantic Ocean (Fig. 18.12). Both groups radiated widely. The primates radiated into the distinctive New World monkeys, evolving habits and characters in parallel with gibbons and Old World monkeys. The rodents evolved into forms that include the world's largest rodent (by far): *Phoberomys* from the Miocene of Venezuela weighed 700 kg (about 1500 pounds)! Other members of the Cenozoic South American fauna included more giants, the largest flying birds of all time, the teratorns (Chapter 13). The largest turtle of all time, *Stupendemys*, lived along the north coast close to *Phoberomys*.

This unique ecosystem suffered four tremendous shocks in ten million years and has almost completely disappeared. First, Antarctica froze up, with the result that the Humboldt Current, flowing most of the way up the west coast of South America, became much colder and stronger. Second, tectonic activity along the Pacific coast raised the Andes as a major mountain chain. Together, these two events drastically lowered rainfall over most of the continent, and much of the area turned from forest and well-watered plain to dry steppe. This led, in the later Miocene, to the extinction of many animals, including the terrestrial crocodiles and especially the large-bodied savanna herbivores.

Third, South America drifted northward towards Central and North America (Fig. 18.12). By about 6 Ma, the gap was small enough to allow a few animals to cross it, more or less by accident. North American raccoons and some mice and rats crossed to the south, while two kinds of sloths crossed to the north. The effect of the competition was seen almost immediately. Many borhyaenids were replaced by raccoons, and the largest of them, the bearlike *Proborhyaena*, was replaced by a bear-sized raccoon. Finally,

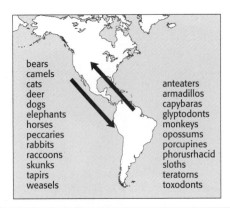

Figure 18.18   The Great American Interchange.

Figure 18.19   Some of the creatures involved in the Great American Interchange. South American creatures that went north, olive green; North American creatures that went south, blue. Your research task is to identify the silhouettes. Diagram by Woudloper and placed into Wikimedia.

at about 3 Ma, the last important sea barrier was bridged, and animals could walk from one continent to the other.

Ecological principles suggest what should happen when an exchange of animals takes place. A larger continent such as North America should contain a larger diversity of animals than its smaller counterpart, and the fossil record confirms that this was true just before the exchange. Therefore, if the same proportion of animals from each continent migrated to the other, one would expect more North American animals to go south than the reverse. If a continent can hold only so many families or genera of animals, then one would predict extinctions on each continent, but more in South America than in North America. The effect would be accentuated because North America was at least intermittently connected with Eurasia, and altogether this huge northern area of temperate open country held a great variety of savanna animals. In contrast, the area of savanna in South America was not as large as one would think, because the continent is widest in equatorial latitudes and narrows significantly to the north and south. South American savanna faunas might have been very vulnerable to invasion from the north.

The "Great American Interchange" happened after 3 Ma (Fig. 18.18, Fig. 18.19). Camels, elephants, bears, deer, peccaries, horses, tapirs, skunks, rabbits, cats, dogs, kangaroo rats, and shrews entered South America. Monkeys, opossums, anteaters, sloths, armadillos, capybaras, toxodonts, porcupines, and glyptodonts migrated north, with the giant birds—a phorusrhacid and a few teratorns.

The South American immigrants to North America flourished there, and so did the successful North American immigrants that moved south. Overall, however, there was a net major extinction of South American groups. The large, native marsupial carnivores and most of the phorusrhacids seem to have been outcompeted by the cats and dogs from the North, and the remaining savanna browsers and grazers were almost all wiped out, perhaps outcompeted by the northern horses and camels, perhaps hunted out by the new predators. The sabertooth marsupial *Thylacosmilus* was replaced by a real sabertooth cat. Even earlier

invaders suffered: the bear-sized raccoon was replaced by a true bear.

Overall, the result was nearly as expected, with the South American animals coming out much the losers. North American invaders survive in strength today in South America, including all the South American cats, the llamas, and dozens of rodents.

The geographical changes that had allowed the interchange also altered the climate of the Atlantic Ocean, and this in turn caused drastic changes in the land ecology of North and South America as the northern ice ages began in earnest around 2.5 Ma. South American faunas suffered another catastrophic extinction in the late Pleistocene. This time, similar extinctions took place in North America too, and we shall examine this in Chapter 21.

Africa

Africa (plus Arabia, so perhaps I should write Africarabia) was part of Gondwana until the Cretaceous, when it broke away from South America on the west and Antarctica and India on the east (Fig. 16.7). From Late Cretaceous times

Figure 18.20 *Moeritherium*, a hippo-like early afrotherian browser from the Oligocene of Egypt. Art by Heinrich Harder, about 1920, now in the public domain.

Figure 18.21 *Arsiniotherium*, a large-bodied browser from the Oligocene of Ethiopia. Image by Trent Schindler for the National Science Foundation, in the public domain.

onward, Africa and South America, and the animals and plants living there, had increasingly different histories.

Africa had dinosaurs much like those of the rest of the world during the Late Cretaceous, but there is no record of any African Cretaceous mammal. This may change, because molecular evidence suggests that there must have been spectacular evolution among the Afrotheria (Chapter 15) on the isolated continent of Africarabia, which had split from South America but was not close to Europe or Asia. Africa lay south of its present position, bounded on the north by the tropical Tethys Ocean.

Our first look at the fossil Cenozoic life of Africa comes from the Eocene rocks of Egypt, laid down on the northern edge of the continent. Shallow warm seas teemed with microorganisms whose shells formed the limestones from which the Pyramids and the Sphinx were carved.

Here we find early whales and sea cows, which probably evolved adaptations for marine life in swamps and deltas along the shores of Tethys. Moeritheres were amphibious animals related to sea cows and to elephants. *Moerithium* itself is an Eocene animal from Africa, and looked like a small, fat elephant with the ecology of a hippo (Fig. 18.20). Other Eocene fossils from Egypt include some early primitive carnivores, the creodonts (also known from other continents).

By Oligocene times, Egypt was the site of lush deltas where luxuriant forest growth housed rodents, primates, and bats, all recent Eurasian immigrants. Piglike anthracotheres had crossed from Eurasia, but there were African groups too. Hyraxes are small- to medium-sized vegetarians that look much like rodents. *Arsinoitherium* was a large-bodied browser (Fig. 18.21).

Eocene and Oligocene African mammals are a mixture of native African groups and a few successful immigrants from Eurasia. Even in the late Oligocene, the large African mammals were still arsinoitheres and a diverse set of elephants. But in the Miocene, Africarabia drifted far enough

Figure 18.22 Africarabia drifted slowly northeast during the Cenozoic. Finally it collided with Western Asia in the Early Miocene along a line that is now the Zagros Mountains. The African continent then rotated slightly clockwise, splitting away from Arabia. At the end of the Miocene the northwest corner of Africa collided with Western Europe to close off the Mediterranean Sea as a vast lake that quickly dried up. Meanwhile the Red Sea opened up as Africa split away from Arabia, and the great African Rift Valley was formed.

north to bring it close to Eurasia, and finally the two continental edges collided around 24 Ma (Fig. 18.22). There were important times of uninterrupted migration between the two land masses. The interchanges affected animal life throughout the Old World, almost on the same scale as the Great American Interchange (Fig. 18.23).

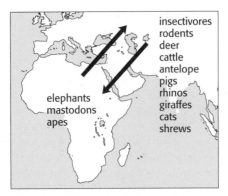

Figure 18.23 The Great Old World Interchange of animals between Africarabia and Eurasia in the Miocene.

Twelve families of small mammals appeared in Africa in the Early Miocene, mostly insectivores and rodents from Eurasia. Early deer, cattle, antelope, and pigs largely replaced the hyraxes at medium sizes, and rhinos and the first giraffes were large invaders. Cats arrived and began to replace the older creodonts. Going the other way, elephants walked out of Africa into Eurasia, in at least two major adaptive groups, mastodons and true elephants. Some large creodonts even reinvaded Eurasia from Africa.

In a second exchange around 15 Ma, a new set of African animals, including apes, quickly spread over the woodlands and forests of Eurasia. Hyenas and shrews migrated into Africa.

It is not clear whether the continental collision itself altered world climate, or whether climate was affected more by major events in the Southern Hemisphere. Whatever the cause or causes, the Miocene change from forest to savanna was partly responsible for the success of the large number of grazing animals listed above.

By the end of the Miocene, more immigrants had appeared in Africa: small animals, including many bats, and the three-toed horse *Hipparion*. Meanwhile, hippos evolved in Africa, and the antelope and cattle that had arrived earlier evolved into something close to the incredible diversity we see today in the last few game reserves.

Africa and Eurasia have been connected by land since the Miocene, but this does not automatically imply free exchange of animals. For example, the Mediterranean Sea dried into a huge salty desert like a giant version of Death Valley, around 6 Ma. Only a few animals could have crossed this barrier. Later, the development of desert conditions in the Sahara formed another fearsome barrier to animal migration for most of the past few million years. Today, North African animals are more like those of Eurasia than those of sub-Saharan Africa.

The Early Pleistocene saw a large extinction in Africa, with one-third of the mammals becoming extinct. But they were replaced by newly evolving species, so total diversity remained high. Africa apparently did not feel the effects of the ice ages too drastically. Human hunting activities have affected Africa less than other continents, perhaps because humans evolved gradually there and the animals had time to adjust to them. On other continents human impact was much more sudden and severe. Animals formerly widespread over the world are now confined to Africa or nearly so (rhinos, lions, cheetahs, hyenas, and wild horselike species, zebras). Protected there by the geographical, climatic, and historical events of the Late Cenozoic, many creatures survived relatively successfully in Africa until this century.

Islands and Biogeography

Strict geographic barriers prevent land plants and animals living on islands from moving easily to other land areas, and potential invaders also must cross barriers. This means that island faunas and floras tend to evolve in greater isolation than those with wider and more variable habitats. Of course, this is true at any scale, whether we look at small islands or continent-sized ones. Islands past and present can teach us a great deal about evolution. It is no accident that Darwin was particularly enlightened by his visit to islands like the Galápagos, and Wallace by his years in Indonesia.

We have seen some of the vagaries of continental faunas over a time scale of tens of millions of years, but it is worth looking at cases where smaller-scale events on smaller islands over smaller lengths of time show the rapidity and power of natural selection in isolated populations.

The Raptors of Gargano

In 1969, three Dutch geologists were exploring the Mesozoic limestones of the Gargano Peninsula in southern Italy (Fig. 18.24). Sometime in the Early Cenozoic, this block of land was raised above sea level and caves and fissures formed in the limestone. In Early Miocene times the Gargano area was cut off from the mainland by a rise in sea level to form an island in the Mediterranean Sea of the time. Land animals living there were isolated on the island as the sea rose. Over only a few million years, animals occasionally fell into fissures in the limestone, where they were covered by thin layers of soil and preserved as fossils. Today, the limestone is quarried for marble, and the fossil bones can be found in the pockets of ancient soils exposed in the quarry walls.

No large animals were isolated on Gargano as it was cut off. The only large reptiles were swimmers (turtles and crocodiles) and the only mammalian carnivore was also a swimmer, a large otter with rather blunt teeth that probably ate shellfish most of the time and would not have hunted on land.

Because there were no land carnivores, small mammals evolved quickly into spectacular forms. Small rabbit-like pikas were abundant, and gigantic dormice evolved on the

Figure 18.24 The Gargano Peninsula in southern Italy. It was a small rocky offshore island in Miocene times.

island. Giant hamsters were eventually outcompeted by true rats and mice. Some of the Gargano mice grew to giant size, with skulls 10 cm (4 inches) long, and many evolved fast-growing teeth as complex as those of beavers. They probably chewed very tough material. *Hoplitomeryx* is a small deer which evolved horns instead of antlers.

If there were no cats, dogs, or other terrestrial carnivores, how were the rodents kept under control? By disease and starvation? And why did *Hoplitomeryx* evolve spectacular horns, if there were no carnivores to fight off? The horns were too lethal to have been used for fighting between individuals of the species.

The answer to these questions seems to have been raptors—birds of prey. A giant buzzard, *Garganoaetus*, was as large or larger than a golden eagle. Presumably it hunted by day. It would have been perfectly capable of taking a small or young *Hoplitomeryx*, and the horns may have evolved to protect the back of the neck against raptors (horns are year-round while antlers are shed seasonally). Normally, small deer hide in vegetation, but Gargano was a bare, limestone island, with no cover by day. At night, the owls took over: the largest barn owl of all time evolved on Gargano.

Giant Pleistocene Birds on Cuba

Cuba had a strange set of animals isolated on it during the ice ages. Pleistocene mammals have been found in enormous numbers in limestone cave deposits, and we have a reasonable idea of the unusual ecology the island must have had. In particular, there were enormous numbers of ground sloths and rodents, and insectivores were very common. Tens of thousands of mouse jaws have been found in one cave, and another site yielded over 200 ground sloths. In addition, large numbers of fossil vampire bats imply that there were large numbers of warm blooded animals for them to prey on. Similar but less spectacular fossils have also been found on Puerto Rico and Hispaniola.

There are practically no carnivorous mammals in these deposits, and as at Gargano, we are forced to wonder what kept the animal populations in check. The answer here too seems to be raptors. In the caves with the animal bones there are also great numbers of the bones of small birds. This suggests that the cave deposits are mainly the accumulations of owl pellets and bat colonies. But the size of the bones indicates that the owls were producing pellets much larger than normal owls do.

In 1954 a gigantic fossil owl was discovered, large enough to have preyed upon baby ground sloths. Later that year, a fossil eagle bigger than any living species was found. A fossil vulture as large as a condor, and a fossil barn owl as large as the species at Gargano, fill out a picture of a set of predators quite alien to our experience today.

The gigantic owl *Ornimegalonyx* must have stood a meter high It may not have been a powerful flyer, because its breastbone looks weak relative to the rest of the skeleton. But with its tremendous beak and claws, it could have preyed successfully on rodents and young sloths. By day the giant eagle would have performed the same function—it is larger than the monkey-eating eagle of the tropical forest today. Presumably the giant vulture fed from the carcasses of giant ground sloths, and the other large owls added to the flying nocturnal predators.

The whole ecosystem became extinct towards the end of the Pleistocene on Cuba and on all the other Caribbean islands. We don't know enough of the geological history of Cuba to suggest that human intervention caused these extinctions.

Further Reading

Archer, M. et al. 1989. Fossil mammals of Riversleigh, Northwestern Queensland: preliminary overview of biostratigraphy, correlation and environmental change. *Australian Zoologist* 25: 29–65. Available at http://www.create.unsw.edu.au/research/files/Archer%20et%20al%20(1989)%20Fossil%20Mammals%20of%20Riversleigh,%20Northwestern%20Queensland.PDF

Diamond, J. M. 1990. Biological effects of ghosts. *Nature* 345: 769–770. [The prehistoric ecology of New Zealand.]

Flannery, T. F. 1995. *The Future Eaters: An Ecological History of the Australasian Lands and People.* New York: George Braziller. [Part I is the story of Australasian life before the arrival of humans.]

Kappelman, J. et al. 2003. Oligocene mammals from Ethiopia and faunal exchange between Afro-Arabia and Eurasia. *Nature* 426: 549–552, and comment, pp. 509–511.

Marshall, L. G. 1988. Land mammals and the Great American Interchange. *American Scientist* 76: 380–388. Available at http://www.eebweb.arizona.edu/Courses/Ecol485_585/Readings/Marshal_1988.pdf

Marshall, L. G. 1994. The terror birds of South America. *Scientific American* 270 (2): 90–95.

Pascual, R. et al. 1992. First discovery of monotremes in South America. *Nature* 356: 704–706.

Paterson, A. B. "Banjo". 1933. Old Man Platypus, in *The Animals Noah Forgot*. Sydney, Australia: Endeavour Press. Available at http://www.middlemiss.org/lit/authors/patersonab/animalsnoahforgot/oldmanplatypus.html

Rich, P. V. and G. F. van Tets. (eds.) 1985. *Kadimakara*. Victoria, Australia: Pioneer Design Studio. [The extinct animals of Australia.]

Richardson, K. C. et al. 1986. Adaptations to a diet of nectar and pollen in the marsupial Tarsipes rostratus (Marsupialia: Tarsipedidae). *Journal of Zoology, London A* 208: 285–297.

Simpson, G. G. 1980. *Splendid Isolation*. New Haven: Yale University Press. [South American mammals.]

Springer, M. S. et al. 1997. Endemic African mammals shake the phylogenetic tree. *Nature* 368: 61–64.

Webb, S. D. 2006. The Great American Biotic Interchange: patterns and processes. *Annals of the Missouri Botanical Garden* 93: 245–257.

Wroe, S. et al. 2005. Bite club: comparative bite force in big biting mammals and the prediction of predatory behaviour in fossil taxa. *Proceedings of the Royal Society B* 272: 619–625. [Thylacoleo.] Available at http://www.ncbi.nlm.nih.gov/pmc/articles/PMC1564077/

Questions for Thought, Study, and Discussion

1. Find three examples where mammals did manage to cross ocean barriers and colonize another continents. Think of ways that ocean crossing could have happened, and try to think of characters that such pioneering animals had in common.

2. Choose a good example of parallel evolution in two mammals or mammal groups in differet continents. Describe the similarities, try to identify differences, and explain how the similarities are related to their way of life.

NINETEEN
Primates

In This Chapter

We are particularly interested in our own ancestry. After all, the recent evolution of primates has produced humans, the most widespread, powerful, and destructive biological agent on Earth. It has always seemed reasonable that the earliest primates would have been small tree-dwellers, but that has now been confirmed by multiple fossil discoveries. It doesn't mean the problems have gone away, because those many small early primates radiated into all living ones, and we have not yet untangled the evolutionary pathways. Living lemurs and lorises are African, and probably always have been. Living tarsiers are probably survivors of the early primate branch that includes anthropoids (us, other apes, and monkeys). That early branch eventually radiated in Eocene and Oligocene times in Africa, where ancestral monkeys and apes have been found. One branch managed to cross the Atlantic to South America to become the South American monkeys. The other branch was dominated at first by African apes, especially in the Miocene of East Africa. Late in the Miocene, some African apes migrated to Asia and were the ancestors of the living orangutan. The apes that remained in Africa were the ancestors of living African apes, and of us: the hominoids.

Primate Characters

Most living primates are small, tropical, tree-dwelling animals that eat high calorie food, mainly insects. This is particularly true of groups that retain the most primitive primate characters. At face value, this suggests that primate ancestors searched for insects, fruit, seeds, or nectar on small branches, high in trees, and in smaller bushes. Evolutionary evidence supports this scenario, because primates are most closely related to two other mammal groups that also live in trees. This group of small arboreal mammals probably invaded forest habitats after the K-T extinction. The surviving members of that clade are primates, tree shrews, and the one surviving species of dermopteran, the colugo or "flying lemur" of the Indonesian rain forest (Fig. 19.1). All these animals were small and probably ate nectar, gum, pollen, seeds, insects, and fruit in the canopy forest.

History of Life, Fifth Edition. Richard Cowen.
© 2013 by Richard Cowen. Published 2013 by Blackwell Publishing Ltd.

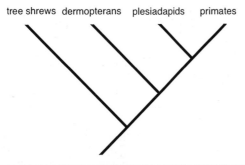

tree shrews dermopterans plesiadapids primates

Figure 19.1 Cladogram of living tree-dwelling mammals close to primates, plus the extinct plesiadapids. This cladogram, one of several possibilities, shows primates and plesiadapids as closest allies.

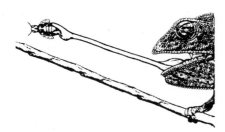

Figure 19.2 Catching small agile prey without moving the body. The chameleon solves the problem in a different way than primates do.

Living primates have large eyes, turned forward to give excellent stereoscopic vision. The combination of large eyes and stereoscopic vision may have evolved in primates—as also in cats, owls, and fruit bats—to help search for food by sight rather than smell, because it allows the animal to judge the distance of a food item without moving its head. Stereoscopic vision promotes agility and coordination, especially when an animal has hands and feet adapted for grasping and fine manipulation, with pads and nails rather than paws and claws. Grasping feet and hands allow primates to forage along narrow branches, and live prey or other food can be reached or seized by a hand or hands rather than by a lunge with the whole body and head. Compare the coiled strike of snakes and the tongue strike of chameleons, frogs, and toads, which all do the same thing in different ways (Fig. 19.2).

Primate fetuses show rapid growth of the brain relative to the body, so they are born with relatively larger brains than other mammals. Gestation time is long for body size, and primates have small litters of young that develop slowly and live a long time. Primates evolved high learning capacity, complex social interactions, and unusual curiosity. The evolution of curiosity is useful in searching for food, and high learning capacity, memory, and intelligence help individuals to make correct responses in a complex, ever-changing environment.

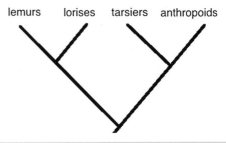

lemurs   lorises   tarsiers   anthropoids

Figure 19.3 Cladogram of the major groups of living primates.

Living primates are often divided into two groups: small-brained, small-bodied animals called prosimians, and the relatively large-brained anthropoids (monkeys and apes). However, prosimians contain two clades, each with a long evolutionary history. The lorises and lemurs are an African group of primates that forms one clade (Strepsirrhini or "wet noses," while tarsiers form another, linked with anthropoids in the Haplorhini ("dry noses") (Fig. 19.3). But there are many early primates that muddy our picture of very early primate evolution.

The Living Prosimians

Living lemurs are confined to Madagascar, and must have reached that island from Africa. Molecular evidence suggests that the ancestors of lemurs reached Madagascar in Paleocene times, but no other primates arrived there until humans did about 2000 years ago. Lemurs flourished in their island refuge. The actual fossil record of lemurs in Madagascar goes back only as far as the Miocene, but that is enough to document a startling radiation into at least 45 species of lemurs on the island, adapted to a great variety of life styles.

Living lemurs are specialists at vertical clinging and leaping, in which the front limbs are used for manipulating, grasping, and swinging, while the hind limbs are powerful for pushing off. Most lemurs are medium-sized (weighing a few pounds), and are omnivorous, eating fruits and leaves. A few lemurs are small: *Microlemur* the mouse lemur (Fig. 19.4) weighs only 50 grams or so (about 2 ounces). The largest lemur, *Archaeoindris*, reached around 200 kg, the size of a gorilla (Fig. 19.5), and became extinct only recently; as an adult it must have been a ground dweller. The recently discovered extinct lemur *Palaeopropithecus* was adapted for moving slowly in the forest canopy in the same way as the South American sloth, while *Megaladapis* was probably rather like the Australian koala in its ecology.

Lorises and bushbabies are small, slow-moving, nocturnal hunters of insects. Lorises live in African and Southeast Asian tropical forests, while bushbabies are exclusively African.

Tarsiers, in comparison, are small, agile primates, adapted to eating small animals and insects (Fig. 19.6). Today they

Figure 19.4 The grey mouse lemur, *Microlemur* from Madagascar. Photograph by Gabriella Skollar, edited by Rebecca Lewis, and placed into Wikimedia.

Figure 19.6 Tarsier from the Philippines. Photograph by Roberto Verzo and used under Creative Commons license.

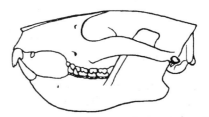

Figure 19.7 The skull of a typical plesiadapid, one of a radiation of early primatelike animals that included animals with an ecology like rodents. (After Gingerich and Krause.)

Figure 19.5 The largest lemur, *Archaeoindris* from the Pleistocene of Madagascar, now extinct. Art by Smokeybjb, based on a reconstruction by Stephen Nash, and placed into Wikimedia.

Earliest Primates

The plesiadapids, an important group of animals found mainly in the Paleocene of North America and Europe, are the extinct sister group of primates (Fig. 19.1). Larger plesiadapids were rather heavy in build, ecologically like squirrels or marmots, small brained and rather small-eyed, with teeth adapted for cropping vegetation. Because plesiadapids looked and probably lived like large rodents (Fig. 19.7), they may have competed to some extent with Paleocene multituberculates (Chapter 15).

Some of the last plesiadapids evolved some characters that are found also in primates: the plesiadapid *Carpolestes* (Fig. 19.8) had grasping hands and feet, and the big toe had a nail (like a primate) while the other toes had claws (like other plesiadapids). But *Carpolestes* did not have stereoscopic vision, and apparently did not leap. It is most likely that *Carpolestes* evolved its "advanced" characters in parallel with true primates (Bloch and Boyer 2002).

Recently research has begun to focus on China in the search for the early primate radiation, and the ancestry of

survive only in Southeast Asia, but were much more widespread in the past. Essentially, tarsiers are living fossils, with possible ancestors in the early Cenozoic that seem to have had much the same anatomy and way of life. They diverged from anthropoids so long ago that the two groups share little similarity today.

Figure 19.8 The advanced plesiadapid *Carpolestes* from the Eocene of North America (see text). Image by Sisyphos23, based on a drawing by Mateus Zica, and placed in Wikimedia.

Figure 19.9 Life reconstruction of *Darwinius* by Nobu Tamura, and placed into Wikimedia.

Figure 19.10 Life restoration of *Tetonius*, a small Eocene omomyid from North America, as an alert tarsierlike animal. By L. Kibiuk under the supervision of K. D. Rose. Courtesy of Kenneth D. Rose of The Johns Hopkins University.

small, alert, active nocturnal insect eaters in the forest (Fig. 19.10). Ecologically they were probably like tarsiers, and in terms of evolution they were probably near the base of the anthropoid/tarsier clade (Fig. 19.1).

The northern continents slowly cooled as they drifted northward during the Eocene. Finally, at the end of the Eocene, primates disappeared from northern latitudes. Refugee adapids reached Southeast Asia in the Late Eocene, but "anthropoid" characters that some of them have probably evolved independently in this region, and they seem to have died out without descendants. By Oligocene times there were practically no primates left in northern continents.

The Origin of Anthropoids

The living higher primates, or anthropoids (monkeys and apes), have evolved into a variety of life styles and habitats that extends from the huge herbivorous gorillas to the tiny gum-chewing marmosets of South America. Various Eocene primates were most likely adept at four-footed climbing and leaping from branch to branch in three dimensions, using the full grasp of hands and feet for catching and holding small branches. All the different ways

anthropoids. Anthropoids may have evolved in East Asia, but at the moment that is only the best guess among several alternatives.

The very warm PETM event at the end of the Paleocene (Chapter 18) allowed Asian mammals to reach North America. Among the primates that arrived were omomyids and adapids. Adapids include *Diacronus*, from the Paleocene of South China, a plausible ancestor for the Eocene adapids in western North America. Adapids look like small lemurs in limb structure, and probably moved in the same way. Many adapids evolved toward larger body size and turned to a diet that included much more plant material as well as animal prey, with some eating fruit and others leaves. *Darwinius* (Fig. 19.9, Fig. 19.11) is the best-preserved adapid, from the Eocene of Germany. Omomyids were

Figure 19.11 The beautifully preserved adapid *Darwinius* from the Eocene of Germany. Photograph from Franzen et al. (2009), published in PLoS ONE, http://www.plosone.org/article/info%3Adoi%2F10.1371%2Fjournal.pone.0005723;jsessionid=E8154D7406947B36A39470C790A4F08C Publication in PLoS ONE automatically places the image into Wikimedia.

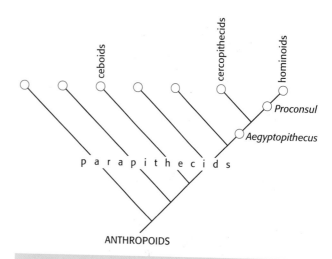

Figure 19.12 Phylogram of the higher primates, including evidence from the fossils of the Fayum. A miscellaneous group of Fayum anthropoids, the parapithecids, include the ancestors of ceboids (New World monkeys), and of Aegyptopithecus, which has characters that allow it to have been the ancestor of both Old World monkeys (cercopithecids) and the hominoids. *Proconsul* is a possible ancestor of all hominoids. This scheme is consistent with biogeographic evidence that suggests that the ceboids diverged from Old World primates around the end of the Eocene.

in which lemurs, monkeys, gibbons, great apes, and humans move could have evolved from this generalized style shared by early primates. Of course, that does not help us find which Eocene primates were the anthropoid ancestors.

The arm-swinging or brachiating of gibbons could have arisen by emphasizing the arms in movement. The careful, multilimbed climbing of orangutans in trees, the agility of monkeys, the four-footed scrambling and shambling of heavy apes, and the trotting of baboons on the ground could each have evolved by using all the limbs equally. The bipedal walking and running of australopithecines and humans could have been achieved by accentuating the role of the hind limbs in powerful pushing and of the fore limbs in grasping and handling.

Living anthropoids are divided into three evolutionary groups: cercopithecoids (Old World monkeys), ceboids (New World monkeys), and hominoids (Fig. 19.12), which include gibbons, apes, and humans. Most evidence points to the Eocene of Africa as the time and place for the radiation of anthropoids, and may imply they originated there too.

The Late Eocene Primates of Egypt

In Late Eocene and Oligocene times, the Fayum district of Egypt, not far from Cairo, lay on the northern shore of Africarabia as it drifted slowly northeastward (Chapter 18). Thousands of fossilized tree trunks, some of them more than 30 meters (100 feet) long, show that tropical forests of mangroves, palms, and lianas grew along the levees of a lush, swampy delta. Water birds such as storks, cormorants, ospreys, and herons were abundant, as they are today around the big lakes of central Africa. Fishes, turtles, sea snakes, and crocodiles lived in or around the water, and early relatives of elephants and hyraxes foraged among the rich vegetation. The primates presumably ate fruit in the trees. The same fauna has been discovered as far south as Angola, so the Fayum animals were widespread around the coasts of Africa in Eocene and Oligocene times.

More than 2000 specimens of 19 species of fossil primates have now been collected from the Fayum deposits, most of them on expeditions led by Elwyn Simons. The primates include tarsiers, lorises, and bushbabies. But the others are anthropoids, all of which look like tree-climbing fruit and insect eaters.

Many of the Fayum primates have some advanced characters, but do not look like the direct ancestors of monkeys or apes. They are placed into a basal stem-anthropoid group called parapithecids (Fig. 19.12). Parapithecids are small, weighing only up to 3 kg (7 pounds). Their skulls are rather like those of Old World monkeys, but the rest of the skeleton looks primitive, more like that of South American monkeys. This miscellaneous group probably had a basic style of primate ecology, eating fruit in the trees. *Apidium*, for example, seems to have been adapted for leaping and grasping in trees.

Those Fayum anthropoids that are well enough known to compare individual sizes are sexually dimorphic. Males are larger and had much larger canine teeth than females, implying that males displayed or fought for rank, and that the animals had a complex social life that included groups of females dominated by a single male. Among living primates, it is generally the larger-bodied species that have these characters, especially in the Old World. However, the Fayum anthropoids show that size is not important in evolving these sex-linked characters, and they also suggest that these traits may well be basic to anthropoids. Simons and his colleagues suggest that they arose when anthropoids became active in daylight: group defense may be linked with the social structure.

Aegyptopithecus (Fig. 19.13) is the best known of the Fayum anthropoids. It is a larger, monkey-sized primate with an adult weight of 3 to 6 kg (7 to 14 pounds). Its heavy limb bones suggest that it was a powerfully muscled, slow-moving tree climber, ecologically like the living howler monkey of South America. It had many primitive characters, but its advanced features were more like those of apes than monkeys. Its brain was large for its body size, for example, and its foot bones were like those of Miocene apes. It had powerful jaws for its size, too. It may well be

Figure 19.13 One of the fine casts on display at the Museum of Anthropology at the University of Zürich. This is *Aegyptopithecus* from the Fayum deposits of Egypt. Photograph by Nicholas Guérin, and used under the Creative Commons Attribution-Share Alike 3.0 Unported license.

the common ancestor of all higher primates in the Old World: the cercopithecoids, or Old World monkeys, and *Proconsul* and the line leading to hominids (Fig. 19.12).

The New World Monkeys

Primates reached South America by Oligocene times, and evolved there in isolation, never again influenced by exchange and contact with other primate groups. No prosimian or apelike primate has ever been found in South America. Instead, the New World primates evolved to fill the ecological niches that monkeys and gibbons occupy in Old World forests.

New World primates, the ceboids, probably evolved from African immigrants that crossed the widening Atlantic in early Oligocene times. For want of better information I have shown them as diverging from early Fayum parapithecids (Fig. 19.12). But ceboids have some unique characters: prehensile tails that can be used as a fifth limb, and four more teeth than Old World primates, cercopithecoids. Ceboids are related to cercopithecoids only in that they are both anthropoids (Fig. 19.12); many of their monkey-like characters evolved independently. Ceboid color vision uses different nerve pathways than that of cercopithecoids and apes, for example, and it evolved once only, very early.

The earliest ceboids, *Dolichocebus* from Patagonia and *Branisella* from Bolivia, are both Late Oligocene in age, perhaps 26 or 27 Ma. *Dolichocebus* is very much like the living squirrel monkey, and is an ideal ancestor for it. In the same way, some Miocene South American monkeys are much like living spider monkeys and howler monkeys. A genuinely modern-looking owl monkey is known from Miocene rocks of Colombia at 12 to 15 Ma. Many of today's South American monkeys therefore qualify as living fossils. Either they evolved early and rapidly, or they have a longer fossil record still to be discovered.

All of them are tree dwellers. South American primates did not evolve into terrestrial ways of life as Old World primates did, even though there have been extensive savannas in South America since the Miocene.

The Old World Monkeys

We know more about the emergence of Old World monkeys. The small fossil *Victoriapithecus* from the Miocene of East Africa is an ideal ancestor for this group. Interestingly, it is small even for a monkey at 3–5 kg (7 to 11 pounds), and seems to have had a semi-terrestrial ecology rather than the tree-dwelling habit that one might have expected.

Emergence of the Hominoids

Living hominoids include hylobatids (gibbons); pongids or Asian apes (only orangutans survive); panids, the African apes (chimps and gorillas); and hominids (only humans survive). The physical, molecular, and genetic structure of

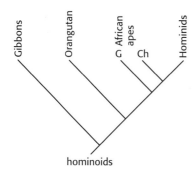

Gibbons
Orangutan
G African apes
Ch
Hominids

hominoids

Figure 19.14 The cladogram of living hominoids that is suggested by molecular and fossil evidence. G, gorillas, Ch, chimps. The timing of the branching points is uncertain, and is not likely to be settled soon. Extinct groups are not shown on this cladogram.

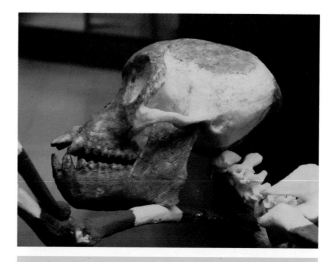

Figure 19.15 One of the fine casts on display at the Museum of Anthropology at the University of Zürich. This is a skull of *Proconsul* from the Miocene of East Africa. Photograph by Nicholas Guérin, and used under the Creative Commons Attribution-Share Alike 3.0 Unported license.

living hominoids has been studied closely. Humans, gorillas, and chimps are very similar in genetic makeup and in protein chemistry, much closer than they are in body structure, but the orangutan differs significantly, and gibbons even more.

Hominoids almost certainly evolved from some African genus like *Aegyptopithecus*. The DNA clock suggests that the common ancestor of all hominoids split from monkeys about 33 Ma. The gibbons split off about 22 Ma, followed by the orangutan lineage at about 16 Ma. Finally the various living lineages of panids and hominids diverged from one another between 10 Ma and 6 Ma (Fig. 19.14). Protein clocks suggest more recent branching points.

Miocene Hominoids

About 20 Ma, Africarabia formed a single land mass that lay south of Eurasia and was separated from it by the last remnant of the Tethys Ocean. African animals were evolving largely in isolation from the rest of the world, and some groups, including the hominoids, were confined to Africarabia at this time, though they were widespread across it.

Early Miocene faunas of Africa were dominated by elephants and rhinos at large body size, primitive deer and hyraxes at medium size, and insectivores common at small sizes. The environment was forest, broken by open grassland and woodland. Primates of all kinds flourished, although it is difficult to describe their ecology and habits because body skeletons are not as well known as skulls. But prosimians and monkeys were rare, while hominoids were diverse and abundant. We have over 1000 hominoid fossils from the Early Miocene of Africa, most dating from 19 to 17 Ma and most from East Africa.

The dominant hominoids were the apelike dryomorphs. Like living African apes, they had relatively small cheek teeth with thin enamel, implying a soft diet of fruits and leaves, and a way of life foraging and browsing in trees like

most living monkeys. (True monkeys were scarce at this time, remember.) Dryomorphs varied in weight from a large species of *Proconsul* in which males weighed about 37 kg (80 pounds) down to *Micropithecus* at about 4 kg (9 pounds), and their locomotion varied accordingly. *Micropithecus* and *Dendropithecus* were not as well adapted for arm-swinging as living gibbons, but they were lightly built and relied more on brachiating than did other Miocene primates.

The best-known and most important dryomorph is *Proconsul* (Fig. 19.15). There were several species of this animal by 18 Ma. The most complete specimens are from a small species that weighed only about 9 kg (20 pounds) but had a baboon-sized brain; that is, its brain was larger relative to body size than that of living monkeys. Its skeleton was a mixture of primitive characters that are also found in monkeys, gibbons, and chimps; altogether, these characters indicate a rather basic quadrupedal, tree-climbing, fruit-eating primate (Fig. 19.16) that could be the ancestor of all later hominoids. *Proconsul* had advanced hominoid characters of the head and jaws, though most of its body skeleton remained unspecialized. A large *Proconsul* may have spent a lot of time in deliberate climbing or on the ground, like a living chimp. Like them, it was probably versatile in its movements, and capable of occasional upright behavior.

Morotopithecus is a large ape from Uganda, probably as old as 20 Ma. Although we do not have a skeleton as complete as *Proconsul*, *Morotopithecus* is clearly large (40 to 50 kg, or 100 pounds), and the pieces we have are more advanced than the same pieces of *Proconsul*. In other words, *Morotopithecus* is probably close to the direct ancestry of

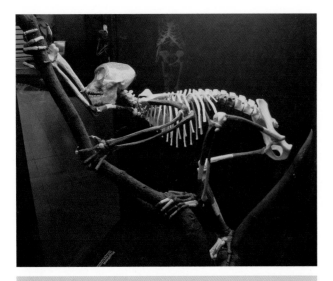

Figure 19.16   One of the fine casts on display at the Museum of Anthropology at the University of Zürich. This is a reconstructed skeleton of *Proconsul* from the Miocene of East Africa. Photograph by Nicholas Guérin, and used under the Creative Commons Attribution-Share Alike 3.0 Unported license.

all later hominoids. It was probably a rather heavy slow climber, hanging in trees and eating fruit.

Africarabia drifted northward during the Miocene (Chapter 18) and finally collided with Eurasia to form an irregular mountain belt from Iran to Turkey. The collision interrupted tropical oceanic circulation and set off climatic changes. Temperatures cooled in East Africa, and almost all the northern continents experienced dramatic changes in faunas and floras. Forests became much more open, and grasses evolved to form wide expanses of savanna.

An exchange of animals between Africarabia and Eurasia added to the ecological turmoil of the times (Chapter 18). In that process, African hominoids successfully invaded Eurasian plains and woodlands.

In Africa the dryomorphs remained in the forests, which were thinned or diminished by cooling temperatures. They came under increasing pressure from the evolving monkeys. Monkeys have increased in abundance and diversity so that today they rather than apes dominate the remaining forests of Africa and Asia.

Some late dryomorphs reached Eurasia, but they were apparently numerous only in Europe. *Dryopithecus* is a European fossil, known from Spain to Hungary. Perhaps because it lived in a cool region, *Dryopithecus* was bigger and stronger than most dryomorphs. It shows adaptations for branch-swinging with the trunk more or less vertical, which gave it some of the characters of the living orang. Its skull is not like that of orangutans, however, and the position of *Dryopithecus* is closer to African hominoids.

Sivapithecids

At about 14 Ma, new hominoids appeared alongside the dryomorphs: the sivapithecids were the dominant group in East Africa. The earliest sivapithecid is *Kenyapithecus*, dated about 14 Ma. It is generalized enough to be a descendant of *Proconsul*, and to be the ancestor of all later large apes: sivapithecids + pongids on one hand, and panids + hominids on the other.

Sivapithecids are apes, known from a wide area that stretches from East Africa to Central Europe, and eastward as far as China. *Sivapithecus* itself was an Asian ape. We have a great number of sivapithecid fossils, but they are mostly jaws, skulls, and isolated teeth; few body or limb bones are well known. Thus we can reconstruct sivapithecid heads rather well, but we know little about their body anatomy, posture, or locomotion.

Many different names have been applied to sivapithecids in their various countries of discovery. Hungarian, Turkish, Kenyan, Indian, Chinese, and Greek specimens were all given different names, for example. Part of the problem of naming sivapithecids is that there is a good deal of variation between individuals. As in orangutans, the skulls of males are much larger and broader than those of females.

All sivapithecids had thick tooth enamel and powerful jaws, suggesting that their diet required prolonged chewing and great compressive forces on the teeth. In living primates with thick enamel, such as orangutans or mangabeys, teeth and jaws like these are correlated with a diet of nuts, or fruits with hard rinds. One can hear an orangutan cracking nuts a hundred meters away! Perhaps sivapithecids diverged from the dryomorph diet of soft leaves and fruits to exploit a food source that had so far been available only to pigs, rodents, and bears.

Nut eating can be an activity of tree or ground dwellers, or creatures making the evolutionary and ecological transition from woodland to open ground. We cannot yet tell whether sivapithecids were foraging for fruit and nuts in the trees (with an arboreal life like that of orangutans) or under the trees (with adaptations for ground living).

One late sivapithecid was adapted to live entirely on the ground. The huge ape *Gigantopithecus* lived in southern and eastern Asia from about 7 Ma well into the Pleistocene. It had huge grinding teeth and weighed several hundred pounds. It probably lived on very coarse vegetation, as an ecological equivalent of the giant ground sloth of the American Pleistocene, or the Asian giant panda, or the African mountain gorilla. *Gigantopithecus* survived in Asia as recently as 300,000 or 250,000 years ago. It was certainly contemporaneous with *Homo* in Eastern Asia, and its bones, teeth, and jaws may be responsible for Himalayan folklore about the abominable snowman, or yeti (Fig. 19.17).

It is clear now that sivapithecids have nothing to do with human ancestry but are instead ancestors of the living Asian ape, the orangutan. In Africa, dryomorphs evolved toward hominids (Fig. 19.18). The molecular clock sug-

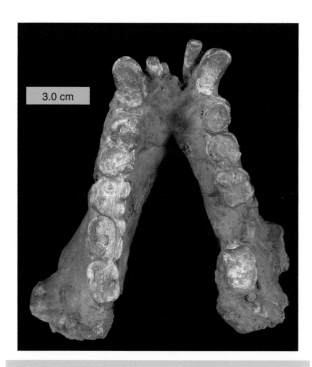

Figure 19.17 The massive lower jaw and teeth of the huge sivapithecid *Gigantopithecus*. Photograph by Dr. Mark Wilson of the College of Wooster, and placed into Wikimedia.

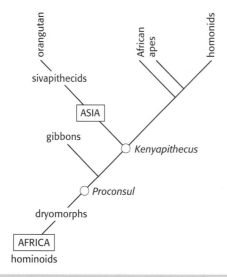

Figure 19.18 A phylogram that shows the sivapithecids fitting into hominoid evolution as ancestors of the orangutan.

gests 17 Ma for the divergence, and new Kenyan fossils seem to agree with that estimate.

After about 11 Ma, migration between Africa and Eurasia was essentially cut off. The hominoid groups evolved independently in Eurasia and Africa, eventually leaving the sivapithecids in Asia and the hominids in Africa. The African fossil record of hominids is horribly incomplete during the critical time after 11 Ma when they radiated to become separate lineages: we simply haven't found these fossils yet.

Between 8 and 5 Ma, the climate of Eurasia slowly changed to encourage even more open grasslands instead of woodland and forest. Then the history of Eurasian apes became one of struggling survival rather than innovation and evolution. European dryomorphs disappeared around 8 Ma, and the only remaining sivapithecids were the East Asian animals that led to *Gigantopithecus* and the orangutan. This means that Eurasia is not the continent in which to search for direct human ancestry (Fig. 19.18). It is the African story that we must now follow.

By 7 to 5 Ma, the African forest had become dominated by monkeys, who displaced the dryomorphs ecologically and presumably restricted all the surviving forest apes. This is the time interval in which we can look for dramatic finds in the near future.

Further Reading

Bloch, J. I. et al. 2007. New Paleocene skeletons and the relationship of plesiadapiforms to crown-clade primates. *PNAS* 104: 1159–1164. Available at http://www.ncbi.nlm.nih.gov/pmc/articles/PMC1783133/

Bloch, J. I. and D. M. Boyer 2002. Grasping primate origins. *Science* 298: 1606–1610, and comment, pp. 1564–1565; arguments, *Science* 300: 741. The plesiadapid *Carpolestes* evolved some primate-like characters, probably independently. Available at http://pages.nycep.org/boyer/data/publications/Bloch&Boyer_science_2002.pdf

Boissinot, S. et al. 1998. Origins and antiquity of X-linked triallelic color vision systems in New World monkeys. *PNAS* 95: 13749–13754. The unusual color vision of New World monkeys evolved only once. See Sumner and Mollon 2000. Available at http://www.pnas.org/content/95/23/13749.full

Ciochon, R. et al. 1996. Dated co-occurrence of *Homo erectus* and *Gigantopithecus* from Tham Khuyen cave, Vietnam. *PNAS* 93: 3016–3020. Available at http://www.pnas.org/content/93/7/3016.full.pdf

Maiolino, S. et al. 2012. Evidence for a grooming claw in a North American adapiform primate: implications for anthropoid origins. *PLoS ONE* 7 (1): e29135. Available at http://www.plosone.org/article/info%3Adoi%2F10.1371%2Fjournal.pone.0029135

Moyá-Solá, S. and M. Köhler 1996. A *Dryopithecus* skeleton and the origins of great-ape locomotion. *Nature* 379: 156–159, and comment, pp. 123–124.

Simons, E. L. et al. 1999. Canine sexual dimorphism in Egyptian Eocene anthropoid primates: *Catopithecus* and *Proteopithecus*. *PNAS* 96: 2559–2562. Available at http://www.pnas.org/content/96/5/2559.full

Tattersall, I. 1993. Madagascar's lemurs. *Scientific American* 268 (1): 110–117.

Walker, A. and M. Teaford 1989. The hunt for *Proconsul*. *Scientific American* 260 (1): 76–82.

Williams, B. A. et al. 2010. New perspectives on anthropoid origins. *PNAS* 107: 4797–4804. Available at http://www.pnas.org/content/107/11/4797.full

Question for Thought, Study, and Discussion

If humans are so wonderful, how is it that dozens of other primate species also flourish all across the warm latitudes of the world? Choose a few varied primate species and find out how they make a living.

TWENTY

Evolving Toward Humans

In This Chapter

The African clade containing humans (*Homo*) and the earlier genus *Australopithecus* is the hominids. The emergence of hominids is not known very well because the fossil record is poor in number of specimens and quality of preservation. The earliest well-known hominid is *Ardipithecus*, dated from 5.5 Ma to 4.4 Ma in East Africa. There is little doubt that it is a direct ancestor of *Australopithecus*. Australopithecines include maybe a dozen species, all of them with large brains (compared with apes) and walking upright. Some species were clearly omnivorous, but some had huge teeth and jaws and probably ate tough vegetation for most of their diet. Beginning about 2.5 Ma, we begin to find primitive stone tools associated with australopithecines, and there is little doubt that they made and used such tools for butchering corpses of game animals. The corpses were probably stolen from big carnivores, so these australopithecines were acting more like jackals than hunters. From about 2 Ma, the earliest *Homo* seems to have behaved in much the same way. But about 1.5 Ma the species *Homo erectus* is associated with much more sophisticated tools that could well have been used for hunting: and the make-up of the African plains ecosystem changes at that time. *Homo erectus* and its close relatives spread all over South Asia, and finally as far north as Beijing. A branch of hominids migrated into Europe and evolved there for nearly a million years, eventually giving rise to the Neanderthals, *Homo neanderthalensis*. They were formidable Ice-Age hunters. *Homo sapiens* evolved in Africa perhaps 200,000 years ago (200 ka), and about 50 ka a wave of *sapiens* migrated out of Africa to take over most of the Old World. Older species became extinct during this time, though there is clear evidence of a small amount of interbreeding with Neanderthals. Natural selection continues to produce evolutionary change in *Homo sapiens*.

The Earliest Hominids

We know practically nothing of the evolution of the hominoid lineages that led to gorillas and chimps. Molecular and genetic evidence suggests that our closest living relatives are chimps, with gorillas a little further away. Our own lineage, the hominids, probably separated from that of chimps around 7 to 10 Ma (White et al. 2009). Even so, our DNA

History of Life, Fifth Edition. Richard Cowen.
© 2013 by Richard Cowen. Published 2013 by Blackwell Publishing Ltd.

is more than 95% identical to that of chimps. Obviously the 5% that is different reflects very important evolutionary changes in our bodies, brains, and behavior.

Over time, there have been perhaps a dozen species of hominids, but we, as *Homo sapiens*, are the only surviving one. As many as six earlier species of *Homo*, ranging back to about 2 Ma, have become extinct, and another six hominid species are placed in a group called **australopithecines**, which ranges back to about 5.5 Ma and contains the ancestor of *Homo*.

This simple picture is changing rapidly as we find more fossils. Be warned that almost everything I have written in this chapter is being argued over by paleoanthropologists. I have tried, as usual, to select what I think are the most likely hypotheses.

Sahelanthropus is dated close to the base of the hominids, at 6 Ma or 7 Ma (Fig. 19.18). It may *be* the first hominid. It is from Chad, far to the west of the "classical" East African sites. It is known from skull pieces, and shows a puzzling combination of very "primitive" characters (small brain, for example), with "advanced" characters such as eyebrow ridges (Brunet et al. 2002). However, the single skull was badly crushed (Fig. 20.1), and a different reconstruction of it might allow a different interpretation. As usual, we need more fossils!

Orrorin dates from perhaps 6 Ma in East Africa. It is known mostly from a few pieces of limb bones, so is difficult to place. Rivals of the discovery team are dropping hints that it may in fact be an ancestral gorilla or chimp rather than a hominid.

Since we cannot yet give a reasonable story of very early hominid evolution, we move to the australopithecines, the hominid species that predate *Homo*. They lived in Africa, south of the Sahara Desert, from perhaps 5.5 Ma to about 1.4 Ma or a little later. Overall, we are reasonably sure of the position of australopithecines in hominid evolution (Fig. 19.18) and have enough evidence to reconstruct a vivid picture of australopithecine life.

Early Australopithecines

The earliest two species of australopithecines are *Ardipithecus kadabba* from rocks in Ethiopia dated at 5.5 Ma and *A. ramidus*, also from Ethiopia and dated at 4.3–4.4 Ma (Fig. 20.2; White et al. 2009). *Ardipithecus* is the most primitive (that is, ape-like) australopithecine yet found, and was probably a forest dweller. It looks as if it was a careful climber in trees, and could move bipedally on the ground.

The slightly younger *Australopithecus anamensis*, from rocks in Kenya dated at 4.1–4.2 Ma, has a jaw that is even more ape-like, but its arm and leg bones suggest fully upright (bipedal) posture and locomotion. It would make a good ancestor for later *Australopithecus*: in fact there is some doubt whether *A. anamensis* is really a different species from the later species *A. afarensis*. *A. anamensis* is rather large, perhaps 50 kg, or 110 pounds. New specimens are still being cleaned, and new information will pour out in the next few years. However, it is fair to say that there are no major evolutionary surprises (yet) in these new fossils.

Footprints at Laetoli

The East African Rift splits East Africa from Ethiopia to Zambia and Malawi. Among its unusual geological features

Figure 20.1 A cast of *Sahelanthropus*, from the Miocene of Chad. Photograph by Didier Descouens, and placed into Wikimedia.

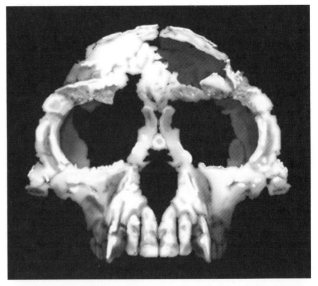

Figure 20.2 *Ardipithecus ramidus*, an early australopithecine from Ethiopia. Art by T. Michael Keesey, and placed into Wikimedia. Based on research by Gen Suwa et al. (see White et al. 2009).

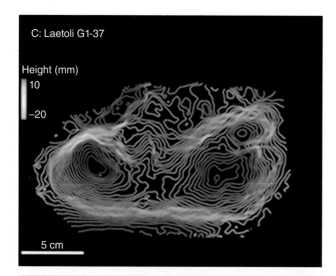

C: Laetoli G1-37

Height (mm)

10

−20

5 cm

Figure 20.3 One of the australopithecine footprints from Laetoli, laser scanned. You can see the heel and toe prints, with a separated big toe, and a distinct arch to the foot. All these are features of an efficient, erect walking action. From Figure 1 of Raichlen et al. (2010), published in PLoS ONE at http://www.plosone.org/article/info%3Adoi%2F10.1371%2Fjournal.pone.0009769, and thereby placed into Wikimedia.

are volcanoes that sometimes erupt carbonatite ash, which is composed largely of a bizarre mixture of calcium carbonate and sodium carbonate. One of these volcanoes, Sadiman, stood near the Serengeti Plain, in northern Tanzania. After carbonatite ash is erupted, the sodium carbonate in it dissolves in the next rain, and as it dries out the ash sets as a natural cement. Any animals moving over the damp surface in the critical few hours while it is drying will leave footprints that can be preserved very well. As long as the footprints are covered up quickly (for example, by another ash fall), rainwater percolating through the ash will react with the carbonate to make a permanent record.

Sadiman erupted one day about 3.6 Ma, towards the end of the dry season. Ash fell on the plains near Laetoli, 35 km (20 miles) away, and hominids walked across it, leaving their footprints along with those of other creatures. The vital point about the tracks is that the hominids were walking fully erect (Fig. 20.3), long before hominid jaws, teeth, skull, and brain reached human proportions, shape, or function.

Why would a hominid become bipedal? Most suggestions are related to carrying things with the hands and arms (infants, weapons, tools, food), to food gathering (seeing longer distances, foraging over greater ranges, climbing vertically, reaching high without climbing at all), to defense (seeing longer distances, throwing stones, carrying weapons), to better resistance to heat stress (less sweat loss and better cooling), or to staying within reach of rich food resources by migrating with the great plains animals (carrying helpless young over long distances). These are all reasonable suggestions, but all are difficult to test.

The footprints at Laetoli were made by australopithecines that were walking upright. All australopithecines are similar below the neck, apart from size differences, so they all probably moved in much the same way. Their movements were probably not exactly like ours, but their leg and hip bones indicate that they walked and ran efficiently. At the same time, the limb joints and toes suggest that they spent a lot of time climbing in trees as well as walking upright on the ground.

Probably the trend toward the use of the fore limbs for gathering food and the hind limbs for locomotion began among tree-dwelling primates long before *Australopithecus*. But this thought is based mainly on my own experience in picking and eating fruit, and the realization that fore limbs are more effective for that job than teeth and jaws alone. The final achievement of erect bipedality on the ground was probably an extension of previous locomotion and behavior, rather than something completely different.

Australopithecines were smaller than most modern people. They varied around 40 kg (90 pounds) as adults, but their bones were strongly built for their size. The skull was even stronger, and very different from ours. The relative brain size was about half of ours, even allowing for the smaller body size of *Australopithecus*, but the jaw was heavy and the teeth, especially the cheek teeth, were enormous for the body size. The canine teeth were large and projecting. The whole structure of the jaws and teeth suggests strength.

The small size of the brain and the thickness of the skull may be linked with another feature that separates us from *Australopithecus*. The birth canal in the pelvis of australopithecines is wide from side to side, but narrow from front to back, so that there may have been a special mode of delivery for even the small-brained babies that australopithecines had. In *Homo* the birth canal is rounder, presumably to accommodate the passage of a baby with a very large head (and brain). If so, a larger brain was important enough that this visible difference in skeletal pelvic anatomy was evolved in *Homo*. And it seems to have evolved in *Homo erectus*, long before *Homo sapiens* (Fig. 20.4).

Like ourselves, australopithecines were built for trotting endurance rather than blinding speed, an adaptation that would be better suited to foraging widely in open woodland than to skulking in forests, provided that *Australopithecus* was not easily picked off by large sprinting carnivores. (Reasons may have been related to group defense, which allows baboon troops to roam freely on the ground, or to early possession and use of defensive tools.) *Australopithecus* had long arms and fingers that were capable of sensitive motor control. In a tall biped walking upright, the arms would have been free for carrying, throwing, and manipulating.

Australopithecus Afarensis

The best-known collections of early australopithecines are from Laetoli and from Hadar in Ethiopia. Each district has produced spectacular finds. At Laetoli there are the footprints, plus remains of at least 22 individuals; at Hadar, bone fragments from at least 35 individuals are preserved

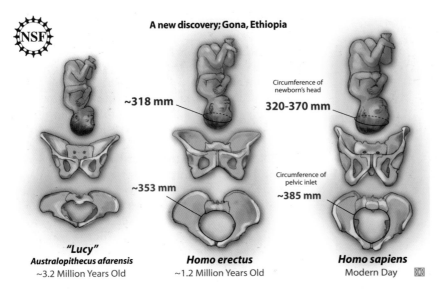

A new discovery; Gona, Ethiopia

~318 mm

Circumference of
newborn's head

320-370 mm

~353 mm

Circumference of
pelvic inlet

~385 mm

"Lucy"
Australopithecus afarensis
~3.2 Million Years Old

Homo erectus
~1.2 Million Years Old

Homo sapiens
Modern Day

Figure 20.4 Australopithecines have a pelvis that is relatively narrow from front to back. Modern humans, and *Homo erectus* from more than a million years ago, have a rounded pelvis, probably related to giving birth to a large-brained baby. Art by Zina Daretsky of the National Science Foundation, in the public domain.

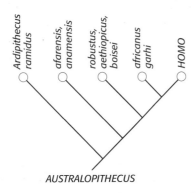

Ardipithecus ramidus

afarensis, anamensis

robustus, aethiopicus, boisei

africanus garhi

HOMO

AUSTRALOPITHECUS

Figure 20.5 Simplified cladogram of australopithecines. I have grouped rather similar species together into clades: the early primitive species *afarensis* and *anamensis*, for example; the "robust" australopithecines (they may belong to only one species, which would have the name *robustus*); and the "gracile" *africanus* and *garhi*. Note that if this cladogram is correct, *Australopithecus* is not a clade unless you include *Homo* in it. This "problem" is not really a problem: the aim of cladograms is to portray evolution, the naming schemes are simply convenient. *Ardipithecus* is an australopithecine, but a separate genus.

rather better. All the specimens belong to one species, *Australopithecus afarensis*, which was very closely related to *A. anamensis*, and was probably ancestral to all the later species of *Australopithecus* and to *Homo* as well (Fig. 20.5).

Hadar lies in the Afar depression, a vast arid wilderness in northeast Ethiopia. At 3 to 4 Ma it was the site of a lake fed by rivers tumbling out of winter snowfields high on the plateau of Ethiopia. The australopithecines lived and are fossilized along the lake edges. Delicately preserved fossils such as crab claws and crocodile and turtle eggs suggest that the australopithecines had rich protein foods available

to them, and skeletons of hippos and elephants suggest that there was rich vegetation in and around the lake edges. All the Hadar specimens are dated at about 3.2 Ma, so they are considerably later than the Laetoli australopithecines.

The best-preserved Hadar skeleton is the famous **Lucy**. Lucy was small by our standards, a little over 1 meter (42 inches) in height. She was full-grown, old enough to have had arthritis. Her brain was small at about 385 cc, compared with 1300 cc for an average human. Her large molar teeth suggest that *A. afarensis* was a forager and collector eating tough fibrous material.

A. afarensis was no more dimorphic than modern humans: males were bigger than females, but not to the extent seen in baboons, for example. Across primates, extreme male size is correlated with intense physical competition between males for females; monogamy is associated with low levels of dimorphism. It would be interesting if it turns out that all hominids (australopithecines and *Homo*) have typically had this sort of social structure.

Baboons sleep in high places—trees or high rocks—and are great opportunists in taking whatever food is available. They live and forage in troops and have a cohesive social structure that gives them effective protection from predators even though they are fairly small as individuals. But *Australopithecus* walked upright, whereas baboons trot on four limbs. Ecologically (but perhaps not socially!), *Australopithecus* may have been a super-baboon. Walking upright, with its arms free for carrying, it may have been a more effective forager than a baboon, which can carry only what it can put into its mouth and stomach. Perhaps the requirements and advantages of efficient troop foraging and defense encouraged tight social cohesion among australopithecines, long before tools permitted technological advances.

Australopithecus *in South Africa*

Isolated caves scattered over the high plains of South Africa are mined for limestone, and hominid fossils have been

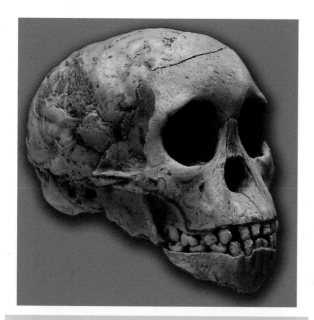

Figure 20.6 The first *Australopithecus* ever found was the fossil of a child, *Australopithecus africanus*. Photograph by José-Manuel Benito Álvarez and placed into Wikimedia.

Figure 20.7 A recent discovery: *Australopithecus sediba* from South Africa. Photograph by Brett Eloff and placed in Wikimedia courtesy of Professor Lee Berger and the University of the Witwatersrand.

found encased in the limestone. But cave deposits are difficult to interpret and date accurately. Roof falls and mineralization by percolating water have disturbed the original sediments, and few of the radioactive minerals in cave deposits allow absolute dating. Thus there have been problems in relating South African hominid fossils to their well-dated East African counterparts.

New research will soon change that. A specimen found at Sterkfontein in 1998 has been claimed to be about 4 Ma in age. This is not outrageous, but it is very early, and the claim will no doubt be examined very carefully.

To date, the best-known early australopithecine from South Africa is *Australopithecus africanus* (Fig. 20.6). Although it was about the same body weight as *A. afarensis*, *A. africanus* was taller but more lightly built and had a larger brain, perhaps 450 cc. The teeth and jaws continued to be large and strong, with molars twice as large as chimpanzee molars, suggesting that the diet remained mainly vegetarian. However, new evidence from isotopes in the teeth suggest that *A. africanus* ate vegetarian animals as well, possibly catching small animals, or scavenging meat from carcasses. Tooth wear suggests an average life span for *A. africanus* of perhaps 20 years, maybe with a maximum of 40 years, about the same as a gorilla or chimpanzee. The arms were relatively long compared with *A. afarensis*, suggesting that *A. africanus*, though perfectly erect and able to walk and run on the ground, spent a good deal of time in trees.

A new species from South Africa, *A. sediba* (Fig. 20.7), is contemporary with the earliest *Homo*. It has a strange mixture of characters, and is still being studied.

Robust Australopithecines

Australopithecines with heavily built skulls are called **robust** to distinguish them from those with lightly built skulls (such as *A. africanus*) which are called **gracile**. The best example of a robust skull is the oldest one, the so-called Black Skull (Fig. 20.8) from the Turkana Basin of northern Kenya dating from about 2.5 Ma. The Black Skull is usually called *Australopithecus aethiopicus*. It has a skull much heavier and stronger than *A. afarensis*, although the brain was no larger and the body was not very different. The jaw extended further forward, the face was broad and dish-shaped, and there was a large crest on the top of the skull for attaching very strong jaw muscles (Fig. 20.8). The molar teeth of the Black Skull are as large as any hominid teeth known, about four or five times the size of ours. Yet the front teeth of robust australopithecines are small.

Later robust forms have been found all over East and South Africa between 2.5 Ma and 1.4 Ma. In South Africa they are called *A. robustus* (Fig. 20.9), and in East Africa they are usually called *A. boisei* (the famous Zinj of Louis Leakey) (Fig. 20.10). There are enough fossils to suggest that robust australopithecines changed over the million years of their history, evolving a larger brain (perhaps 500 cc rather than 400 cc) and a flatter face.

The robust australopithecines are certainly similar ecologically. The large jaw and the huge molars, with their very thick tooth enamel, were adaptations that indicate great chewing power and a diet of coarse fiber. However, almost all the characters that are used to define robust australopithecines are connected with the huge teeth, and the modifications of the jaws and the face during growth that are required to accommodate the teeth. So any

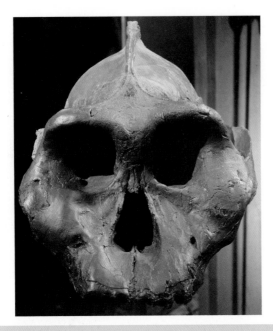

Figure 20.8   One of the fine casts on display at the Museum of Anthropology at the University of Zürich. This is the Black Skull, *Australopithecus aethiopicus*, from East Africa. Photograph by Nicholas Guérin, and used under the Creative Commons Attribution-Share Alike 3.0 Unported license.

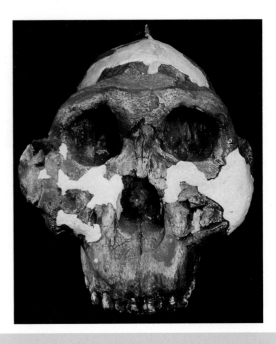

Figure 20.10   One of the fine casts on display at the Museum of Anthropology at the University of Zürich. This is *Australopithecus boisei* from East Africa. Photograph by Nicholas Guérin, and used under the Creative Commons Attribution-Share Alike 3.0 Unported license.

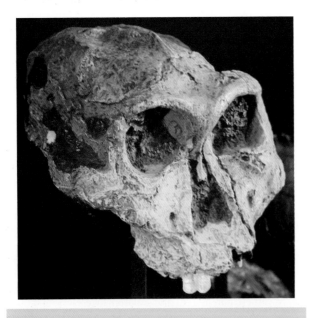

Figure 20.9   One of the fine casts on display at the Museum of Anthropology at the University of Zürich. This is *Australopithecus robustus* from South Africa. Photograph by Nicholas Guérin, and used under the Creative Commons Attribution-Share Alike 3.0 Unported license.

australopithecine population that evolved huge teeth would have come to look "robust." Therefore the robust australopithecines may not be an evolutionary clade. They may be three separately evolved species; they may be three related species; or they may be variants of the same species (which would have to be called *robustus*). Some specialists give robust australopithecines their own generic name, *Paranthropus*, but I have not used this name. All the robust australopithecines could easily have evolved from *A. afarensis*.

Australopithecus Garhi, *and Butchering Tools*

An astonishing find was reported in 1999. Rocks in Ethiopia dated at 2.5 Ma yielded enough pieces of two or three skeletons to allow the description of a new species, *Australopithecus garhi*. Since then, beds of the same age have yielded evidence of the use of stone tools for butchering meat and smashing bones.

   A. garhi is a normal gracile australopithecine, except that it has very large teeth for the size of its jaw and skull. The skull is far too primitive for it to belong to *Homo*, and its brain size is only about 450 cc. But given its age, location, and the features of its skeleton, *A. garhi* would be a reasonable ancestor for *Homo*.

   Some of the animal bones in the same rock bed had been sliced and hammered in ways that betray intelligent butchering. Most likely, the butchers used their tools carefully, because there were no suitable rocks nearby, and all tools had to be carried in (and carried out for further use).

Before 1999, it had generally been thought that the defining characters of *Homo* versus *Australopithecus* included a larger brain and the use of tools. The new evidence suggests that *A. garhi* was making, carrying, and using tools effectively. Perhaps the great ecological advantages gained by the invention of butchering tools encouraged exactly those changes in the *Australopithecus garhi* lineage that led quickly to increased brain size, reduced tooth size, and the status of first *Homo*.

Once again, apparently major transitions disappear as we collect more fossils: we have seen this for the transition between birds and dinosaurs, between cynodonts and mammals, and now between australopithecines and *Homo*. As we see it now, perhaps as early as 2.4 Ma, hominids with increased brain size and reduced teeth and jaws appeared in Africa. They are sufficiently like ourselves in jaws, teeth, skull, and brain size to be classed as *Homo*. But because one genus always evolves from another, there is always room to argue just where to draw the line, and this is happening as we try to decide which species actually was the first *Homo*. Increasingly, we realize that there is a great difference between early, transitional forms, and later species that everyone agrees as belonging to *Homo*. I will continue to call the transitional forms *Homo* until there is more of a consensus.

Transitional Species That May or May Not Remain in *Homo*

The most familiar transitional species is *Homo habilis* (Fig. 20.11). *Homo rudolfensis* is known from East Africa around 2 Ma, largely from skulls, and has been given a separate name. However, a new find at Olduvai Gorge seems to suggest that *habilis* and *rudolfensis* are the same species. This story is bound to change as we find more fossils. Meanwhile I will call them "early *Homo*" or *Homo habilis*.

Early *Homo* was small by modern standards, perhaps just over a meter (4 feet) tall, but was at least as heavy as contemporary robust australopithecines at about 30–50 kg (65–110 pounds). The difference in brain size is striking, however. The brain size was about 650 cc, considerably larger than the brain of an australopithecine. Perhaps, then, early *Homo* is marked by a new level of brain organization.

We have only a few sets of bones of *H. habilis*, but there is enough evidence from hands, legs, and feet to suggest that it spent a lot of its time climbing in trees.

We have a good record of the tools that were used by *Homo habilis* (and probably *Australopithecus garhi* before it). They are called Oldowan tools because they were first identified by the Leakeys in Olduvai Gorge. They are often large and clumsy-looking objects with simple shapes, and not all of them were useful tools in themselves. Instead, many objects may be the discarded centers (cores) of larger stones from which useful scraping and cutting

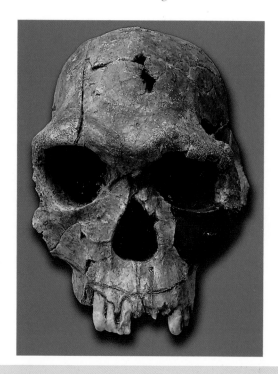

Figure 20.11 Replica of the skull of *Homo habilis* from East Africa. Photograph by José-Manuel Benito Álvarez and placed into Wikimedia.

flakes had been removed by hammering with other stones (Fig. 20.12).

Oldowan tools demonstrate the use of stone in a deliberate, intelligent way, and the flakes were probably made and used for cutting up food items. For example, an excavation in the Turkana Basin turned up the skeleton of a hippopotamus lying near an ancient river bed. Cobbles naturally occurring close by on a gravel bank in the river had been broken to produce simple tools. Marks on the hippo bones showed that they had been scraped, and that the tendons and ligaments had been cut, to allow meat to be taken from the carcass. There was no indication that the hippo had been killed by the tools.

Nicholas Toth has reproduced and used Oldowan-style artifacts from East African rock types. He showed that the toolmaker was sophisticated in selecting appropriate rocks and making the most of them. Toth's experiments on fresh carcasses of East African animals show that Oldowan axes, flakes, and cores are excellent tools for slitting hides, butchering carcasses, and breaking bones for marrow. Toth was also able to determine that *Homo habilis* was right-handed. (Schick and Toth 1993).

Some Oldowan sites were visited many times. They contain accumulations of bones, stones, and tools, brought to the site over periods of years. Flakes were made on site from stones that had been carried there. This may not indicate a systematic return to a homesite, but it does indicate an intelligent return to sites that perhaps were particularly suitable for food processing and tool making.

Figure 20.12 Oldowan tools from Ethiopia, probably cores, about 1.7 Ma. Photograph by Didier Descouens and placed into Wikimedia.

From Super-Baboon to Super-Jackal

Was early *Homo* a hunter or a scavenger? This may be a nonquestion, because all hunters will eat a fresh carcass, and all scavengers will cheerfully kill a helpless prey if they can. Evidence from Turkana and Olduvai suggests that early *Homo* was a scavenger on large carcasses but hunted small- and medium-sized prey. Thus early *Homo* may have had the ecology of a super-jackal, foraging in groups over long distances in search of large, fresh carcasses killed by other predators. Rhinos, hippos, and elephants have thick and leathery hides, difficult for vultures, jackals, and hyenas to pierce, but stone tools allowed *Homo* to make short work of dismembering a large carcass. Between carcass finds of large animals, early *Homo* may have foraged for leopard kills of medium-sized animals, left hanging in trees. Early *Homo* may also have been an opportunistic hunter of small- to medium-sized prey that was brought to central sites for butchering, and also a forager searching for fruits, berries, grains, roots, grubs, locusts, and lizards. *Australopithecus africanus* may have eaten small animals as well. Early *Homo* (and possibly *A. garhi)* used tools to make that opportunistic way of life more efficient.

The concept is exciting. A new ecological niche opened up, or became much more profitable, with the invention of tools and the ability to use them intelligently. Visiting American anthropologists with no previous experience in the African bush were able to learn quickly how to find large carcasses of animals killed in woodlands and smaller carcasses cached in trees by leopards; it is perfectly reasonable to expect that early Homo could have done so too (Blumenschine and Cavallo 1992). Simultaneous or consequent changes in diet, brain size, and possibly even social structure are consistent with the apparently rapid advances in skull characters, but not body anatomy, in early *Homo*,

and its replacement of gracile australopithecines. Perhaps early *Homo* did not compete ecologically with the surviving australopithecines (the robust forms). Certainly robust australopithecines and *Homo* co-existed for over a million years in the same environments.

One can imagine how early *Homo* could have improved its competitive ability by exploring and exploiting the possibilities of tool use. Weaponry would naturally follow from tool use during scavenging. Food and infants could be transported safely from place to place with carrying devices. Increasing behavioral complexity would probably act to increase the value of brain growth and learning ability, and perhaps we may speculate (but not too wildly) about the increasing value of, or need for, sophisticated communication within and among social groups of humans.

Hominids and Cats in South Africa

Most hominid fossils found in South Africa have come from caves, many of which had steep or vertical entrances. It is unlikely that the hominids lived in the caves. Instead, the piles of bones there probably fell into the caves from above. In addition to hominid skeletons (mostly australopithecines), the fossils include bones of rodents, hyraxes, antelopes, baboons, two species of hyenas, leopards, and three extinct species of stalking sabertooth cats, one as big as a lion and two the size of a leopard.

C. K. Brain realized that the hyrax skulls in the cave deposits are all damaged in a peculiar way. Leopards always eat hyraxes completely, except for the fur, the gut, and the skull and jaws, and as they get at the brain and tongue they leave characteristic tooth marks on the skull, just like those on the fossil hyraxes. Cheetahs today can eat the backbones of baboons but not those of antelopes. Fossils from Swartkrans cave include many antelope vertebrae but none from baboons. Only baboon skulls are found. Furthermore, it looks as if the cave fossils were selected by size. There are very few juvenile baboon skeletons at Swartkrans, but many juvenile australopithecines. Some of the primate skulls show teeth marks that look exactly like those made today by leopard canines. Some of the fossil antelopes are bigger than those killed today by leopards, and sabertooth cats may have been responsible for them (Brain 1981).

Leopards today like to carry their prey up trees (Fig. 20.13). On the bare plains of South Africa, a cave entrance is one of the few places that seedlings can find safe rooting away from browsers, fire, and winter frost, and trees can grow. Brain suggests that Pliocene and Pleistocene sabertooth cats killed prey and carried them to safe places to eat, undisturbed by jackals and hyenas, in trees growing at the entrances of caves such as Swartkrans and Sterkfontein. Uneaten parts of the carcasses fell into the caves, away from the hyenas, and were buried and preserved as soil, debris, and limestone deposits filled the cave.

Hominoids may have been a preferred meal for sabertooths for a long time. Many well-preserved fossils of sivapithecids and gibbons in South China, at about 6 Ma, consist

Figure 20.13 This leopard has carried its prey up a tree, to eat without disturbance from hyenas or jackals. Photograph taken by Raphael Melnick in a game reserve in South Africa and placed into Wikimedia.

Figure 20.14 An early *Homo erectus* from East Africa (sometimes called *Homo ergaster*). Photograph by Luna04, and placed into Wikimedia.

almost entirely of skulls and skull fragments, with few other bones of the skeleton, and there are large and impressive sabertooth canines in the same beds.

The large cats were the dominant carnivores in South Africa when the cave deposits were formed, and we can imagine them stalking and killing fairly large prey animals, including *Australopithecus robustus*. But there are relatively few fossils of early *Homo* at Swartkrans, Sterkfontein, or among other early cave deposits, suggesting that early *Homo* was either rare or comparatively safe from big cats by virtue of habits, intelligence, or defensive methods and weapons. Early *Homo* did not have to be immune from big cat predation, just well defended enough that the big cats hunted other prey most of the time. *Homo*, of course, eventually replaced *Australopithecus* in South Africa.

There are many bones of *Australopithecus africanus* in the rock bed Member 4 at Sterkfontein, but no tools. Member 5, which overlies it, contains many tools, including choppers and diggers, animal bones with cut marks on them, and a few fossils of *Homo habilis*. The contrast between these two levels is striking in every aspect of their fossil record. As Brain sees it, the replacement of big cats by *Homo* as the dominant predators in South Africa was a major step toward human control over nature, and the beginning of our rise to dominance over the planet.

Homo Erectus: the First "Real" *Homo*?

Some extraordinary changes took place in the African plains ecosystem, beginning about 1.5 Ma. It is tempting to associate them with the appearance of new species of human, *Homo erectus*. ("*Homo ergaster*" is a name for African fossils that are probably the same species as the later fossils from Eurasia that are called *Homo erectus*.) This is still not clear, but I follow the claim here and call all of them *erectus*. (Even readers of *National Geographic* have to deal with messy alternative schemes, each of which has enthusiastic and often intolerant proponents.)

An excellent specimen of *H. erectus* was discovered in 1984 west of Lake Turkana in Kenya, in sediment dated about 1.5 Ma. The body had been trampled by animals, so the bones were broken and spread over 6 or 7 meters, but careful collecting recovered an almost complete skeleton. The skeleton came from a boy (the "Turkana Boy") who was 11 or 12 years old and stood 1.6 meters (65 inches) high. Adult males were probably close to 1.8 meters (6 feet). The nose was large and projecting, as in modern humans but unlike australopithecines or *H. habilis*. This character suggests that *H. erectus* was adapting to greater exposure to dry air, for longer times and during greater activity.

Homo erectus was strongly built, and was a specialized walker and runner with large hip and back joints capable of taking the stresses of a full running stride. There is less evidence of tree-climbing ability than there is in early *Homo*, though *H. erectus* would have been no worse at it than we are. *H. erectus* is also advanced in skull characters (Fig. 20.14). The skull is thick and heavy by our standards, but brain size had increased to around 900 cc.

Quite suddenly, at about 1.4–1.5 Ma, all over East Africa, *Homo erectus* is found associated with a completely new set of stone tools. The **Acheulean** tool kit is much more effective than the older Oldowan, but experiments by Nicholas Toth show that Acheulean tools required much greater strength and precision to make and use than Oldowan

Figure 20.15 An Acheulean biface (hand-axe) made of flint. This is actually from the locality of St. Acheul, in France. Photograph by Didier Descouens of specimen from the Museum of Toulouse, as part of Projet Phoenix, and placed into Wikimedia.

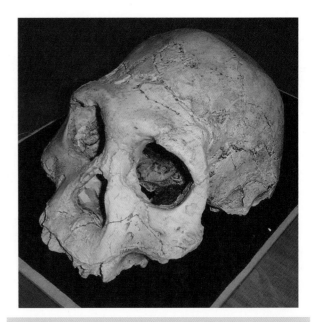

Figure 20.16 Replica of a *Homo erectus* skull from Dmanisi, Georgia, dated about 1.8 Ma. Photograph by Gerbil, and placed into Wikimedia.

tools. Acheulean craftsmen shaped their stone cores into heavy axes (Fig. 20.15) and cleavers at the same time as they flaked off smaller cutting and scraping tools. Most Acheulean tools are well explained as heavy-duty butchering tools. And around this time, robust australopithecines, *A. africanus*, and sabertooth cats all became extinct in South Africa. All other species of early *Homo* were already gone from Africa, and by 1 Ma, the last robust australopithecines and the last two species of sabertooth cats were gone too.

It is tempting to correlate all these events with the achievement of some dramatically new level of intellectual, physical, and technical ability in *Homo erectus*. *H. erectus* was much bigger than any preceding human. Most paleontologists believe that the evidence from anatomy, from tools, and from animal remains found with *H. erectus* suggests that this was the first effective human hunter of large animals. Alan Walker suggested that the entire ecosystem of the African savanna was re-organized as *Homo erectus* came to be a dominant predator instead of a forager, scavenger, and small-scale hunter. African kill sites with butchered animals suggest a sophisticated level of achievement.

The physical stature and ecological impact of this new species of *Homo* is the reason some experts suggest we should re-define the origin of *Homo* to the appearance of *Homo erectus*, perhaps placing *Homo habilis* and the other transitional forms into a genus of its own that would not be called *Homo*.

The first fossils to be named *Homo erectus* were in fact collected a hundred years ago on the island of Java, in Indonesia. The earliest of the Java specimens of *H. erectus*

may be as old as 1.8 Ma, though this date is contested. However, specimens of *H. erectus* have been discovered in the southern Caucasus, in Georgia (Fig. 20.16), and they date to around 1.7–1.8 Ma. It seems likely that the emigration of *H. erectus* from Africa to Asia occurred soon after the species evolved. It was rapid, and it extended across the warm regions of Southern Asia from the Middle East to Indonesia: and it happened before the Acheulean tool kit was invented in Africa.

The migration of *Homo erectus* left a corridor of humanity that stretched from South Africa to eastern Asia. All these populations evolved larger body size and more advanced skull characters, and all made new tool kits. Given the mobility of humans, there was no necessary dramatic or long-lived separation between these pantropical populations. There was one founding, and dominant early human species, *Homo erectus*, with locally variable anatomy and culture, just like *Homo sapiens* today. At least, that's a hypothesis that can stand until more new evidence turns up.

Other specimens of *Homo erectus* from China are compatible with this story. The Chinese specimens have an age around 1.0 Ma and younger. *H. erectus* may have reached as far east as the island of Flores, in Indonesia, before 750,000 BP (years before present), a feat which involved two sea crossings, of 15 and 12 miles. There are no fossils, only a few tools, but the story fits with the fact that three major animals became extinct quite suddenly on Flores around 900,000 BP: a pygmy elephant, a giant tortoise, and a giant lizard related to the Komodo dragon.

Homo erectus from Java had a brain size just under 1000 cc, but brain size had reached 1100 cc by the time of

"Peking Man," who occupied caves outside Beijing between 500,000 and 300,000 BP. The successful long-term occupation of North China by these people indicates that they had solved the problems of surviving a challenging northern winter. The Asian populations of *H. erectus* had their own versions of stone tool making styles. Some Asian *erectus* made tools out of rhino teeth, since they were living in an area without good tool-making stone.

It looks as if *Homo erectus* was the first species to control fire. There is very good evidence for fires in a South African cave dating from at least 1 Ma (Berna et al. 2012), and *H. erectus* lived in the cold north of China around 1.4 Ma.

We know from the shape of the pelvis that *H. erectus* babies were born as helpless as modern human babies are (Fig. 20.4), and it is clear that the brain grew a lot after birth, as our brains do. This implies a long period of care for a baby that probably could not walk for several months. That is an enormous price to pay for a larger brain, and would only have been evolutionarily worthwhile (selected for) if there was a large pay-off of learning and intelligence.

All these lines of evidence imply a complex and stable social structure for *Homo erectus*, though details are certainly not available. I will make one comment of my own. The cooperation required to build, start, control, maintain, and transport a fire is very high. It is difficult (for me) to imagine a campfire without conversation. But as soon as any hominid evolved words, and then language, that would have begun the novel process of transferring abstract information and knowledge directly and immediately from individual to individual, replacing indirect methods such as taking and showing, demonstrating and copying, or sharing the same real experience. The ability to short-circuit the processes of teaching and learning must have allowed much more knowledge to be transmitted, absorbed, and retained in a society, with obvious advantages to all its members.

Later modifications in educational techniques have served mainly to accelerate the transfer of information and to make it possible at a distance, by the invention of writing and reading, numbers and alphabets, schools, printing, telephones, broadcasting, computers, and so on.

Many experts believe that language is a very recent invention, essentially by *Homo sapiens*. That may be true for the complex activity that all modern humans are so good at. But language must have evolved, like every other characteristic of modern humans. I suspect that the information explosion set off by modern electronics is only the latest in a series that started around a campfire a million years ago. This book and the Macintosh I wrote it on are direct continuations of that tradition.

After *Homo Erectus*

After *Homo erectus*, the story of human evolution becomes very messy as anthropologists argue about the origin of modern humans. As fossil and molecular evidence has accumulated, anthropologists have accepted finer and finer

subdivisions for species. As I write this chapter we have a majority opinion that several species of *Homo* have evolved during the last million years, with all but one becoming extinct. As always, you are free to construct your own interpretation of the story, and as always, new data will cause us all to re-think and re-interpret in the future.

According to the current majority story, some local populations of *Homo erectus* evolved separately and significantly. The name *Homo antecessor* was proposed for a group of well-preserved specimens found in Spain, dated between 1.2 Ma and 800,000 BP, and living on what was then the fringes of humanity. Obviously, *H. antecessor* must have had ancestors somewhere in Africa (*Homo erectus*?). *H. antecessor* left evidence for cannibalism: some *antecessor* bodies were processed—by human tools—to provide convenient chunks of meat, leaving characteristic cut marks on the skeletons (Fernández-Jalvo et al. 1999).

Perhaps dating as far back as 500,000 BP, a population, perhaps descended from *H. antecessor* or *H. erectus* and known from central and Western Europe, is called *Homo heidelbergensis* (Fig. 20.17). Around 400,000 BP, a number of *H. heidelbergensis* skeletons were deposited in the same caves in Spain associated with *H. antecessor* 400,000 years before. In an act of complex behavior, a *heidelbergensis* placed a beautiful pink quartzite hand axe on a pile of skeletons in the cave. At about the same time, *H. heidelbergensis* in Germany were making beautifully crafted wooden hunting spears from spruce and pine. These are throwing spears, up to 2.5 m long (8 feet), carved to angle through the air like modern javelins, and the spears are associated with butchered horses and other bones from elephant,

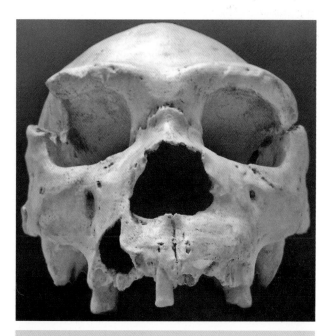

Figure 20.17 *Homo heidelbergensis*. This skull "Miguelon"(!) from Spain is about 400,000 years old. Photograph by José-Manuel Benito Álvarez and placed into Wikipedia.

rhino, deer, and bear. Modern athletes have thrown replicas of these spears 70 meters (over 200 yards).

Homo heidelbergensis in turn seems to have evolved into the Neanderthal people, *Homo neanderthalensis*. They were strongly adapted for life in cold climates along the fringes of the Ice-Age tundra from Spain to Central Asia.

We have only patchy evidence from Africa during this time, but we assume that *Homo erectus* continued to live across that continent. Meanwhile, *Homo erectus* thrived in Eastern Asia, and we have populations from Indonesia ("Java Man") and China ("Peking Man") in the age range 1 Ma to 500,000 Ma.

The Origin of *Homo Sapiens*

For some years now, the majority story has been that around 200,000 BP an African *Homo* species (*erectus/ ergaster/heidelbergensis/rhodesiensis/antecessor/archaic sapiens*) (all these names have been used, so we need more specimens to decide precisely what to call them) gradually developed a distinctive new type of stone tool technology which we call Middle Stone Age/Middle Paleolithic technology: MSA for short. And gradually, one of these African populations evolved into fully modern *Homo sapiens*.

This story was strengthened when a new set of fossils was described from Ethiopia in 2003. Three skulls, dating to about 160,000 BP, were identified as *Homo sapiens*. They still have some "archaic" features such as big eyebrows that would allow us to recognize them as "different" even if they were wearing blue jeans and T-shirts, so they are called *Homo sapiens idaltu*, a separate subspecies. Their tools were on the boundary between MSA and Acheulean. This time was in the middle of a very cold Ice Age, when other populations of *Homo* outside the tropics may have been stressed and fragmented. It seems that *Homo sapiens* had some feature or features that allowed it to expand and compete effectively against other *Homo* species.

During the last interglacial (around 120,000 BP) modern humans, *Homo sapiens sapiens*, are found in South Africa. And sometime after that, a small population of pioneering or refugee *Homo sapiens* left Africa for the first time and expanded into and over the Old World, presumably through the Middle East, which was not such a severe desert as it is now. Modern humans may have traversed around the shores of the Indian Ocean to Southeast Asia and Australia (Fig. 20.18), but they were apparently not yet able to sweep aside other species easily. *Homo neanderthalensis* was firmly entrenched in the colder areas of Europe and Western Asia, and *Homo erectus* in Eastern

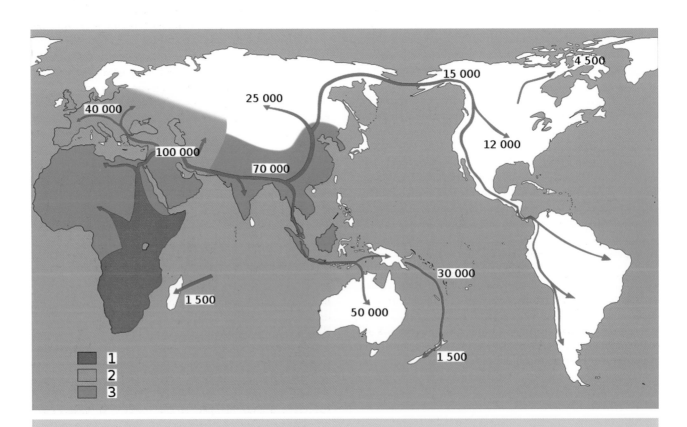

Figure 20.18 The final spread of *Homo sapiens* out of Africa, with estimates of years BP. Red, *Homo sapiens*. Green, earlier *Homo*. Olive, Neanderthals. Art by Magasjukur2 and placed in Wikimedia.

Asia. As the interglacial ended and the climate turned cool again, Neanderthals re-occupied the Middle East around 70,000 BP, replacing modern humans.

All this changed, beginning around 45,000 BP. Modern humans now swept all competing species from the Old World: the Neanderthals from Europe, and *Homo erectus* from Asia. This was overwhelming competition, perhaps even ethnic cleansing: the fossil record suggests there were no survivors. And, of course, *Homo sapiens* also spread to parts of the world previously not occupied by any *Homo*: the Americas, and Polynesia. All living humans are thus descended from what must have been a comparatively small original population of *H. sapiens* that emigrated from Africa (Fig. 20.18). This scenario was first favored in the late 1980s, when it was called the **Out-of-Africa hypothesis**. It is now strongly supported by masses of new data.

This fundamental scenario is debated, and it is important. There have been claims of transitional skeletons between *erectus* and *sapiens* in Indonesia, China, Africa, and the Middle East.

All these events are so recent that their effects can still be seen in, and interpreted from, the genetics of *Homo sapiens*. And those genetic data overwhelmingly support the out-of-Africa model. Even if genetic clocks do not run accurately, the story would not change: only the dates would.

The DNA of living humans shows a distinct division into African and non-African types. Furthermore, the variation in modern human DNA is very restricted. A single breeding group of chimpanzees in the Taï forest in Africa has more variability than does the entire human race today. This suggests very strongly that all modern humans are descended from an ancestral population that was not only small—say 10,000 or so—but was small for a long time.

If this estimate is correct, there is no way that 10,000 people could have interbred and inhabited more than a relatively small area, even if they were wandering hunter-gatherers. The only possible conclusion (if the assumptions and calculations are correct) is that indeed all living humans are descended from a small ancestral *sapiens* population who evolved in a restricted region in Africa, and spread from there throughout the Old World.

The difference that remains today between "African" and "non-African" DNA is explained if a small **founder population** left Africa, carrying with them only a small sample of the genetic variation that had by then evolved across Africa. These founder populations expanded as they occupied Eurasia, growing into a large population with a distinctly non-African DNA structure. Once again, a small subset of East Asian humans crossed the Bering Strait and populated the Americas with people who had even less genetic variation.

The Neanderthals

The people we call Neanderthals lived between 90,000 and 30,000 BP in Europe and along the mountain slopes on the northern edges of the Middle East, as far east as Iraq. They

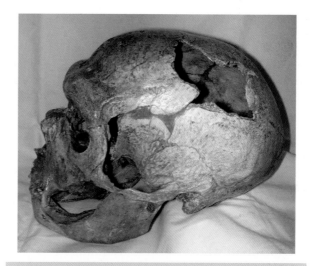

Figure 20.19 *Homo neanderthalensis*. This skull is from La Chapelle aux Saintes in France. Image published in PLoS Biology at http://www.plosbiology.org/article/info%3Adoi%2F10.1371%2Fjournal.pbio.0020080, which places the image into Wikimedia.

are named after a site in the Neander Valley in Germany. Neanderthals had a way of life that was distinctly sophisticated in living sites, tools, and behavior.

Neanderthals differ from living humans in having big faces with large noses, large front teeth, and little or no chin (Fig. 20.19). These characters are connected: Neanderthal front teeth show heavy wear, as if they used their incisors for something that demanded constant powerful pressure (perhaps softening hides by chewing, as Eskimos used to do?). Human faces are plastic, especially in early growth, and either by use or by genetic fixation, the frontal chewing of Neanderthals seems to have encouraged growth of the facial bones to support the front teeth against the skull. The nasal area was essentially swung outward from the face, so that the nose was even bigger and more projecting than it is in most hominids.

Neanderthal brain size, at 1450 to 1500 cc, was at least equal to that of living humans and sometimes greater. Another special Neanderthal character was a very strong, stocky body with very robust bones, which may have helped conserve body heat in a cold climate and/or may reflect a life style that required great physical strength. Most Neanderthal fossils are found in deposits laid down in the harsh climates of the next-to-last ice age.

Most Neanderthal tools are made in a style called **Mousterian**. They include scrapers, spear points, and cutting and boring tools (Fig. 20.20) made from flakes carefully chipped off a stone core. Marks on Neanderthal teeth suggest that they stripped animal sinews to make useful fibers by passing them through clenched teeth, just as Australian aboriginals do. But perhaps the most enlightening Neanderthal finds are their ceremonial burials. Bodies were carefully buried, with grave offerings of tools and food.

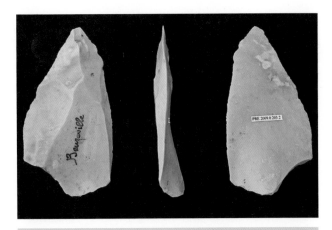

Figure 20.20   Mousterian tools made from flint, from Beuzeville, France. Photograph by Didier Descouens of specimens from the Museum of Toulouse, as part of Projet Phoenix, and placed into Wikimedia.

Enormous quantities of pollen were found with the body of Shanidar IV, a Neanderthal man buried in Iraq. The pollen came from seven plant species in particular. All seven have brightly colored flowers, all seven bloom together in the area in late April, and all have powerful medicinal properties. It is difficult to avoid the conclusion that Shanidar IV was carefully buried with garlands of healing herbs chosen from early summer flowers, suggesting an intense concern for the abstract world.

Neanderthals became adapted to life in the cold climate along the edges of the ice sheets from Western Europe to central Asia, by evolving characters of their own. The more geographically isolated they were, the more extreme their Neanderthal characters became, until they became visibly different from both *heidelbergensis* and from the *sapiens* populations that were evolving in Africa. The reasonable interpretation of these facts is that Neanderthals were a separate and extinct species, *Homo neanderthalensis*.

In the Middle East, Neanderthals seem to have alternated with *Homo sapiens*, with a fluctuating border between them. Both peoples in the region were making the same Mousterian tools, which have been identified as far south as the Sudan, but none of the fossil skulls are intermediate, suggesting that the two did not interbreed. Neanderthals lived in the Middle East in cooler, wetter times, while *Homo sapiens* lived there in hotter, drier times. Each was fitted to a particular climatic zone in which the other could not compete; neither was "superior" during those tens of thousands of years.

Neanderthals disappeared from the Middle East about 45,000 BP, then from Eastern and Central Europe, and finally from northwest Europe (France and northern Spain). The last dated surviving Neanderthals survived in upland France until about 38,000 BP and in southern Spain and Portugal until about 30,000 BP.

European Neanderthal sites typically contain less standardized tools, made only from local stone and flint, but the last Neanderthals in Western Europe showed a distinctly more advanced culture, with some similarities to the Aurignacian tools that the newly arrived, fully modern humans (the CroMagnon people), were using at about the same time in Western Europe. The last Neanderthal sites in France also contain simple ornaments, and it is tempting to suggest that Neanderthals may have copied some of the CroMagnon technology and art.

The skeletons at Châtelperronian sites are Neanderthal. A beautifully preserved skull of a late Neanderthal baby from France shows a structure of the inner ear that is quite different from that in modern humans. Neanderthal DNA (recovered from the original fossil from the Neander Valley) is distinctly different from that of any living human population, and also different from two samples of CroMagnon DNA.

Dramatic new genetic evidence (the first complete genome from a Neanderthal) shows there was some limited interbreeding between *Homo sapiens* and Neanderthals. It turns out that some Neanderthal gene sequences are found (at low frequency) in European and Asian genomes, but not in African genomes. The only way that could have happened is that the ancestral (African) human lineage had no Neanderthal genes. But Eurasian *Homo sapiens* that mated with Neanderthals and produced viable offspring thereby introduced what must have been a few particularly favorable Neanderthal genes into the *sapiens* population.

It looks as if somewhere between 2% and 5% of the genes of Europeans are Neanderthal in origin. Computer models confirm that the interbreeding could only have been minimal, or Neanderthal genes would have been more widespread. This explains why we have not found more than two fossils that could have been hybrids between *Homo sapiens* and Neanderthals, but both of them are disputed.

CroMagnons (Fig. 20.21) and Neanderthals were quite different socially. But Californians are quite different socially from Amish and Papua New Guineans: that means nothing genetically. We need a better explanation for the extinction of the Neanderthals. What superiority did *Homo sapiens* have? Was it weaponry? or social cohesion? It was not language as such: Neanderthals could certainly speak.

As CroMagnon people took over the cold and forbidding European peninsula, they were only a local population on the northwest fringe of the human species, but they are important because they have yielded us the best-studied set of fossils, tools, and works of art from the depths of the last Ice Age (Fig. 20.22).

CroMagnon sites yield richer and more sophisticated art and sets of tools, and more complex structures and burials than Neanderthals. In particular, CroMagnons made stone tips for projectiles. Neanderthals may have had arrows and spears, but if they did, they were not stone tipped. Mousterian tools are more frequently wood-working tools than those of CroMagnons, who worked more with bone, antler, and stone.

CroMagnons and their contemporaries in Eastern Europe, the Gravettians, left evidence of a capacity for

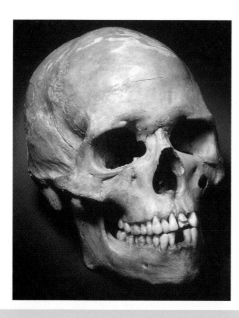

Figure 20.21 *Homo sapiens* (CroMagnon skull from France). Photograph by Laténium and placed into Wikipedia.

Figure 20.22 A bison, painted on the wall of the cave at Altamira, Spain, well over 10,000 years ago. Photograph by Ramessos and placed into Wikipedia.

habitat destruction that is typically modern in style. From Russia to France, sites contain the remains of thousands of horses and hundreds of woolly mammoths. CroMagnons were also responsible for the magnificent cave paintings of extinct Ice Age animals (Fig. 20.22), drawn by people who saw them alive, and they made (and presumably played) bone flutes. CroMagnons were painting on cave walls at 30,000 BP, and Gravettians were firing small terracotta figurines in kilns more than 25,000 years ago. CroMagnons and their contemporaries had a tremendous ecological impact on the world (Chapter 21).

So is *Homo sapiens* collectively guilty of ethnic cleansing? All other species of *Homo* disappeared. It is unlikely that some natural catastrophe affected *Homo* but not the other surviving apes. It is much more likely that the branches of human descent were pruned by other humans.

Migration-with-ethnic-cleansing may not be one's favorite image of the human race, but we have to deal with evidence. Certainly genocide occurred in the past and still occurs. Horrific examples are found in the history of Soviet Russia and Nazi Germany, and in modern times in Syria, Sierra Leone, Liberia, Bosnia, Rwanda, Kurdistan, Uganda, and the Congo. Within living memory, cannibal villagers in Papua New Guinea would try to kill off all members of a target community, because survivors would be likely to retaliate.

Evolution among Humans Today

Given the vastly different biological and ecological environments of the species of *Homo* since 2.4 Ma, it's likely that the selective pressures on soft-part anatomy and behavior have been as intense as those on skeletal features. There is clear evidence among living humans of regional evolution to suit the particular environment; for example, some of the characters of soft anatomy, body proportions, and even parts of the mitochondrial genome are strongly linked to climate in many human groups. Nose shape is strongly correlated with humidity. Eskimos are endomorphic to resist body cooling. Peoples living at high altitude in Asia, Ethiopia, and South America have adapted physiologically to low oxygen levels. There are extremely ectomorphic, dark-skinned tropical people, pink-skinned people in northern Europe, and so on. Testis size varies markedly among human groups, as does the frequency of twinning.

Such features must have evolved under intense regional selection, combined with the slow spread and mixing of genes at a time when human groups could not travel great distances. Such differences are visibly diminishing in certain modern populations (Hawaii, Brazil, California, and London come to mind).

The evolution of behavior cannot be assessed very well from the fossil record, but the variation in social structure within hominoid species is very great and suggests radical behavioral evolution, at least over the past 10 m.y. New research on mating patterns among animals suggests that much of human sexual anatomy and sexual behavior may be linked with the evolutionary breakthrough that began with early *Homo* and the use of tools to achieve ecological dominance.

It is sometimes claimed that natural selection no longer acts on modern humans because our surroundings are so artificial. Most people are now more insulated against diseases, environmental fluctuations, and accidents than humans were only a few centuries ago. Yet selection still operates strongly even in the most "advanced" societies. The genes for sickle-cell anemia are now generally harmful, instead of being favored in malarial regions. The genes that predispose non-Europeans to diabetes and gall-bladder cancer are more easily triggered into action on a Westernized ("coca-colonized") diet, whereas they were beneficial in their original cultural environment (Diamond 2003). Among the Micronesian population of Nauru, nearly two-thirds of the population have diet-induced diabetes by the age of 55, and 50% of all Pima Indians of Arizona have diabetes.

Finally, it appears that although the maximum human lifespan has not changed very much, average life expectancy has. At least some human societies now depend strongly on physical, social, and intellectual nurturing of children long past physical maturity. Characteristics that are normally juvenile attributes in primates, such as imagination, curiosity, play, and learning, are now encouraged in early adult years. The trend is presently social. There is no evidence yet of any evolutionary feedback creating delayed physical maturity or increased mental capacity. It is not yet clear whether this will occur and alter human biology as well as human culture. Potentially, an increased learning period could have enormous consequences for us and for every living thing on Earth.

Further Reading

Aiello, L. C. and J. C. K. Wells. 2002. Energetics and the evolution of the genus *Homo. Annual Reviews in Anthropology* 31: 323–338.

Appenzeller, T. 1998. Art: evolution or revolution? *Science* 282: 1451–1454.

Asfaw, B. et al. 1999. *Australopithecus garhi*: a new species of early hominid from Ethiopia. *Science* 284: 629–635, and comment, pp. 572–573. See also companion paper by J. de Heinzelin et al. on environment and behavior of *A. garhi*, pp. 625–629.

Bahn, P. G. and J. Vertut 1998. *Journey Through the Ice Age*. Berkeley, California: University of California Press. The art work of the Cro-Magnons.

Berger, L. R. et al. 2010. *Australopithecus sediba*: a new species of Homo-like australopith from South Africa. *Science* 328: 195–204. Available at https://evolutionaryanthropology. duke.edu/uploads/assets/Sebiba%20Berger%20et%20al .pdf

Bermudez de Castro, J. M. et al. 1997. A hominid from the Lower Pleistocene of Atapuerca, Spain: possible ancestor to Neandertals and modern humans. *Science* 276: 1392–1395, and comment, pp. 1331–1333. [*Homo antecessor*.]

Berna, F. et al. 2012. Microstratigraphic evidence of in situ fire in the Acheulean strata of Wonderwerk Cave, Northern Cape province, South Africa. *PNAS*, in press.

Blumenschine, R. J. and J. A. Cavallo 1992. Scavenging and human evolution. *Scientific American* 267 (4): 90–96.

Brain, C. K. 1981. *The Hunters or the Hunted? An Introduction to African Cave Taphonomy*. Chicago: University of Chicago Press.

Brunet, M. et al. 2002. A new hominid from the Upper Miocene of Chad, Central Africa. *Nature* 418: 145–151, and comment, pp. 133–135. [*Sahelanthropus*. See also companion paper on the age and environment, pp. 152–155.]

Carmody, R. and R. W. Wrangham 2001. The energetic significance of cooking. *Journal of Human Evolution* 57: 379–391. Available at http://www.anthro.utah.edu/PDFs/ CarmodyWrangham09cookingHumEv.pdf

Churchill, S. E. 1998. Cold adaptation, heterochrony, and Neandertals. *Evolutionary Anthropology* 7: 46–61.

Diamond, J. M. 2003. The double puzzle of diabetes. *Nature* 423: 599–602. Why are many people predisposed to get diabetes, and why are Europeans not among them? Available at http://www.wsrn.sccs.swarthmore.edu/users/08/ wnekoba/DoubleDiabetes.pdf

Diamond, J. M. and J. I. Rotter 1987. Observing the founder effect in human evolution. *Nature* 329: 105–106. South African Afrikaner genetics.

Fernández-Jalvo, Y. et al. 1999. Human cannibalism in the Early Pleistocene of Europe (Gran Dolina, Sierra de Atapuerca, Burgos, Spain). *Journal of Human Evolution* 37: 591–622. Someone ate *Homo antecessor*, and it looks as if it was *Homo antecessor*.

Fifer, F. C. 1987. The adoption of bipedalism by the hominids: a new hypothesis. *Human Evolution* 2: 135–147. Stone throwing.

Gabunia, L. et al. 2001. Dmanisi and dispersal. *Evolutionary Anthropology* 10: 158–170. Very early *Homo erectus* in Georgia.

Klein, R. G. 2009. *The Human Career: Human Biological and Cultural Origins*. 3d ed. Chicago: University of Chicago Press. Comprehensive text with hundreds of references.

Nesse, R. M. and G. C. Williams 1995. *Why We Get Sick: An Introduction to Darwinian Medicine*. New York: Times Books. [Read this, and then ask whether natural selection has ended! Paperback edition, Vintage Books, 1999.]

Pinker, S. 1994. *The Language Instinct*. New York: William Morrow and Company. Chapter 11 deals with the origin of language among humans.

Raichlen, D. A. et al. 2010. Laetoli footprints preserve earliest direct evidence of human-like bipedal biomechanics. *PLoS ONE* 5 (3): e9769. Available at http://www.plosone.org/ article/info%3Adoi%2F10.1371%2Fjournal.pone.0009769

Reno, P. L. et al. 2003. Sexual dimorphism in *Australopithecus afarensis* was similar to that of modern humans. *PNAS* 100: 9404–9409, and comment, pp. 9103–9104. This could be important because social structure such as monogamy is associated in primate biology with low levels of dimorphism. No proof, but reasonable inference. Available at http://www.pnas.org/content/100/16/9404.full

Ruxton, G. D. and D. M. Wilkinson 2011. Avoidance of overheating and selection for both hair loss and bipedality in hominins. *PNAS* 108: 20965–20969.

Schick, K. D. and N. Toth 1993. *Making Silent Stones Speak*. New York: Simon and Schuster.

Semaw, S. et al. 2003. 2.6-million-year-old stone tools and associated bones from OGS-6 and OGS-7, Gona, Afar, Ethiopia. *Journal of Human Evolution* 45: 169–177.

Suwa, G. et al. 1997. The first skull of *Australopithecus boisei*. *Nature* 389: 489–492, and comment, pp. 445–446. *A. boisei* and *A. robustus* may be the same.

Swisher, C. C. et al. 1996. Latest *Homo erectus* of Java: potential contemporaneity with *Homo sapiens* in Southeast Asia. *Science* 274: 1870–1874.

Thieme, H. 1997. Lower Palaeolithic hunting spears from Germany. *Nature* 385: 807–810, and comment, pp. 767–768. [Throwing spears from 400,000 BP.]

Toth, N. et al. 1992. The last stone ax makers. *Scientific American* 267 (1): 88–93.

Vekua, A. et al. 2002. A new skull of early *Homo* from Dmanisi, Georgia. *Science* 297: 85–89, and comment, p. 26–27.

Walker, A. and P. Shipman 1996. *The Wisdom of the Bones: in Search of Human Origins*. New York: Alfred A. Knopf. Superb account of the way an outstanding scientist works and thinks, with a vivid reconstruction of *H. erectus*.

White, T. D. et al. 2009. *Ardipithecus ramidus* and the paleobiology of early hominids. *Science* 326: 75–86. *Ardipithecus*: this and companion papers made up an entire issue of *Science*. Available at http://nekhbet.com/ardi.pdf

Zollikofer, C. P. E. et al. 2002. Evidence for interpersonal violence in the St. Césaire Neanderthal. *PNAS* 99: 6444–6448. St. Césaire survived a vicious blow to the head, so he must have been cared for. Available at http://www.pnas.org/content/99/9/6444.full

Questions for Thought, Study, and Discussion

1. The more hominid fossils we find, the more we find intermediate specimens and species along the various branches of hominid evolution. Explain how this makes it easier for a paleontology instructor to describe human evolution, and how at the same time it makes it harder to teach. Summarize your arguments in a few short questions that need to be answered. (They would all start with "Why?")

2. The following limerick was composed for a newly discovered species. Who was it? Now give a scientific discussion of the paradox described in the "poem".

   He butchered and hammered the dead
   We assume he was very well fed
   　　More astonishing still
   　　He learned this new skill
   With a very small brain in his head.

3. Summarize the evidence that modern humans evolved in Africa and spread across the world from there.

4. There is a growing consensus that there was some interbreeding between Homo sapiens (modern man) and Homo neanderthalensis (Neanderthal man). Do some research on recent articles, and explain the evidence for and against the claim. Should we "believe" it (=accept it as a working hypothesis for now)?

TWENTY-ONE
Life in the Ice Age

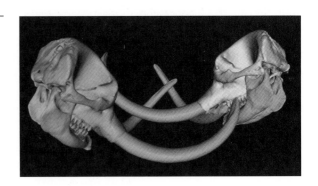

In This Chapter

The last 2.5 million years of human evolution took place against a time of climate change known as the Ice Age. First, I describe how ice ages can happen, and in particular how the recent one occurred. Climate change caused plants and animals to move north and south to avoid too much temperature change, but only in rare circumstances were there extinctions or radiations associated with the climate change. Geography changed too as ice was frozen out of the seas to form ice caps, but again these were not catastrophic changes from an evolutionary standpoint. Even so, there has recently been an astonishing extinction of life on land, with the dis-

appearance of such iconic animals as sabertooth cats, mammoths and mastodons, giant armadillos and sloths, and many more. But their extinctions do not coincide with climate change. Instead, they coincide with the arrival of *Homo sapiens* in their ecosystem. There is no doubt that the extinctions were caused by humans, directly or indirectly. Worse still, the extinction continues, as human populations increase and ecosystems are increasingly devastated. Given that the evidence is so clear, it is a terrible judgment on the human race that we continue to ravage our planet, causing irreversible damage with every decade.

Ice Ages and Climatic Change

Climate is one of the most important environmental factors for all organisms, and climatic changes have been major factors affecting the evolution of life. Plate tectonic movements can change oceanic and continental geography, and those geographic changes can modify seasonal climatic patterns and affect the ecology and evolution of

organisms in major global events (Chapter 6). There were major effects on life as world geography changed with the breakup of Pangea in the Late Mesozoic and Early Cenozoic. Some effects resulted directly from geographic isolation (Chapter 18), but others were mediated through the indirect effects of geography on climate. Many puzzles of Mesozoic evolution may be resolved when we can reconstruct paleoclimates more accurately. Certainly this has

History of Life, Fifth Edition. Richard Cowen.
© 2013 by Richard Cowen. Published 2013 by Blackwell Publishing Ltd.

been the result of concentrated research programs on Cenozoic paleoclimates.

We are living through an ice age now, and have been for the past 2.5 million years or so. We happen to live during a warm stage in it, but there is no sign that it is over. Great ice sheets expanded and covered much of the northern continents, then retreated again. They have done so at least 17 times in the past two million years. How did the present ice age affect life?

Ice ages are not common events in Earth's history. There was a widespread ice age toward the end of the Precambrian around 600 Ma: Snowball or Slushball Earth (Chapter 4). In Late Ordovician times, when northern Gondwana was over the South Pole, a great ice sheet spread over most of North Africa and probably further, triggering enough changes in marine life to mark the end of the Ordovician and the beginning of the Silurian. Gondwana drifted across the South Pole during the rest of the Paleozoic, with a particularly important glacial period in South America at the end of the Devonian. A small ice sheet lay over South Africa in the Early Carboniferous, but large-scale glaciation once again spread over most of Gondwana in the Late Carboniferous and Early Permian. Traces of this event, in the form of scratched rock surfaces and piles of glacial rock debris, are widespread in South Africa, South America, Australia, India, and Antarctica (Chapter 10). But afterward there was no major ice age for 250 m.y., until the present one began. Paleoclimatic evidence suggests that the Earth's surface cooled over the past 60 m.y., until finally the planet dropped into the present ice age.

The only external factors that could generate major climate change are astronomical processes—**changes in Earth's orbit** or **changes in solar radiation**. Such changes occur, but they are probably too small to generate major climate change by themselves. They cause fluctuations in climate, however. An **asteroid impact** could conceivably trigger a climate change, but only for a short time and only if conditions were already just right to start and maintain a change over considerable time.

It seems that we must look for mechanisms here on Earth for major climatic changes. Two processes can affect the amount of solar radiation that Earth retains. Some solar radiation is reflected back into space (the **albedo effect**), and a change in the amount of heat reflected would cool or warm the Earth. Gases in the atmosphere, especially carbon dioxide and methane, are very effective in absorbing solar radiation (the **greenhouse effect**), and changes in the amounts of these gases could strengthen or weaken the effect of solar radiation, or override it completely.

The basic preconditions for climate change on Earth are simple. For an ice age, there must be a lot of snowfall in areas where it will build up rather than remelt. Such a situation can occur if Earth's global geography is arranged in the right way. An ice age, or any other climate change, can be encouraged or discouraged by geographic changes that result from plate tectonic movements. Changes in geography also act to vary the albedo of the Earth, the scale and

activity of ocean currents, and the distribution of heat to different regions, all of which affect climate.

In general, ice ages require large areas of land in high latitudes. The poles should be isolated from warm water. Finally, to lock the Earth into a long glacial period, there must be room for large continental ice sheets to spread out and provide high reflectivity over large regions. Geography thus controls whether or not Earth's heat is well distributed, and whether or not polar ice sheets can form. Plate tectonics controls continental distributions, and the necessary conditions for generating an ice age may arise from time to time just by the motions of the plates.

Earth's major climate changes include distinct fluctuations. Dramatic advances and retreats of ice can occur even while Earth is locked into an ice age. Huge areas of the northern continents are still covered by debris dropped by ice sheets during 40 or so glacial advances and retreats during the past two million years. Large temperature fluctuations are recorded by microfossils in seafloor sediments. World sea level has fluctuated up and down by more than 70 m (220 feet) as 5% of Earth's water has alternately been frozen into ice sheets and melted away. Such changes in sea level are recorded worldwide in sedimentary deposits far from the ice sheets. For example, islands and atolls in the Atlantic and Pacific Oceans have been repeatedly exposed and reflooded.

Ancient rocks also show evidence of frequent and important change in sea level. For example, regular cycles of limestone, sandstone, and coal formation in the Carboniferous rocks of North America and Europe resulted from

Box 21.1 The Components of The Milankovitch Theory

Tilt. Increased or decreased tilting of the axis (Fig. 21.1a), which varies between 22° and 24.5°, increases or decreases the effect of the seasons in a cycle of about 41,000 years (Fig. 21.2).

Precession. Earth's orbit is not a circle but an ellipse, with the sun at one focus (Fig. 21.2). One pole is closer to the sun in its winter, while the other is closer in its summer. Thus, at any time, one pole has warm winters and cool summers, while the other pole has warm summers and cool winters. However, the slow rotation or precession of the Earth's orbit around the sun alternates the effect between the two poles in cycles of 19,000 or 23,000 years (Fig. 21.1c).

Eccentricity. Earth's orbit varies so that it is more elliptical at some times than at others, strengthening or weakening the precession effect (Fig. 21.1b). This variation in orbital eccentricity affects climate in cycles of about 100,000 years and about 400,000 years. Of course, when eccentricity is low (when the orbit is closer to being circular), the precession effect is much lessened.

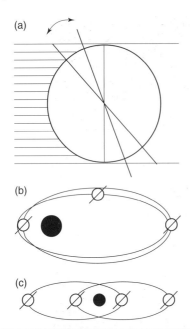

Figure 21.1 Some parameters of Earth's orbit can change over time, and as they do so they affect Earth's climate. a) greater or lesser tilting of Earth's axis causes stronger or weaker seasons. b) the eccentricity of the Earth's elliptical orbit means that one pole almost always feels greater seasonal effects than the other; changes in eccentricity weaken or strengthen that effect. c) the precession of the elliptical orbit alternates the eccentricity effect between the poles.

Figure 21.2 Earth's axis is tilted. As it orbits around the sun, solar radiation is concentrated on one hemisphere and then the other, giving Earth its seasons. As the seasons go by, we see the Sun gradually moving higher in the sky, then lower.

the cyclic rise and fall of sea level, and those rocks were deposited in tropical latitudes far from the glaciations of Gondwana that indirectly generated them. Many other cases of climatic cycling have been identified, even at times when Earth had no ice sheets, so we should look for a general cause for them, unconnected with ice sheets as such.

The **astronomical theory of ice ages** was suggested more than a century ago. It was worked out by hand by Milutin Milankovitch in the 1920s, and refined by computer calculations in the 1970s. It has been confirmed by evidence from microfossils that record temperature fluctuations in

the oceans. The Milankovitch theory suggests that slight variations in Earth's orbit around the sun and in the tilt of the Earth's axis make significant differences to climate (Fig. 21.1, Fig. 21.2).

Most important, Milankovitch cycles can trigger the advance and retreat of ice sheets, if conditions for an ice age are already present. Computer models of ice advances and retreats agree well with data from the geological record. The models suggest that the present mild climate on Earth is very unusual for our geography. Interglacial periods with reduced northern ice sheets are very short in comparison with glacial periods with large ice sheets.

Now it is time to apply all this theory to Earth's present ice age, the one in which we are living.

The Present Ice Age

Earth has been locked into an ice age since about 2.5 Ma, but its effects have been most marked in the Northern Hemisphere. Thus the northern glaciations that began in the late Pliocene and continued through the Pleistocene were centered on huge new ice sheets, mostly around the North Atlantic Ocean (Fig. 21.3). At the same time there was severe cooling in the Southern Hemisphere.

Once ice sheets built up, they altered climatic patterns in the North Pacific and North Atlantic. Sea surfaces in the North Atlantic froze as far south as New York and Spain. Warm Gulf Stream waters were diverted eastward toward North Africa, instead of bringing warm, moist climate to Western Europe as far north as Scandinavia.

At its maximum somewhere around 20,000 ± 2000 BP, Canadian ice advanced as far south as New York, St. Louis, and Oregon (Fig. 21.3). Ice scour removed great blocks of rock and transported them for hundreds of miles. The North American ice sheets diverted the jet stream and the main storm track southward. The western United States became much wetter than it is today, so that great freshwater lakes formed from increased rainfall and from meltwater along the front of the ice sheet. River channels were blocked by ice to the north, and at the southern edges of the ice sheets, much of the melt water drained south to the Gulf of Mexico down a giant Mississippi River.

As the North American ice sheets began to melt and retreat, water flow down the Mississippi to the Gulf of Mexico must have increased enormously. The water draining from the melting North American ice sheet changed the seawater composition of the Gulf of Mexico as it poured southward down the Mississippi in enormous quantities beginning about 14,000 BP, perhaps at ten times its current flow. Finally, as the edge of the ice sheet retreated, the Great Lakes began to drain to the Atlantic instead, first down the Hudson River, then the St. Lawrence, and finally north to the Mackenzie delta and the Arctic Ocean.

More subtle effects occurred in warmer latitudes. For example, increased rainfall in the Sahara during ice retreat formed great rivers flowing to the Nile from the central Sahara; they were inhabited by crocodiles and turtles, and rich savanna faunas lived along their banks.

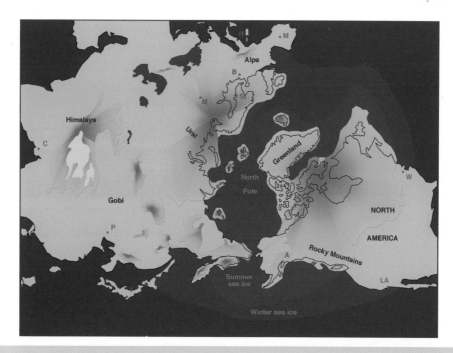

Figure 21.3   Great ice sheets around the Northern Hemisphere were a major influence on climates during the successive ice ages. Map by Hannes Grobe and placed into Wikimedia.

Life and Climate in the Ice Ages

Amazingly, the severe fluctuations of climate do not appear to have affected ice-age plants or animals very much. Glacial advances and retreats were rapid on a geological time scale, yet they were slow enough to allow ecological communities to migrate north and south with the ice sheets and the climatic zones and weather patterns affected by them. Communities close to mountain glaciers were able to adjust to advances and retreats by simply moving up and down in altitude. Tropical rain forests were very much reduced, but the habitat did not disappear, and their fauna and flora survived well. Tropical savannas were more extensive during the drier times that accompanied glaciations.

The most interesting effects were controlled by changes of sea level that occurred with every glacial advance and retreat. Each major glaciation dropped world sea level by 120 meters or so (about 400 feet), exposing much more land area and joining land masses together. Each new melting episode reflooded lowlands to recreate islands.

Most continents carry examples of creatures stranded by flooding and the warming that occurred during and after the last ice retreat. In the Sahara Desert, for example, there are cypress trees perhaps 2000 years old, just a few hundred survivors of their species in the Tassili mountains (Fig. 21.4). Ancient rock paintings of giraffes and antelope confirm the evidence of the cypress trees of a moist climate 2000 years ago (Werner 2007). The giraffes migrated south to the savannas; other cypresses are confined to the north along the Mediterranean coast; and the Sahara is a fearsome barrier to biological exchange.

Figure 21.4   A cypress tree in the middle of the Sahara Desert, a relic of a much moister climate 2000 years ago. Photo by Gruman at Flickr and placed into Wikimedia.

Box 21.2 Major Land Masses, Now and At 18,000 BP

NOW	AT 18,000 BP
1. Eurasia plus Africa	1. One major world continent
2. The Americas	
3. Antarctica	2. Antarctica
	3. Meganesia: Australia + New Guinea
4. Australia	
5. Greenland	4. Madagascar
6. New Guinea	5. New Zealand
7. Borneo	
8. Madagascar	
9. Baffin Island	
10. Sumatra	

Figure 21.5 Woolly mammoths foraging in a European Ice Age winter, among horses, lions and their prey, and a woolly rhino. Painting by Mauricio Antón, published in an article by C. Sedwick in PLoS Biology: http://www.plosbiology.org/article/info:doi/10.1371/journal.pbio.0060099. This places the art into Wikimedia.

A few creatures were trapped in geographical cul-de-sacs and wiped out. Advancing ice sheets, not St. Patrick, wiped out snakes from Ireland, and snakes have not yet been able to cross the Irish Sea to recolonize the island. The Loch Ness Monster is impossible because Loch Ness was frozen under a mile of ice 18,000 years ago. The forests of Western Europe were trapped between ice sheets from Scandinavia and wiped out, except for a few stragglers that hung on near the coast of Norway (Parducci et al. 2012). After the ice sheets melted, Western Europe was largely recolonized by deciduous hardwoods; though elsewhere, in North America, Scandinavia, and Siberia, the great boreal forests are dominated by conifers.

We have good evidence of the plant and animal life of the Pleistocene. Enormous bone deposits in Alaska and Siberia and fossils found in caves, sinkholes, and tar seeps have provided excellent evidence of rich and well-adapted ecosystems on all continents.

There were much greater changes in terrestrial animals and plants during and after the last glaciation than in any previous one, and the effects have often varied with the size of the land area. The land areas sometimes showed dramatic changes. For example, Alaska and Siberia were usually joined across what is now the Bering Strait, and Greenland was joined to North America, to form one giant northern continent (Fig. 21.3). Australia was joined to New Guinea, and Indonesian seas were drained to form a great peninsula jutting from Asia. Box 21.2 compares the major land masses now and at 18,000 BP at the height of the last glaciation; it shows how drastically seawater barriers were removed to join land masses together.

Continental Changes

On major continents, the larger birds and mammals of the Pleistocene were most unlike their modern counterparts. Just before the last ice advance, North America had mastodons, mammoths, giant bison, ground sloths, sabertooth cats, horses, camels, and dozens of other large mammals. Eurasia had most of these, plus giant deer and woolly rhinos (Fig. 21.5). The giant ape *Gigantopithecus* roamed the Himalayan slopes (Chapter 19). The moas of New Zealand and the elephant birds of Madagascar are well known (Chapter 18), but Australia had giant ground birds as large as these and a dozen giant marsupials. All these creatures are now extinct.

The catastrophic extinctions occurred at different times on different continents, but in each case, the mammals and birds were part of flourishing ecosystems. For example, North America has a very good fossil record of large Pleistocene mammals. Twenty genera became extinct in the 2 m.y. before the last ice sheet melted, then 35 genera were lost in less than 3000 years! (Fig. 21.6). Radiocarbon dates for the extinction cluster around 13,000 BP; where the record is good, the extinctions look sudden.

Some ice-age animals, such as the woolly mammoth and the woolly rhinoceros, were specifically adapted to life in cold climates. They were much hairier than their living relatives, and they have been found in areas that were very close to the ice sheets at the time. Woolly mammoths were sometimes killed by falling into ice crevasses. Their bodies have been found still frozen in permafrost in Siberia, preserved well enough to tell us quite a lot about their way of life (Fig. 21.7, Fig. 21.8). Gallons of frozen stomach contents show that woolly mammoths ate sedges and grasses and browsed tundra trees such as alder, birch, and willow (Fig. 21.9). The tusks of adults were well shaped for clearing snow from forage in winter. Woolly mammoths reacted to ice advances and retreats, but the only change was in size. Siberian woolly mammoths were about 20% larger during the warm interglacials than they were at the coldest times.

Large Pleistocene mammals were well able to withstand climatic change as well as climatic severity. Their large body

Figure 21.6 The astounding number of large North American mammals that became extinct about 13,000 BP. Art © 2012 Mike Hansen, and used by permission.

Figure 21.7 The discovery of the Beresovka mammoth, which had been frozen into the permafrost of Russia. The mammoth is now stuffed and on display in St. Petersburg. From a turn-of-the-century "magic lantern" slide.

sizes gave them low metabolic rates, so they could live on rather poor-quality food. As adults, they were largely free from the danger of predation by carnivores. Yet the large mammals and birds became extinct, while smaller species did not suffer as much. The plants the large mammals ate are still living, and so are the small birds, mammals, and insects that lived with them. In the oceans, nothing happened to large marine mammals.

In North and South America, the extinctions took place in a short time toward the end of the ice age, very close to 13,000 BP. This was an unusually cold, dry time, so it has been easy for North American geologists to argue that the extinction was related to climate change.

But that explanation, even if true, covers only some of the American extinctions and does not apply at all to the rest of the world. For example, the giant ground sloths of Arizona were browsers and ate semidesert scrub that was available in the area before, during, and after they died out. Other things being equal, we should prefer another hypothesis if it explains more data more simply.

Figure 21.8   A baby woolly mammoth (called "Dima"), as it was found frozen into the permafrost in Siberia. It is now preserved and on display in St. Petersburg. Photograph by NOAA, in the public domain.

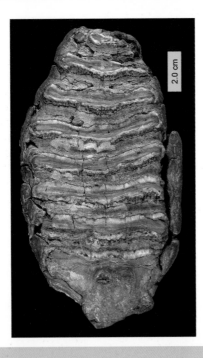

Figure 21.9   A mammoth's molar, well adapted for chewing coarse forage. Photograph by Dr. Mark A. Wilson of the College of Wooster, and placed into Wikimedia.

There is no question that climatic change around 13,000 BP was rapid. Yet the very same species of animals had already survived a dozen or more similar events. There is nothing climatically unique about the last ice retreat. The previous ice retreat, about 125,000 BP, was just as sudden

Box 21.3   Paul Martin's Evidence for the Overkill Hypothesis

1. Large mammals and ground-living birds were affected most. North America lost 35 genera, and South America lost even more.
2. Extinctions occurred in different areas at very different times.
3. Extinct animals were not replaced.
4. Extinctions were closely linked in time and space with human arrival.
5. Large mammals survived best in Africa and Asia. Extinctions were much more severe in the New World (Australasia and the Americas).
6. Where extinctions are well dated, they were sudden: North America and New Zealand are the best examples.
7. There are very few places where mammal remains occur with human remains or human artifacts. This implies that co-existence was brief.

but caused no extinctions. So if climatic change did not result in the extinctions, what did? The problem of Pleistocene extinctions has been debated ever since ice-age animals were discovered, and there is still continuing major disagreement.

The strongest evidence supports an idea put forward in its current form mainly by Paul Martin (for example, Martin 2005). Martin's **overkill hypothesis** gets its name because he stresses one human behavior in particular: hunting. In every case, invading humans were skilled hunters, encountering animals that had never seen humans before. Martin listed seven major lines of evidence (Box 21.3). All the pieces of evidence, argues Martin, are consistent with the idea that the sudden arrival of human invaders in an ecosystem was responsible for the extinctions. Other corollaries of human arrival may play an important part, so Martin's idea should not be judged entirely on the hunting overkill that he stresses most. To test his idea, we look at data from the only three major continents that were colonized suddenly by humans: North and South America, and Australia.

The Americas: Human Arrival

Humans crossed into North America from Siberia at a time when the Bering Strait region was a dry land area, Beringia. In the depths of the Ice Ages, Beringia and the Alaskan lowlands next to it were a frigid plain swept by violent winds blowing dust and sand from the edge of the ice sheet. Yet even then a varied Arctic vegetation supported a fauna of large ice-age mammals, including woolly mammoths, horses, camels, sheep, deer, musk-oxen, and ground sloths.

Beringia was separated from the rest of North America by the ice sheets of the Canadian Shield and the Rocky Mountains, which flowed together in what is now Alberta. An ice-free corridor to the south opened up into the rest of the Americas only as the main Canadian ice sheet retreated eastward. The important event in human migration is not when people reached Beringia, but when they broke past the ice barriers to the temperate and tropical Americas to the south.

Did humans reach the Americas only as the last glacial period ended, or had they done so long before? There is now compelling evidence that humans were living in Monte Verde, in southern Chile, at close to 15,000 BP. Most likely, these people had arrived by boat along the western American coast. When Europeans arrived on the Alaska coast, they found very proficient fishers and hunters there, using a variety of boats for hunting, transportation, and trade along the coast. There is scattered evidence of very early American fisherfolk operating much the same way of life thousands of years earlier, from sites in British Columbia, southern California, and Peru. These people seem to have eaten shellfish, seabirds, and fish. As far as we can tell, they had very little effect on American continental ecosystems, and may not even have ventured inland across the mountain barriers that lie behind the entire west coast. American continental ecosystems did not receive full human impact until the arrival of big game hunters in the interior.

But when did the next great wave of continental colonists spread into the Americas (after the coastal fisherfolk)? For decades, we have been gathering evidence of a short-lived, distinctive tool and weapon culture, the **Clovis culture**, which was widespread across North America from Washington to Mexico. All the dates for Clovis sites in the western United States cluster around 13,000 BP. The trademark of the Clovis culture is a large lethal spear point made of obsidian or chert. These are weapons made to kill large mammals (Fig. 21.10). There is a strong implication that the Clovis people were already skillful hunters of large mammals across the far northern plains of Asia and Beringia, before they reached the open plains of North America with its native population of large herbivores. Yet Clovis spear points have not been found in Beringia.

This problem has been solved by a spectacular find in the State of Washington. Here an adult male mastodon skeleton was found with a long large spear point embedded in a rib (it must have been driven a foot (30 cm) deep into the body to reach this far). But the spear point was not made from obsidian: it had been made from mastodon bone or tusk. The mastodon kill is dated at 13,800 BP, nearly a thousand years before Clovis. In addition, evidence of mammoth hunting around the Great Lakes, also close to 14,000 BP, is likely to have been the work of pre-Clovis hunters who used bone to arm their spears, rather than stone.

This means that the great extinction of large American mammals may have been an event that lasted 1500 years rather than 500 years. It is still true that the extinction followed the arrival of skilled large-animal hunters from the north.

The Americas: Large Animals

Clovis people hunted mammoths and mastodons. There are cut marks on mastodon bones found around the edge of the ice sheet near the Great Lakes, and it seems that humans butchered the carcasses into large chunks and cached them for the winter in the frigid waters under shallow, ice-covered lakes, just as Inuit do today in similar environments. We can tell that the favorite hunting season for mastodons was late summer and fall, whereas natural deaths occurred mainly at the end of winter when the animals were in poor condition. A mammoth skeleton from Naco had eight Clovis points in it. Two juvenile mammoths and seven adults were killed with Clovis tools near

Figure 21.10 A collection of Clovis spear points from a site in Iowa. The skilful preparation of the edges is clear. Once the point was made, two careful blows chipped off a central smooth area on each side of the point so that it could be hafted on to the shaft of the spear. The scale is in centimeters. Photograph by Billwhittaker@en.wikipedia.

Colby, Wyoming, and the way the bones are piled suggests meat-caching there too.

It cannot have been easy to kill these elephants in any direct attack. Mastodons were lethally effective in using their tusks against one another (Fig. 21.11, Fig. 21.12), and that skill would easily have carried over into effective defense against their natural predators. But Clovis people were another story: they were armed with formidable weapons, traps, poisons, and intelligent group hunting tactics, and the American megafauna was devastated. The

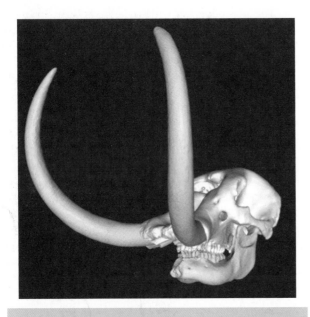

Figure 21.11 Reconstruction of a male American mastodon. Courtesy Daniel Fisher of the University of Michigan. © Daniel C. Fisher].

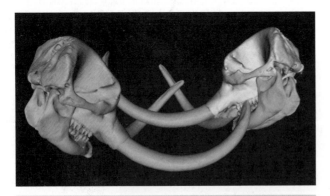

Figure 21.12 Reconstruction of fighting between two male American mastodons. The reconstruction is based on massive damage to some fossil male skulls, which indicates lethal or sublethal strikes on a target area in which the tusk would smash under the cheekbone toward the braincase. Courtesy Daniel Fisher of the University of Michigan. © Daniel C. Fisher.

last mammoths in the Great Lakes area, the last native North American horses, and ground sloths and mountain goats in the Southwest all died out around 13,200 BP. In west Texas, mammoths, horses, and bison are common before 13,000 BP, but after that we find only bison bones. The skeleton of a giant extinct turtle found in Little Salt Spring in southwest Florida had a sharpened wooden stake jammed between the shell and the breastbone. The turtle had apparently been killed and cooked on the spot in its shell. There are ground sloths, bison, and a young elephant at about the same level, but afterward the PaleoIndians ate white-tailed deer.

The Americas: Megaherbivores and Medium-Sized Animals

There was extinction among medium-sized mammals, although one would expect some of them (camels, horses, and deer, for example) to have been resistant to extinction because of their speed, agility, and rapid reproduction, even in the face of expert hunters.

The answer to this puzzle may be found in ecosystems that include **megaherbivores** (very large herbivorous mammals more than 2000 kg [2 tons] in weight). On the plains of Africa today, for example, the largest animals, elephant and rhino, can have drastic effects on vegetation. Elephants destroy trees and turn dense forest into open woodland by opening up clearings in which smaller browsing animals multiply. Eventually elephants turn any local habitat into grassland. They then migrate to another woodland habitat, leaving the trees to recover in a long term ecological cycle that can take decades to complete. White rhinos graze high grass so effectively that they open up large areas of short grassland for smaller grazing animals.

Thus, in the long run, megaherbivores keep open habitats in which smaller plains animals can maintain large populations. Where elephants have been extinct for decades, the growth of dense forest is closing off browsing and grazing areas, and smaller animals are also becoming locally extinct. Many of the problems in African national parks today occur because they are not large enough to allow these cycles of destruction and migration to take place naturally.

But what would happen if megaherbivores were completely removed from an ecosystem—by hunting, for example? Megaherbivores breed slowly and cannot hide. They would be particularly vulnerable to skillful hunters. Norman Owen-Smith proposed that the disappearance of Pleistocene megaherbivores (Fig. 21.13) soon led to the overgrowth of many habitats, reducing their populations of smaller animals too. Thus, even if early hunters hunted or drove out only a few species of megaherbivores, they could have forced ecosystems so far out of balance that extinctions would then have occurred among medium-sized herbivores too, especially if hunters were forced to turn to the latter as prey when the megaherbivores had gone.

Figure 21.13 We don't have to rely on the inferences of paleontologists to describe the megaherbivores in Pleistocene ecosystems. The woolly mammoth and the woolly rhinoceros were observed and illustrated by competent European ecologists of the time.

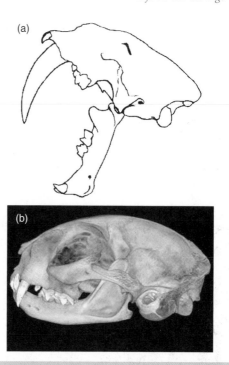

Figure 21.14 a) the North American sabertooth cat *Smilodon*. b) a CAT scan of a cat (sorry, but it's true!). The skull structure is much the same, and the two cats differ mainly in the extravagant size of the upper canine in *Smilodon*. Courtesy Tim Rowe and the Digimorph Project at the University of Texas, www.digimorph.org)

There may be more subtle effects of removing large herbivores. Plants sometimes coevolve with herbivores that disperse their seeds. Very large herbivores are likely to encourage the evolution of large, thick-skinned fruits, and a sudden extinction would leave the fruits without dispersers. Even today, guanacaste trees of Central America produce huge crops of large fruits, most of which lie and rot. Daniel Janzen suggested that these fruits coevolved with large elephants (gomphotheres), which became extinct with the other large American mammals.

The Americas: Predators and Scavengers

Predator species such as the sabertooth cats (Fig. 21.14) and the North American lion could have been reduced to dangerously low levels by the removal of their prey by overkill; there is no need to think in terms of the direct, systematic overkill of predator species that modern humans often carry out. In turn, scavengers may also depend on populations of large mammals to provide the carcasses they feed on. For example, the giant teratorn known from the La Brea tar pits (Fig. 13.23) is extinct, and the so-called "California" condor once nested from the Pacific coast to Florida. Pleistocene caves high on vertical cliffs in the Grand Canyon of Arizona contain bones, feathers, and eggshells of this condor, along with the bones of horses, camels, mammoths, and an extinct mountain goat. The condor vanished from this area at the same time as the large mammals did, presumably because its food supply largely disappeared.

The Americas: Survivors

What about the surviving large mammals in North America? It turns out that many of them were originally Eurasian and crossed into the Americas late in the Pleistocene. Thus bear, moose, musk-oxen, and caribou had been exposed to humans in Eurasia before 13,000 BP, so may have had behaviors that reduced their vulnerability to human hunting. There were no North American extinctions after 8000 BP at the latest, presumably when a new stable balance had evolved. There were separate regional cultures in the Americas by this time, but there were no new significant extinctions even though tools and weapons had improved.

Bison were a special case. They were American natives, and although an immense long-horned bison became extinct, the smaller "American bison" survived in great numbers. Perhaps the removal of larger competitors encouraged its success. Moreover, bison survived in the face of intense and wasteful hunting by PaleoIndians, whose methods were by no means as ecologically sound as their descendants sometimes claim. PaleoIndians, hunting on foot with stone weapons, would stampede whole herds of bison along preplanned routes that led to a cliff edge or buffalo-jump (Fig. 21.15a). The bison would then be finished off and butchered at the cliff base (Fig. 21.15b). The

Figure 21.15   a) the famous Head-Smashed-In buffalo jump site in Alberta, Canada. Photograph by Ken Thomas, and released into the public domain. b) some of the bones being excavated at the Vore buffalo jump site in Wyoming. Photograph by Jeff the quiet, and released into the public domain.

method naturally resulted in the deaths of many young animals. Only about a quarter of the animals killed at the site that is appropriately named Head-Smashed-In (Fig. 21.15a) were full-grown. That site had been used for more than 5000 years when it became obsolete as firearms reached the tribes in the nineteenth century. Given that there are dozens of buffalo-jump sites stretching across the Great Plains from Alberta to Texas, most of them still to be investigated, it seems less surprising that humans would be capable of the overkill that the fossil record suggests, and more surprising that the bison survived so successfully for so long in the face of human predation pressure.

Australia

Australia suffered more severe extinctions than any other continental sized land mass. It lost every terrestrial vertebrate larger than a human. It lost a giant horned turtle as big as a car, and its giant birds, the dromornithids (Chapter 18). It lost its top predators, including *Megalania*, the largest terrestrial lizard that ever evolved, 7 meters (24 feet) long. *Megalania* was closely related to the living Komodo dragon but weighed more than eight times as much (Fig. 21.16). Other predators were a huge terrestrial crocodile, a carnivorous kangaroo, and a 5-meter (16-foot) python. Australia lost about 20 large marsupials, including all the diprotodonts, huge four-footed vegetarians the size of tapirs; a wombat the size of a cow; and the largest kangaroo of all time, *Procoptodon*, a browser 3 meters high that was the ecological equivalent of a tapir or ground sloth. Only a few large animals survived in Australia, but small animals were less affected.

The extinctions are dated to around 45,000 to 50,000 BP, a time that coincides, as far as we can tell, with human arrival in Australia. The slow-running giant dromornithid

Figure 21.16   The skull of *Megalania*, the largest terrestrial lizard of all time, from the Pleistocene of Australia. This skull is 30 inches long, and the largest individuals were over 20 feet long, the top terrestrial carnivores in the ecosystem. Photograph by Stephen G. Johnson, and placed into Wikimedia.

bird *Genyornis* (Fig. 21.17) disappeared from habitats where fast-running emus survived, but there seems to be a memory of it in aboriginal legend as the mihirung. Some of the oldest Aboriginal rock art seems to show *Genyornis*, perhaps the last ones!

The extinctions coincide roughly with a change in vegetation associated with increased burning. This was not a time of climatic change, so the increased burning may have been generated by the early Australians. They were migrating into a dry country ecosystem that was unfamiliar to them because they came from the moister tropical ecosystems of New Guinea. They had to learn slowly how to adapt to drier Australian conditions, just as Europeans had to do tens of thousands of years later.

One of the easiest ways of clearing Australian vegetation is to burn it: burning makes game easier to see and hunt,

Figure 21.17 *Genyornis*, a giant ground-running bird from the Pleistocene of Australia, over 6 feet tall. Art by Nobu Tamura, and placed into Wikimedia.

and Australian aborigines today have complex timetables for extensive seasonal brush burning that has dramatic effects on regional ecology. Most likely, then, the Australian extinctions were the direct result of human invasion, through the introduction of large-scale burning as well as hunting.

On the basis of this evidence from three continents, it looks as if Martin's hypothesis of human impact is stronger than any other. Now we will look at smaller land masses that were subjected to much the same human impact and see how they fared rather differently.

Island Extinctions

Island animals can often evolve into unique sets of creatures, and geographical changes that connect previously isolated areas can have dramatic and damaging effects on species and communities (Chapter 18). Human arrival has often had a catastrophic effect on island faunas. The world may have 25% fewer species than it did a few thousand years ago, and most of those extinctions took place on islands. For example, native Tasmanians killed off a unique penguin sometime after the 13th century, 600 years before they in turn were wiped out by European settlers.

On Madagascar, large lemurs, giant land tortoises, and the huge flightless elephant birds (Chapter 13) disappeared after the arrival of humans somewhere between 0 and 500 AD. Here too, large forest areas were cut back and burned off to become grassland or eroded, barren wasteland. No native terrestrial vertebrate heavier than 12 kg (25 pounds)

survived after 1000 BP. Humans took a long time to penetrate the forest of this large island, and the extinction may have taken 1000 years instead of being sudden. It's clear that human arrival was part of a "recipe for disaster" (as David Burney has called it). The desperate erosion and poverty of much of the countryside of Madagascar today underlines the fact that humans are still involved in self-destructive deforestation, in spite of the evidence all around them of its horrific after-effects.

A panda-sized marsupial lived in New Guinea in the Late Pleistocene, and although it is now extinct, the plants that it ate are still flourishing.

New discoveries of extinct flightless birds in Hawaii suggest that devastating extinctions followed the arrival of Polynesians. The Hawaiian Islands are famous for honeycreepers, which evolved there into many species like Darwin's finches on the Galápagos Islands. But there were 15 more species of honeycreepers before humans arrived. Two-thirds of the land birds on Maui were wiped out by the Polynesians, probably by a combination of hunting, burning, and the arrival of rats. As in New Zealand (Box 21.4), the extinctions that followed the European arrival were severe but not as drastic, probably because the bird fauna was already so depleted.

The same process is recorded on almost all the Pacific islands in Melanesia, Polynesia, and Micronesia. Almost all of them, apparently, had species of flightless birds that were killed off by the arriving humans and/or their accompanying rats, dogs, pigs, and fires. As many as 2000 bird species may have been killed off as human migration spread across the ocean before the arrival of Europeans. It may not be an accident that Darwin was inspired by the diversity of the Galápagos Islands: these were never occupied before European discovery in 1535, and human impact was relatively slight until whalers arrived in strength around 1800.

Several islands in the Mediterranean Sea (Cyprus and Crete are good examples) held fascinating evolutionary experiments after the ice ages. There were pigmy elephants and pigmy hippos, giant rodents, and dwarf deer. These mammals had evolved on these isolated islands in much the same way as did the fauna of Gargano during Miocene times (Chapter 18). Many of the island animals disappeared as Neolithic peoples discovered how to cross wide stretches of sea and colonized the islands several thousand years ago.

Europeans killed off the dodo on Mauritius (Fig. 21.18) and several species of giant tortoises there and in the Galápagos; deliberate burning, and the goats, pigs, and rats they brought, completed a great deal of destruction of native plants, birds, and animals. They killed off the great auk of the North Atlantic, the huia and other small birds of New Zealand (Box 21.4), and an unknown number of species of birds of paradise in New Guinea, all to satisfy the greed of egg and feather collectors. They drove fur-bearing mammals close to extinction worldwide.

Irrespective of race, color, and creed, it seems, human arrival in the midst of a fauna and flora unused to hunting pressure, to extensive burning, or to rats, cats, and pigs, has

Figure 21.18   The dodo, playing a cameo role in Alice in Wonderland, by Lewis Carroll. This famous image by Sir John Tenniel from 1869 shows an outmoded reconstruction of the dodo, perhaps based on overfed captive specimens.

spelled disaster. One common factor among the victims is naiveté. Charles Darwin described the complete lack of fear of humans of the Galápagos animals and birds, an almost universal feature of creatures never exposed to human hunting; and modern observers like Tim Flannery have recorded the same behavior in Papua New Guinea. New Zealand's inhabitants had never seen a land mammal before the Polynesians arrived (Box 21.4).

The Du

Very large bird bones were discovered recently on the Isle of Pines, off New Caledonia. The fossil was named *Sylviornis*, and although it was large, it was not a ratite, but a very large flightless megapod. Megapods are a family of birds that includes the mallee fowl of Australia and ranges through eastern Indonesia and Australasia. No bird anywhere near that size now lives on New Caledonia. *Sylviornis* is therefore extinct, and its remains are dated at about 3500 BP, when humans had already reached its island.

Melanesian folk tales from the Isle of Pines describe a giant red bird, the Du (Fig. 21.20), which did not sit on its single egg to hatch it. Although the Melanesians did not know it at the time, this behavior is unique to megapod birds, which lay their eggs and cover them with rotting vegetation to keep them at an even warm temperature. The male keeps close control of the egg temperature by adjusting the compost heap on an almost hourly basis for weeks at a time. The Du, then, was *Sylviornis*, and the legend shows that it was known (at least briefly) to early man.

The Isle of Pines has large areas covered by large and mysterious mounds, never associated with original human artifacts. The mounds are the right size to be Du hatching mounds, still preserved in enormous numbers. They give some idea of the numbers of the Du, and they illustrate the massive disaster that overtook the bird at a time when there was no significant climatic change in its environment.

Experienced Faunas

We have already seen that many survivors in North America had been used to hunting pressure in Eurasia. Humans developed their hunting skills in the Old World, and although there were extinctions of large mammals there, they were spread out over longer times than the New World extinctions were.

For example, in Africa *Homo erectus/ergaster/antecessor* butchered giant baboons, hippos, and the extinct elephant *Deinotherium*. The remains of 80 giant baboons have been found at one site dated at about 400,000 BP. At Torralba, in Spain, the remains of 30 elephants, 25 horses, 10 wild oxen, 6 rhinos, and 25 deer were found on one site.

It therefore fits the overkill or human-impact hypothesis that the most important local extinctions in the Old World took place in habitats that modern humans were invading in strength for the first time. These invasions took place along the edges of the ice sheets, and even then humans are implicated in the disappearance of only a few large mammals of the northern Eurasian tundra, especially the woolly mammoths and the woolly rhinoceros.

It looks as if mammoths became extinct as advanced hunting techniques allowed humans to range closer to the ice sheets. For example, the advance of ice sheets toward the peak of the last glaciation seems to have driven the Gravettian people (Chapter 20) out of the northern Carpathian Mountains of Central Europe toward the south and the east, where they discovered and invaded the mammoth steppe of Ukraine for the first time. The Gravettians were already using mammoth bones as resources. At Predmost in the Czech Republic, a site dating from just before the coldest period of the last glaciation (28,000–22,000 BP) contains the bones of at least 1000 woolly mammoths. These people routinely buried their dead with mammoth shoulder bones for tombstones.

The pattern in these extinctions was always the same. The large mammals were hunted out of the optimum part of their range, and then the last survivors hung on in the inhospitable (usually northern) parts of their range until newly invading humans or climatic fluctuations killed them off. For example, woolly mammoths, woolly rhinoceroses, and giant deer, along with horses, elk, and reindeer, reinvaded Britain from Europe after the ice sheets began to retreat and birch woodland and parkland spread northward. Mammoths flourished in Britain until 12,800 BP at least, but then human artifacts appeared at 12,000 BP, and the largest animals of the tundra fauna quickly disappeared.

Box 21.4   Case Study: New Zealand

A thousand years ago, New Zealand was an isolated set of islands without land mammals (except for two species of bats). Birds were the dominant vertebrates, and the largest were the moas, huge flightless browsing birds the size of ostriches (Chapter 18) (Fig. 21.19). The moas and other native creatures survived as glacial periods came and went, yet they became extinct within a few hundred years of the arrival of the Polynesian Maori people after 1000 AD.

There seem to have been two main reasons for the extinctions, and all of them are connected with human arrival. First, evidence of hunting is clear and appalling. Midden sites that extend for acres are piled with moa bones, with abundant evidence of wasteful butchering. The bones are so concentrated in some places that they were later mined to be ground up for fertilizer. The middens contain bones of 11 of the 12 extinct species of moas, and they also contain bones of tuataras, very primitive reptiles. Second, the Maori brought rats with them, which ate insects directly, killed off reptiles by eating their young, and exterminated birds by robbing their nests. The tuataras (Chapter 10), the giant flightless wetas (insects that had been the small-bodied vegetarians of New Zealand (Chapter 18), and many flightless birds including the only flightless parrot, the kakapo, were practically wiped out by rats. There were many other more subtle ecological changes. A giant eagle that may have preyed on moas died out with them, for example. And when the moa were gone, the Maori took up serious cannibalism, because humans were the largest remaining protein packets.

Half of the original number of bird species in New Zealand were extinct before Europeans arrived, and the new settlers only acted to increase the changes in New Zealand's landscape and biology. Forests were cleared even faster, and new mammals were introduced. European rats were the worst offenders against the native birds, but cats, dogs, and pigs were also destructive, rabbits destroyed much of their habitat, and deer competed with browsing birds. The tuatara now lives only on a few small, rat-free islands, and the kakapo survives precariously in remote areas where it is threatened by wild cats. Bird populations are still dropping in spite of efforts to save them.

As a microcosm of the problem, consider the Stephen Island wren, the only flightless songbird that has ever evolved. This species had already been exterminated from New Zealand by the Polynesian rat before European arrival. The entire remaining population of this species, which was by then confined to one island, was caught and killed by Tibbles, a cat brought to the island in 1894 by the keeper of a new lighthouse.

A convict colony established by the British wiped out an endemic seabird on Norfolk Island, between Australia and New Zealand. Several small, unique native birds fared better for a while on Lord Howe Island, further north: they lived alongside the early settlers until a shipwreck allowed rats to reach the island in 1918. Within a few years five species had completely disappeared.

Figure 21.19   One of the moas of New Zealand. After Frohawk.

Figure 21.20   *Sylviornis*, the extinct Du of New Caledonia. The bird stood about a meter (3 feet) high. Redrawn from a reconstruction by Poplin and Mourer-Chauviré; the head ornament is based on oral legends of the Melanesians of New Caledonia.

Figure 21.22   The coat of arms of Northern Ireland. Like all such devices, it is redolent in symbolism. The lion and the crown represent Great Britain, and the Red Hand is an ancient Celtic symbol adopted by the O'Neill kings of Ulster. The Irish elk, along with the harp, are perhaps the only pan-Irish symbols in the entire device.

Figure 21.21   *Megaloceros*, reconstruction by Pavel Riha, placed into Wikimedia. This giant deer has several species across Pleistocene Europe, but the largest and last-surviving was the "Irish el."

The giant deer *Megaloceros* (Fig. 21.21) is sometimes called the Irish elk, partly because it is best known from Ireland. It was not an elk but a deer the size of a moose, with the largest antlers ever evolved, more than 3 meters (10 feet) in span. It was adapted for long-range migration and open-country running, and its diet was the high-protein willow vegetation on the edges of the northern tundra. It once ranged from Japan to France, but it did not reach North America, where the moose is an approximate ecological counterpart.

The giant deer disappeared from Eurasia in a sequence that started in Eastern Siberia and proceeded westward. It survived longest in Western Europe and finally, after the ice sheets melted and sea level rose, it was confined to the island of Ireland (Fig. 21.22). The giant deer flourished there in a warm period until about 11,000 BP, but it then died out in a cold period, possibly because it was unable to retreat southward to a better climate. Humans did not reach Ireland from Europe and Britain until the climate warmed again, around 9000 BP.

Mammoths, which had lived much farther south, were confined to the tundra north of the Black Sea by 20,000 BP. We have an interesting record of life around 15,000 BP on the plains of Russia and Ukraine. Several dozen living sites were built on low river terraces by people from the Gravettian culture. Each major site contains the remains of several large buildings whose foundations and lower walls were made entirely from mammoth bones. The buildings were large, 4–7 meters (13–23 feet) across and up to 24 sq m (240 sq ft) in area. The foundation was made of the heaviest bones, carefully aligned. Skulls at the base were followed by jaws and then long bones, with the resulting pattern providing an aesthetic geometric arrangement as well as sound architecture. The roofs were probably lighter structures made from branches, hides, or sod. (Only a thousand years ago, the Inuit of Greenland were using whale skulls and whale ribs in the same way, roofing the dwellings with sod.)

One group of four buildings was built using bones from at least 149 mammoths. It's not clear whether the mammoths had been killed, or whether bones from old skeletons had been collected. Zoia Abramova has remarked that it would have been easier to obtain bones from living mammoths than to dig them out of permafrost. Others disagree, arguing that dwelling sites may even have been chosen because they were close to massive mammoth bone accumulations. Since no one involved in the arguments has completed either one of these tasks, we are not likely to get any agreement soon. Certainly, the sites in this unique area give some idea either of the numbers of mammoths that once roamed the plains, or of the hunting efficiency of these stone-age peoples, or both.

Inside their homes, the Gravettians laid down clear river sand as a floor, built hearths (fuelled by mammoth bone), and remained secure and warm during the winter. They made finishing touches to stone tools (leaving behind their antler hammers and chipped flakes), skinned animals (leaving behind the bones), ground up ochre for dye, and sewed with ivory needles.

The mammoth-bone dwellings are surrounded by pits dug into the permafrost that were probably used to store meat long-term. Lewis Binford made a close and vivid comparison between the whale- and caribou-hunting Inuit of Arctic North America today and the tundra dwellers of the mammoth steppe. Many features of their buildings and the food storage pits are closely similar, implying that the ancient Gravettians were effective hunters even if they also used and re-used old bones.

Older generations of Inuit did not hesitate to attack a 25-ton bowhead whale from flimsy boats, though their modern descendants prefer motor boats and assault rifles. Ice-age hunters may or may not have attacked an 8-ton mammoth directly, but it doesn't take a great deal of imagination to see why man and mammoth could not have coexisted for long in this open steppe country that had been the main range of the species. (The same Gravettian people had built mammoth-bone dwellings in Central Europe several thousand years before, when mammoths still lived there.)

Some woolly mammoths survived in the permafrost areas of northern Siberia until perhaps 10,000 BP, in an area where humans arrived late. Even then, there was still one mammoth refuge left, in the Wrangel Islands, a small group of low lying islands off the north coast of Siberia. Forage was poor, and the last mammoths were small, perhaps 2 tons instead of the 6 tons of their ancestors. The last woolly mammoths died out in the Wrangel Islands only 3000–4000 years ago, at a time when there were large cities in the ancient civilizations of Eurasia and the Egyptian pyramids were already old. We are not sure what killed off these last pitiful survivors of the great mammoths, but humans reached the Wrangel Islands about that time.

There is a myth that primitive peoples live in ecological harmony with the plants and animals around them, and that it has been only with the arrival of modern civilization that major ecological imbalances have arisen. We have seen several examples that explode this myth, and there are many more.

Ancient peoples have destroyed their own civilizations on islands with delicate ecosystems. The Easter Islanders who built their famous enormous stone statues on the island also deforested their fertile, productive land until it became a barren waste and they became a wretched band of refugees surviving by shoreline scavenging and primitive fishing. But sophisticated peoples on large continents have harmed themselves too. The Anasazi Indians, who built a complex civilization on the Colorado Plateau, stripped their environment of trees until the erosion and siltation that followed ruined their irrigation projects and they disappeared as a significant people. (Tim Flannery calls these sorts of self-destructive societies *The Future Eaters.*) But are we doing any better?

The World Today

The Spanish introduced cattle to Argentina in 1556; by around 1700 there were about 48 million head of wild cattle on the plains. By 1750 they had been all but exterminated by a comparatively sparse human population with primitive firearms (muskets). This is even more incredible than the North American slaughter of about 60 million bison a century later with much more effective rifles, and it is more evidence in support of the plausibility of the overkill hypothesis.

Stripping tropical forest from hillsides not only removes the plants and animals that are best adapted to life there, but it results in erosion that removes the few nutrients left in the soil, destroying any agricultural value the land may have. It also results in much greater run-off and downstream flooding, which destroys or silts up rivers, irrigation channels, and fields downstream, harming ecosystems and productivity there too. This scenario has been played out now in Ethiopia, Madagascar, and Haiti in horrific proportions; it is happening throughout the rain forests of Brazil and Indonesia, it is destroying the reservoirs that provide water for the Panama Canal, and yet we do not seem to have learned the lesson.

One can argue that humans at 13,000 BP, perhaps even at 500 BP, did not know enough ecology, did not have enough recorded history, did not know enough archeology or paleontology, and did not have enough of a global perspective to realize the consequences of their impact on an ecosystem. But that is not true today. We have the theory and the data to know exactly what we are doing. We transport species to new continents and islands without proper ecological analysis of their possible impact. We approach our environment sometimes with stupidity, sometimes with greed, but usually with both.

We know very well that the tropical regions of the world are a treasure house of species, many of them valuable to us and many of them undescribed. Yet we deliberately introduce alien carnivorous fish into tropical lakes, ruining fisheries that have been stable for centuries.

We know that clearing tropical forests will quickly destroy the low level of nutrients in the soil and will render those areas useless for plant growth. Yet we go ahead anyway, sometimes for a quick profit on irreplaceable timber, sometimes to achieve a few years' agricultural cropping before the land is exhausted.

We poach gorillas and shoot animals for trophies. Indonesians and Malaysians clear tropical forests to supply the wood for the eleven billion disposable chopsticks used each year in Japan. Africans destroy rhinos to supply Asians with useless medicines and Yemenis with dagger handles; Africans poach elephants for ivory that ends up as ornaments and trivia in Japan, China, Europe, and North America.

Fishing grounds have been plundered worldwide. Tanzanian and Filipino fishermen use dynamite sticks, killing off the reefs their fish depend on and any hope they have of catching anything next year. Filipinos capture tropical fish for American aquariums and Chinese restaurants and "medicine" shops by dosing their reefs with cyanide and catching the few survivors. Japanese, Norwegians, and Icelanders catch whales under the guise of "research" even though they sell the meat: after all, they've already paid for the ships and want to get their money back. Giant clams are poached from marine preserves all over the South Pacific to feed the greed of Chinese "gourmets."

North Americans complain about the destruction of tropical rain forests for export to Japan, while their own lumber companies are felling the last of the old Douglas fir and redwood forests of the northwest for export to Japan. Italians and French shoot millions of little songbirds each year for "sport." All industrial nations continue to destroy forests and lakes with acid rain, though we know how to prevent the pollution that causes it. Ignorance is not the problem in any of these cases: poverty, greed, and arrogance are to blame.

We could live perfectly well—in fact, we could have a vastly increased quality of life—without disturbing the equilibrium of marmosets, gorillas, orangutans, chimps, whales, and all the other endangered species, if we took a grip on our own biology and behavior. What we need is a sense of collective responsibility and enlightened self interest. It's a difficult message to get across because evolution and society, and simple principles of economics, all favor the short-term goals of individuals rather than the long-term welfare of communities or societies.

It is in the interest of everyone, for themselves and for their children, to make our future secure not just for survival but for quality of life. If we don't solve our problems by our own voluntary actions, natural selection will do it for us. If we can learn anything from the fossil record, it is that extinction is the fate of almost every species that has ever lived on this planet. There is no automatic guarantee of success. Every individual in every generation is tested against the environment. We have the power and the knowledge to control our environment on a scale that no other species has ever done. So far, we have used those abilities to remove thousands of other species from the planet. If we destroy our environment to the point where the human species fails the test, becoming either extinct or less than human, it will serve us right.

But the greater tragedy would be our legacy, because we'll destroy much of the world's life along with ourselves. I believe that any rational God would have intervened long ago to prevent the wholesale destruction of so many of His creatures. We have only ourselves and one another to blame and to rely on.

The anthropologist David Pilbeam wrote that we have only just begun to tap the potential of the human brain. He had better be right.

Further Reading

Climate and Ice Ages

Ruddiman, W. F. 2007. *Earth's Climate: Past and Future.* 2nd edition. New York: W. H. Freeman.

Zachos, J. et al. 2001. Trends, rhythms, and aberrations in global climate 65 Ma to present. *Science* 292: 686–693. Review of Cenozoic climate as seen in 2001.

Ice Age Life and the Great Extinction

Burney, D. A. and T. F. Flannery 2005. Fifty millennia of catastrophic extinctions after human contact. *Trends in Ecology & Evolution* 20: 395–401. Available at http://www.anthropology.hawaii.edu/Fieldschools/Kauai/Publications/Publication%204.pdf

Fedje, D. W. and H. Josenhans 2000. Drowned forests and archaeology on the continental shelf of British Columbia. *Geology* 28: 99–102. Migration route for the first Americans?

Fisher, D. C. 1984. Mastodon butchery by North American PaleoIndians. *Nature* 308: 271–272.

Frison, G. C. 1998. Paleoindian large mammal hunters on the plains of North America. *PNAS* 95: 14576–14583. Available at http://www.pnas.org/content/95/24/14576.full

Gladkih, M. I., et al. 1984. Mammoth-bone dwellings on the Russian Plain. *Scientific American* 251 (5): 164–175.

Guthrie, R. D. 1990. *Frozen Fauna of the Mammoth Steppe.* Chicago: University of Chicago Press. [Paperback, excellent reading and excellent science.]

Martin, P. S. 2005. *Twilight of the Mammoths.* Berkeley, California: University of California Press.

Meltzer, D. J. 2009. *First Peoples in a New World: Colonizing Ice Age America.* Berkeley, California: University of California Press.

Owen-Smith, N. 1987. Pleistocene extinctions: the pivotal role of megaherbivores. *Paleobiology* 13: 351–362.

Sandweiss, D. H. et al. 1998. Quebrada Jaguay: early South American maritime adaptations. *Science* 281: 1830–1832, and comment, pp. 1775–1777.

Soffer, O. and N. D. Praslov (eds.) 1993. *From Kostenki to Clovis: Upper Paleolithic—PaleoIndian Adaptations.* New York: Plenum Press. Chapters by Olga Soffer and Lewis Binford.

Sutcliffe, A. J. 1985. *On the Track of Ice Age Mammals.* London: British Museum (Natural History).

Waters, M. R. et al. 2011. Pre-Clovis mastodon hunting 13,800 years ago at the Manis site, Washington. *Science* 334: 351–353, and comment, p. 302. Available at http:/

centerfirstamericans.org/cfsa-publications/Waters-etal-science334-2011.pdf

The Modern World (The Last 5000 Years or So)

Diamond, J. M. 1986. The environmentalist myth. *Nature* 324: 19–20.

Diamond, J. M. 1989. The present, past and future of human-caused extinctions. *Philosophical Transactions of the Royal Society of London B* 325: 469–477.

Diamond, J. M. 1991. Twilight of Hawaiian birds. *Nature* 353: 505–506.

Diamond, J. M. 1992. Twilight of the pygmy hippos. *Nature* 359: 15.

Diamond, J. M. 1995. Easter's end. *Discover* 16 (8): 62–69. The destruction of Easter Island by its own inhabitants. Required reading! Available at http://mrhartansscienceclass.pbworks.com/w/file/fetch/46798073/Easter%20Island's%20End%20by%20Jared%20Diamond.pdf

Diamond, J. M. 2001. Australia's last giants. *Nature* 411: 755–757. Oz megafaunal extinctions tied more firmly to human arrival.

Flannery, T. F. 1995. *The Future Eaters*. New York: George Braziller. The history of Meganesia (Australasia and associated islands). This book has much deeper significance than simply a regional history.

Flannery, T. F. 1996. *Throwim Way Leg: Adventures in the Jungles of New Guinea*. London: Weidenfeld and Nicolson. Enthralling account by an Australian zoologist in New Guinea, showing how easy it is for people to see species go extinct without either noticing or caring very much.

Goldschmidt, T. 1996. *Darwin's Dreampond: Drama in Lake Victoria*. Cambridge, Mass.: MIT Press. The devastation of the astounding biology of Lake Victoria.

Holdaway, R. N. and C. Jacomb 2000. Rapid extinction of the moas (Aves: Dinornithiformes): model, test, and implications. *Science* 287: 2250–2254, and comment by Jared Diamond, pp. 2170–2171. Available at http://www.esf.edu/efb/gibbs/efb413/moa.pdf

Jackson, J. B. C. et al. 2001. Historical overfishing and the recent collapse of coastal ecosystems. *Science* 293: 629–637. Required reading, showing that humans began to wreck marine ecosystems as soon as they learned how to fish. Available at http://jenkins.cos.ucf.edu/courses/biogeography/readings/Jacksonetal2001.pdf

Montevecchi, B. 1994. The great auk cemetery. *Natural History* 103 (8): 6–9.

Parducci, L. et al. 2012. Glacial survival of boreal trees in northern Scandinavia. *Science* 335: 1083–1086.

Quammen, D. 1996. *The Song of the Dodo*. New York: Scribner. [Highly recommended.]

Reeves, B. O. K. 1983. Six milleniums of buffalo kills. *Scientific American* 249 (4): 120–135.

Steadman, D. W. 1995. Prehistoric extinctions of Pacific island birds: biodiversity meets zooarchaeology. *Science* 267: 1123–1131. Available at http://web2.uwindsor.ca/courses/biology/weis/55-437/Steadman.pdf

Steadman, D. W. et al. 2002. Rapid prehistoric extinctions of iguanas and birds in Polynesia. *PNAS* 99: 3673–3677. You can reach other papers from the references. Available at http://www.pnas.org/content/99/6/3673.full

Stiner, M. C. et al. 1999. Paleolithic population growth pulses evidenced by small animal exploitation. *Science* 283: 190–194. Overhunting around the ancient Mediterranean. Available at http://www.u.arizona.edu/~mstiner/pdf/Stiner_etal1999.pdf

Werner, L. 2007. A cypress in the Sahara. *Saudi Aramco World* 58 (5): 32–39. Available at http://www.saudiaramcoworld.com/200705/a.cypress.in.the.Sahara.htm

Worthy, T. H. and Holdaway, R. N. 2002. *The Lost World of the Moa: Prehistoric Life of New Zealand*. Bloomington, Indiana: Indiana University Press.

Questions for Thought, Study, and Discussion

1. Starting from the present day, explain how the Earth will drop into its next ice age (in the absence of human activity).
2. As you know, bison are quite large animals, and before they were slaughtered by modern rifles in the 1800s, they travelled in huge herds. Read again what "megaherbivores" are, and discuss whether bison really are megaherbivores that survived the hunting of ancient Americans.
3. Elephants and rhinoceroses really are megaherbivores. How did they survive in Africa when megaherbivores were killed off on other continents?
4. Describe the science behind this limerick:

   They fall from the branches to wait
   But they're 12,000 summers too late
       You can smell them for miles
       They're rotting in piles
   The fruits that the gomphotheres ate

5. Describe the science behind this limerick:

   The morning was hardly propitious
   When sailors discovered Mauritius
       They killed off the lot
       Stewed them up in a pot
   And pronounced them extinct, but delicious.

Index

Page numbers in italics refer to figures

History of Life, Fifth Edition. Richard Cowen.
© 2013 by Richard Cowen. Published 2013 by Blackwell Publishing Ltd.

ASKHAM BRYAN
COLLEGE
LEARNING RESOURCES